Peter Richards

IET PROFESSIONAL AP ES 9

Model-Based Requirements Engineering

Other volumes in this series:

Volume 1	**Knowledge discovery and data mining** M.A. Bramer (Editor)
Volume 3	**Troubled IT projects: prevention and turnaround** J.M. Smith
Volume 4	**UML for systems engineering: watching the wheels, 2nd edition** J. Holt
Volume 5	**Intelligent distributed video surveillance systems** S.A. Velastin and P. Remagnino (Editors)
Volume 6	**Trusted computing** C. Mitchell (Editor)
Volume 7	**SysML for systems engineering** J. Holt and S. Perry
Volume 8	**Modelling enterprise architectures** J. Holt and S. Perry

Model-Based Requirements Engineering

Jon Holt, Simon A. Perry
and Mike Brownsword

The Institution of Engineering and Technology

Published by The Institution of Engineering and Technology, London, United Kingdom

The Institution of Engineering and Technology is registered as a Charity in England & Wales (no. 211014) and Scotland (no. SC038698).

First published 2012

The Institution of Engineering and Technology
Michael Faraday House
Six Hills Way, Stevenage
Herts, SG1 2AY, United Kingdom

www.theiet.org

British Library Cataloguing in Publication Data
A catalogue record for this product is available from the British Library

ISBN 978-1-84919-487-7 (paperback)
ISBN 978-1-84919-488-4 (PDF)

Typeset in India by MPS Ltd, a Macmillan Company
Printed and bound by CPI Group (UK) Ltd, Croydon, CR0 4YY

This book is dedicated to Jude, Eliza and Roo.

JDH

To Hannah, Austin, Jacob, Isabella, Beatrice, Grace and Lillian.
With love from Uncle Jolly.

SAP

To my wonderful wife, Jean.

MJB

Contents

Acknowledgement

> Wise men talk because they have something to say. Fools talk because they have to say something.
>
> *Plato*

As I write more books I find that the list of people who should be acknowledged grows longer, whilst the list of people that I actually mention becomes shorter.

The information in this book is the result of more than 12 years of work, from when I first realized what a use case actually *was*, rather than what everyone else thought it was. There have been so many people who contributed to this work over the years that thanking them seems impossible, but there are a few stalwarts who deserve a mention: Duncan and his colleagues, Dipesh and Mick, Rick and his people at the Defence Academy, all at INCOSE, Pemby and his analogy, all at the people from the IET PN for systems engineering (Ian, Andrew, Doug, Steve, Michael, and the rest of the crew), my BCS colleagues and, of course, all (rather some) of my colleagues at Atego. Thanks to Kay, who drew the original Brontosaurus picture, based on my rubbish sketch, that we still use to this day.

Special thanks go to Lisa and her team at IET publishing for putting up with us for 10 years now!

As ever, I cannot thank Mike and Sue enough for all that they have done for me over the years, both professionally and personally. Rather worryingly, I think that I am the same age now as Mike was when I first met him!

I could not have achieved what I have today without two very special individuals – my co-authors and cohorts for many years now Simon A. Perry and Mike Brownsword. People say that you can judge a person by the company that they keep and I feel truly honoured to share the front cover with both of them today and, hopefully, for many years to come.

Finally, the reason why I do all of this, I must thank my beautiful wife Rebecca and my three children Jude, Eliza and Roo for all of their love and support over the years. Olive and Betty also deserve a mention for their love and devotion to whoever controls the cat food.

Jon Holt, July 2011

There have been many people who have contributed to the ideas captured in this book and Jon has done an excellent job of naming the most influential. Rather than adding to this list, I would like to say a big thank you to all who have attended training courses that I have been lucky enough to deliver and who have challenged, questioned and contributed to the often lively discussions that have taken place over the last 6 years. Your thoughts and opinions have been invaluable.

Without turning this into a love-fest, I must take this opportunity to thank Jon and Mike for being great friends, as well as colleagues and coauthors, for more than 10 and 5 years, respectively. I hope that we have many more years working together.

To conclude, Motley, the cat, is still, just about, with us and continues to serve as an excellent feline alarm clock, if one is with no sense of time.

Finally, as ever, I must thank my wife Sally for all the love and encouragement she has given me over the last 20 years. I am, without doubt, a better person for having met her.

Simon A. Perry, July 2011

In writing my first book I find myself in awe about the number of people who have contributed, through comment, discussion, model and criticism, whether on site, during a course or in the office. I would like to thank them all. However, I will simply echo Jon and Simon's thanks and extend it a little to the iFEST partners who have been exposed to some of these ideas more recently.

To complete the triangle, I will take this opportunity to thank Jon for letting me hang him upside down from a crane, and Simon for reviewing all my work. They have made writing this book, as with most of my work, an enjoyable task. I hope they will let me work with them on more books in the future.

Finally, I thank my wonderful wife Jean and my two energetic boys Samuel and Daniel for all the love and laughter they give. Also to continue the theme of cats, I would like to thank Shelly, our little cat, who never fails to remind me that it's late at night and I really should stop modelling or writing to feed her.

Mike Brownsword, July 2011

Chapter 1
Introduction

What you want and what you get are two different things.

Christine Holt

1.1 Introduction to requirements

The whole area of requirements is one that is fraught with conflicts, misunderstandings and many contradictory views and opinions.

It is widely acknowledged that getting the requirements 'right' is crucial to the success of any project or to realising any form of successful system, but what exactly is meant by this?

It is clear that a good understanding of requirements is important, so why do so few people actually apply any sort of rigour to their requirements engineering activities?

This section discusses these two issues at a high level by looking at exactly what we mean by 'requirements' and 'requirements engineering' and then looking at some key concepts that are required to ensure good requirements and rigorous requirements engineering activities.

1.1.1 The need for requirements engineering

To understand the requirements for a system, it is not enough to simply state the needs of a group of users, but it is necessary to *engineer* the requirements of all relevant stakeholders.

'Requirements engineering', therefore, is the discipline that enables this understanding and is important for a number of reasons:

- *Systems approach.* All systems approaches identify a need to understand the requirements properly. This relates to all types of systems whether they are technical, social, financial etc.
- *Quality.* There are many definitions of quality, but the two that are considered here are from the International Standards Organisation (ISO) [1]. This first definition, which is arguably the most common definition of quality used in the world, is 'fitness for purpose'. Fitness for purpose means that the system does what it is supposed to do or, to put it another way, the system satisfies its

original requirements. The second definition that is cited here is 'conformance to requirements' – no explanation needed.

- *Requirements drive the project.* Every aspect of the project should be traceable back to the source requirements. The requirements should drive everything that is done in the project, and if they are not, then some serious questions need to be asked.
- *Benchmark for acceptance.* Acceptance testing – the tests that provide the customer with the confidence that the system is fit for purpose are based solely on requirements. Acceptance tests are not based on the design or the implementation techniques that are applied (unless they are constraints) but are based entirely on the requirements set for the system. Therefore, if the requirements are not fully understood, how on earth can the final system be accepted?
- *For increased confidence.* One of the least tangible, yet most powerful benefits of applying an effective systems approach is one of confidence. When the requirements of the system are understood, the system can be demonstrated and accepted. When the requirements are understood, they can be agreed by the customer somewhere in the initial, or conceptual, stage of a project. This confidence refers to the confidence of both the customer and the supplier sides of the relationships. In other words, the confidence of all the stakeholders will be increased. Although difficult to measure, confidence is an immensely valuable attribute for any relationship.

The points raised here make the case for understanding requirements. To understand the requirements, it is necessary to engineer them, and this gives rise to the discipline known as 'requirements engineering'. An effective approach to requirements engineering will result in a concise, consistent and lucid definition of the requirements of any system and will yield the following benefits:

- *Increased probability of optimum solution.* If the requirements for any system are unknown or badly defined, then there is no way that the system under development can meet them. There is also no way that the system can be validated, as validation is based solely on the requirements.
- *Full traceability.* It is important that any aspect of the system can be traced back to its source requirement(s). Not only is traceability essential for effective verification and validation of the system, but it is also a crucial part of quality assurance.
- *Requirements that are independent of the solution.* A good set of requirements should be independent of any specific solution. In the real world, this is rarely completely achievable, as there will almost always be some aspect of the requirements where the solution, part of the solution or the technology used by the solution has already been decided. When and where this does occur, such requirements are considered to be 'constraints' because they will limit, in some way, how that requirement may be realised. In some cases this may refer to a solution or partial solution. In others, there may be a quality constraint, such as meeting a particular standard or following a specific process. Otherwise, the

constraint may come from a specific technology or approach being adopted on the development. There are, of course, many more types of constraints that exist, but this brief list should provide an insight into the kind of requirements that may be considered as constraints.

The points mentioned above provide an overview of good reasons to perform effective requirements engineering. Requirements engineering is an entire subject matter in itself, so this chapter can only hope to provide a very high-level view of the rationale behind requirements engineering [2–5].

From a systems engineering point of view, getting the requirements right is crucial. From the point of view of this book, there are five main aspects of getting it right, which are the following:

- *Capture*. It is important that an initial set of requirements can be captured in some way, whether this is by someone producing an initial requirements specification, or whether the requirements need to be extracted from somewhere, such as existing documentation.
- *Specification*. By the term 'specification' here, we mean to state specific features of the requirement in an unambiguous manner. This is very important but is only useful when it is done alongside the other four items in this list. As discussed later, specifying individual requirements, although necessary, is only a small part of the overall requirements engineering process.
- *Analysis*. By the term 'analysis' we mean *understanding*. By the term understanding, we do not mean understanding the sentence that describes the requirement or the individual words, but true understanding of the nature of the requirement, where it came from, how important it is, how it can be tested and so on.
- *Organisation*. The requirements themselves must be organised in some way. This may include grouping the requirements into related categories, defining different contexts, relating requirements together and so on.
- *Management*. Once the requirement have been analysed, organised and specified, it is important that they can be managed effectively. Requirements form a complex, living entity that will evolve over time and, hence, they must be properly controlled. Invariably, this will be made easier by the use of tools, whether it is within a modelling tool or whether a requirements-specific tool is used.

The requirements engineering must be carried out within the context of a project life cycle and it is usual for this to happen close to the beginning of a project. Any life cycle will be made up of a number of stages, or phases, one of which will represent the first stage in a project and have a name like 'conception', 'initial', 'requirements' or 'inception'. It should be kept in mind, however, that the requirements set is a living entity and, as such, it will evolve and change as the project progresses through its life cycle. It is crucial, therefore, that the requirements set is managed and that it can be revisited at different points in the project to validate that the project is being developed correctly.

The approach described in this book covers all of these aspects and may be used during any life cycle stage. The approach itself is model-based, in that the model forms the heart of all requirements engineering activities, and uses the idea of a context, or viewpoint, as a basis for the model and its realisation.

1.2 Concepts and terms

This section looks at a number of key concepts and terms that are used in this book. This is deliberately kept at a high level, as all of these concepts and terms are discussed in far more detail in the subsequent sections of the book.

1.2.1 Defining 'requirement'

1.2.1.1 Dictionary definition

This section will look at several definitions of the term 'requirement' and then define exactly what is meant by this term for the purposes of this book.

The basic definition of a 'requirement' in the Oxford English Dictionary is:

> *(noun) a thing that is needed or wanted* [6]

This definition, albeit a very high-level one, is very interesting for two main reasons:

- A requirement is basically defined as a 'thing' – not the most unambiguous of definitions.
- This 'thing' is either 'needed' or 'wanted'. This is particularly interesting because there is often a big difference between what stakeholders want and what they actually need. Indeed, part of any requirements analysis activity should concern itself with the difference between these two terms.

In conclusion, it is, therefore, a nice, if somewhat ambiguous, definition of the term.

1.2.1.2 INCOSE definition

The next definition that is considered is taken from the world of systems engineering. The whole area of systems engineering advocates that requirements must be understood, so much so that the areas of requirements engineering and systems engineering are often confused. The next definition, therefore, is taken from the International Council on Systems Engineering (INCOSE):

> *A statement that identifies a system, product or process characteristic or constraint, which is unambiguous, clear, unique, consistent, stand-alone (not grouped), and verifiable, and is deemed necessary for stakeholder acceptability.* [7]

This definition is more complex than the previous one, although it also shares some characteristics with it. Consider the following:

- For the first time, this definition states explicitly that a requirement takes the form of a 'statement'.

- This definition then states that a requirement is a 'system, product or process characteristic or constraint'. This is, essentially, putting a better definition to the term 'thing' that was used in the first definition.
- The definition then goes on to qualify this statement, by stating that it must be 'unambiguous, clear, unique, consistent, stand-alone (not grouped) and verifiable'. This set of basic characteristics is discussed throughout this book, but, suffice to say, there exists a set of desirable characteristics for any requirement.
- The final part of the INCOSE definition states that the requirement is 'deemed necessary for stakeholder acceptability'. This is very much in the 'needed' rather than 'wanted' camp when compared with the OED definition.

In conclusion, therefore, a fuller definition of the term that adds more detail when compared with the OED definition.

1.2.1.3 Modelling definition

The next definition of the term 'requirement' is taken from the world of modelling. This is an important definition because modelling is core to the whole approach taken in this book. This definition is taken from the Unified Modelling Language definition as is:

> *a desired feature, property or behaviour of a system* [8]

This definition is certainly simpler than the INCOSE one and has a number of noteworthy aspects:

- The term 'desired' is used here, which really lies somewhere between 'wanted' and 'needed'.
- The term 'requirement' is defined in the context of being a property of a system rather than the context of the stakeholder.

In conclusion, therefore, another simple definition and, again, one that is very similar to both the INCOSE and OED definitions.

1.2.1.4 Standards definition

The next definition that is considered is taken from the standards community and, more specifically, the ISO:

> *need or expectation that is stated, generally implied or obligatory* [1]

Again, this may appear to be a simple definition, but there are some interesting points to be made here:

- For the first time, the term 'expectation' is used here as an alternative to 'need'. The main difference between these two terms is that one is more specific than the other. A need may need to be stated, whereas an expectation may be viewed as more assumed than stated.
- Following directly on from the previous point, this definition provides three qualifiers on the term 'requirement'. It states that a requirement may be 'stated' explicitly, 'generally implied' implicitly or 'obligatory', in that it is mandatory.

In conclusion, therefore, this definition has similarities with the previous definitions but adds more detail with the nature of the requirement.

1.2.1.5 Definition for this book

This list of definitions could go on indefinitely, but the list so far includes examples from a good cross-section of the systems engineering community. Therefore, based on these definitions, the definition that will be used for the purposes of this book is the following:

> *a requirement is the definition of a property of a system that is either needed or wanted by a stakeholder*

A few points to consider about this definition:

- The term 'definition' is used here rather than 'statement' because the term 'statement' may imply a text-only description of a requirement.
- The term 'property of a system' is used rather than 'thing' or a list of what it may or may not be. This was chosen because it leaves the definition of a property open and ties this property to the 'system'.
- The terms 'needed' and 'wanted' were kept as part of this definition, as both are seen as very important and are discussed later in this book.
- Everything about a requirement must be related back to a stakeholder, hence the use of the term here.

This definition has been deliberately held at a high level, but is one that will be expanded upon later in this chapter and throughout the remainder of this book.

1.2.2 Types of requirements

1.2.2.1 Basic types of requirement

This section looks at some of the different types of requirement that exist. There is no definitive list of different types of requirement, but there are some basic generic types that exist in most organisations.

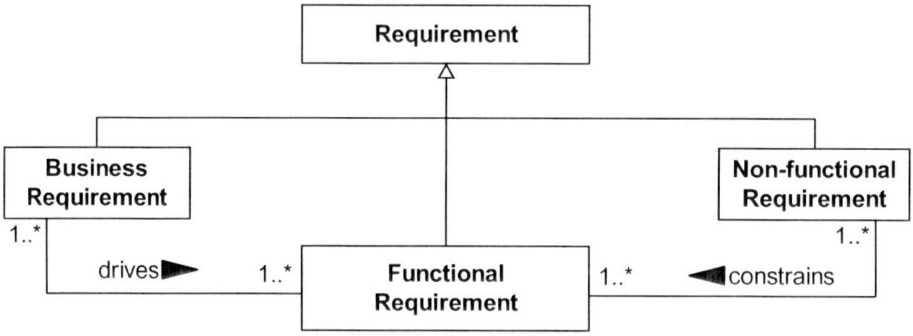

Figure 1.1 Different types of requirement

Figure 1.1 shows that there are three types of 'requirement': 'Business Requirement', 'Functional Requirement' and 'Non-functional Requirement'.

The 'business requirements' drive the 'functional requirements', which are themselves constrained by 'non-functional requirements'. The exact nature of these three main types of requirement is described in more detail in the following sections.

1.2.2.2 Business requirements

A 'business requirement' is a high-level requirement that encapsulates, in a concise manner, one single requirement that describes what the business is setting out to do. Business requirements drive the organisation, and everything that is done in the organisation should be traceable, either directly or indirectly, back to the business requirements.

Business requirements represent the fundamental nature of the business and may be based on the following:

- *Strategy, such as technical, expansion and planning for change strategies.* In today's business world, organisations are constantly evolving and mergers and acquisitions are becoming more common. As a direct result of these business changes, the strategy of the organisation may very well change and, hence, the business requirements will also change.
- *Markets, such as current and future.* Again, the world is forever turning and the markets associated with our businesses are constantly changing. This could be due to many reasons, such as the rapid advances made in technologies, changes in buying approaches or patterns because of recessions or even the success or failure of another business.
- *Mission, the business requirements should reflect the overall mission of the business.* In some cases, where the organisation has an explicit mission statement, it may be desirable to represent the mission as a business requirement. Mission statements can be very useful and very powerful when used appropriately and when they are based in the real world. Unfortunately, mission statements generally have become the butt of many jokes in business. This is a reputation that is, on the large, well deserved, as people will write any sort of nonsense, laminate it, attach it to a wall and call it a mission statement.

Business requirements always exist but must be identified and classified accordingly. It will be seen later how the use of 'contexts' will facilitate this.

1.2.2.3 Functional requirements

The concept of 'functional requirement' is one that is mainly intended when people use the generic term 'requirement' or 'user requirement'. A functional requirement is one that has some sort of perceivable impact on a stakeholder. Functional requirements often have verbs that imply some sort of strong activity, such as 'produce', 'perform', 'provide' and so on.

To many people, the main focus of a requirements exercise will be to capture the functional requirements and only the functional requirements. Functional requirements must be analysed in conjunction with the stakeholders and the non-functional requirements in order to provide a complete picture of the requirements and their context.

1.2.2.4 Non-functional requirements

The 'non-functional requirements', or constraints, in a system represent any requirement that will limit or constrain another requirement in some way. As has already been discussed, examples of non-functional requirements or constraints include quality requirements, implementation requirements, solution-specific requirements.

1.2.2.5 Differentiating between types of requirement

There is a lot of difference of opinion over which is the more important – functional or non-functional requirements. At the end of the day, this will depend on application or project. In some projects, the functional requirements may be very straightforward but there may be many very complex constraints that apply to them. Conversely, the system may have many complex functional requirements yet have few constraints. Whichever type of requirement is considered the more important is irrelevant and can be perceived as a waste of time and effort, but what is absolutely crucial is that the two of them are considered together. In many cases, functional and non-functional requirements are separated and treated independently and, in some situations, dealt with using different teams of people. It is essential that functional and non-functional requirements are related to each other, and this is one of the key aspects of requirements modelling that is discussed in this chapter.

It is important to be able to differentiate between the different types of requirement and looking through a list of generic requirements and simply asking which of the three broad types of requirements classifies each one can be a very worthwhile exercise. In many cases, the answer will be clear but try to do the same exercise with more than one person in the room, particularly if each person is a different stakeholder in the system. Each stakeholder may potentially have a set of requirements that is different from others, and the stakeholders may also interpret the same requirement in a different way, depending on their role.

In the situation where everyone agrees on the requirement classification, all is well and good. However, in some cases there will be a difference of opinion. In many of these cases it will indicate either of the following:

- The requirement is very ambiguous and means something completely different to each stakeholder.
- There are two separate types of requirement that exist.

As well as classifying requirements, it is also necessary to define or specify each requirement. Before we look at specifying requirements, however, it is necessary to look in more detail at the concept of a 'stakeholder'.

1.2.3 Stakeholders

One of the most important concepts to understand when dealing with requirements is that of the 'stakeholder'. A stakeholder is defined, for the purposes of this book, as:

> *the role of any person, organisation or thing, that has an interest in, or is influenced by, the system or project under development*

The concept of a stakeholder, and the different types of stakeholder, is fundamental to the approach described in this book, and there will be several discussions in later chapters concerning stakeholders.

To provide a quick overview of the various types of stakeholders, which may exist, consider the diagram in Figure 1.2.

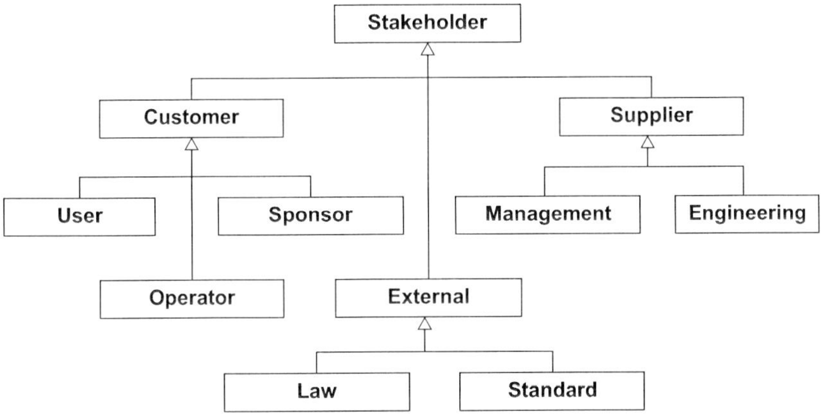

Figure 1.2 A simple stakeholder view

Figure 1.2 shows a simple classification hierarchy, or 'taxonomy' of different types of stakeholders. Notice that the first level of the taxonomy shows that stakeholders can be divided into three main categories:

- 'Customer' that describes all roles that may make use of the system or service or that the supplier needs to keep satisfied. In many cases, the supplier may have some influence over the customer and be in a position to discuss issues and agree compromises with them.
- 'External' that describes the roles that must be satisfied by the supplier but that cannot be negotiated with in any way. In other words, the supplier has no influence on external stakeholder roles.
- 'Supplier' that describes all the roles that are involved with the development and delivery of the product or service.

Each of the stakeholder roles will have one or more names associated with it, which describe who or what realises this role.

To discuss the nature of these roles, consider the example of a passenger airline.

For the 'Customer' stakeholder roles, the following will happen:

- The 'User' role will be realised by any number of passengers who are the end users of the airline. In almost all cases, this will be a list of names that can be specified to realise a single role. In other words, a single stakeholder role is not limited to a single name and may have any number of names associated with it in order to realise that role.
- The 'Operator' role will encompass all the employees or associates of the airline, ranging from ticket sellers, to the pilots and stewards, to the air-traffic controllers. Again, this is an example of many names that will realise a single role. Notice here that the role does not imply in any way how the role is realised. For example, the role of ticket sellers may be realised by a person (or people) or by an automated system. The stakeholder role does not care how it is realised and is only concerned with the definition of the role.
- The 'Sponsor' role will represent all the people or businesses that have a financial stake in the airline enterprise. Again, this will be realised by many names in reality. Notice here, however, that some of the names may represent organisations or businesses, rather than individual names.

This quite nicely makes the case for a single stakeholder role having any number of names associated with it. It is also possible, however, to have a single person that takes on any number of roles. Imagine a person who is a pilot for his or her main job. When working, this person takes on the role of a pilot, which will be a type of 'Operator'. When the person is not working, he or she may go on holiday and may travel on an aeroplane. In this situation, the same person that usually takes on the role of pilot (when working) will take on the role of passenger (which will be a special type of 'User'). Also, this person may have shares in the company, which means that the same person will sometimes take the role of shareholder, which will be a type of 'Sponsor'. In this example, it is possible for a single person to take on multiple roles, which is why it is essential to avoid individual or company names when it comes to specifying stakeholder roles.

For external roles, the following will happen:

- The 'Law' role will include all legal roles that may exist. Examples of this role include individual piece of legislation, treaties or agreements, national or international laws or any industry-specific mandatory requirements. This will be realised by all the laws and legislation that apply to the world of air travel.
- The 'Standard' role will encompass all the standards that may apply to the airline. This will range from international standards, to country-specific standards, to industry standards, right down to in-house processes, guidelines and work instructions.

Notice in the example here that none of the roles discussed in this section are realised by people, but are 'things'. In some cases these things may be pieces of paper (documents) or may even be virtual documents (such as work instructions on the internet).

For supplier stakeholder roles, the following will happen:

- The 'Management' roles will be realised by all the managers involved in the project. This will include, for example, project managers, operational managers, configuration managers.
- The 'Engineering' role will be realised by all the engineers involved in the project, such as software engineers, hardware engineers, electrical engineers, mechanical engineers.

In the example provided here, almost all of these roles will be taken by actual people rather than things.

It is important that the various stakeholder roles involved with the project can be identified properly and accurately because the requirements will differ for each stakeholder and they may even conflict with other stakeholder's requirements. To put it another way, each stakeholder has, potentially, its own individual set of requirements. A set of requirements from a particular stakeholder's point of view is known as a 'context', which must be understood for any effective requirements engineering. It is this idea of the 'context' that forms the basis for the approach described in this book.

1.2.4 Context modelling

Context modelling is a concept that is fundamental to requirements engineering, yet one that is often glossed over or, more worryingly, not considered at all. Understanding the context of a system is essential for understanding the requirements.

A context may be thought of, when using modelling, as the highest level view of the requirements of a system. The context identifies the boundary of the system and separates the outside world from the capabilities or requirements of the system. Everything inside the system boundary is represented as a 'use case' of the system. A use case, for the purposes of this book, is defined as:

a requirement that exists in a specific context

These use cases should consist of verb constructs. Everything outside the system boundary may be thought of as an external system, person or anything that represents the roles of something that has an interest in the system. As a point of interest, these are often, but not always, referred to as 'enabling systems'. For the purposes of this book, we will always use the term 'stakeholder'.

Each use case inside the system boundary must have some sort of relationship with an enabling system. By this, it is meant that the use case must yield some sort of perceived result to a stakeholder – of course, this perceivable result may be desirable, or in the case of the snail daters, wholly undesirable. This may be a direct relationship between the use case to the enabling system, or it may be an indirect relationship that is traceable through several use cases that are closely related together.

Each time one of these relationships exists, the relationship must cross the system boundary. Each time the system boundary is crossed, a 'system interface' is identified. This is a good way to identify the high-level system interfaces by using a context.

The point to be stressed here is that any system can be looked at from a number of points of view (contexts) and that invariably there will be conflicts between these different points of view. To look at a system from only a single point of view is quite simply a folly.

The idea of a context, however, is not just related to points of view based on stakeholders but may be based on other definitions, for example:

- *The level of hierarchy in a system.* It is possible to base the contexts of a system on the level of hierarchy, such as subsystem, assembly level, element level.
- *The life cycle stage.* The context of system may very well depend on which life stage the system is at. For example, the use cases may be very different depending on whether the system is in its 'concept' stage or 'disposal' stage of its life cycle – particularly where chemicals or radioactive substances are concerned!

The context is also crucial from the point of view of validating requirements. It is essential that each and every requirement can be validated, but this validation will differ depending on the context of the requirement. Therefore, the validation of requirements is achieved by validating the use cases in each context and, hence, the requirements may be demonstrated to be validated.

The main philosophy of this book is to consider a number of contexts and then to use them as a basis for truly understanding the requirements of a system.

1.2.4.1 Realising model-based systems engineering

The approach described in this book forms part of an overall model-based systems engineering (MBSE) approach. Systems engineering is concerned with realising successful systems [7] across the whole life cycle whilst keeping all of the stakeholder requirements satisfied. MBSE is an approach to systems engineering where the model forms the basis of all of the system artefacts – see Chapter 2 for more discussion on this point. This results in a number of benefits, such as the following:

- All system documents may be generated automatically from the model, resulting in simpler document maintenance, more consistent document content and drastically reduced documentation effort and time.
- A true model results in consistent and coherent views across the whole system architecture.
- When the model is correct, then traceability between all the system artefacts, across all life cycle stages, is inherent in the model.
- There is a true centralised 'master' model of the system that will exist, potentially, from all requirements through to low-level design, implementation and, indeed, cover the whole life cycle. The alternative is that the knowledge of the system is spread across multiple sources, such as heterogeneous models, spread sheets, documents.

- The whole issue of the maintenance of documentation is also made far simpler, as only the model need be maintained. All documentation may be generated directly from the model; therefore, there are no multiple sources of information to complicate configuration control.

To realise such an MBSE approach, it is necessary to address three fundamental aspects of MBSE, which are shown in Figure 1.3.

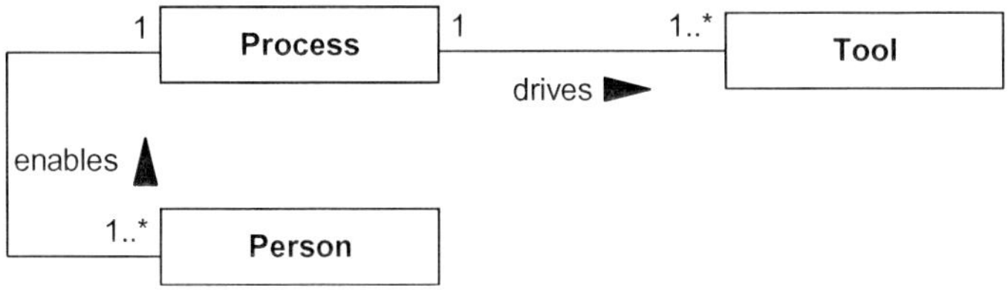

Figure 1.3 Fundamentals of MBSE

Figure 1.3 shows three main concepts that drive everything that is described in this book, which are the following:

- *Person, or people*. By this, we mean competent people who hold the necessary skills and experience to enable the process.
- *Process, or processes*. By this, we mean the approach taken to MBSE, which will be realised by process and methodology. This is enabled by the people and drives the tools.
- *Tool, or tools*. By this, we mean anything that can be employed to make the people enable the process more efficiently or effectively. This may include software-based tools, techniques, notations, standards, frameworks and even pen-and-paper systems!

These fundamentals will be described in great detail throughout this book, and the diagram in Figure 1.3 will also be seen again in subsequent chapters.

1.3 Overview of the book

This book is made up of a number of chapters that should provide a good overview of every aspect of the approach to context-based requirements engineering (ACRE) described in this book. After this introductory chapter, the rest of the book is made up of the following parts and chapters:

Part one is mainly concerned with the generic concepts of modelling and the notation that are used in the book.

- Chapter 2 introduces the concept of modelling including the basic types of modelling and how it applies to MBSE.
- Chapter 3 describes the notation that is used to realise the modelling concepts that were introduced in the chapter 2. The notation described there is kept at a

high level and only includes sections of the modelling language that are relevant to this book.

Part two is aimed at introducing the basic approach that forms ACRE.

- Chapter 4 introduces the concept of the ACRE ontology that describes the concepts and terminology that are used in the book.
- Chapter 5 introduces the ACRE framework that describes a number of views that are necessary for effective requirement modelling.

Part three of the book is aimed at realising the ACRE approach.

- Chapter 6 provides an in-depth case study that looks at examples of how all of the ACRE views may be realised.
- Chapter 7 looks at the pragmatic issues associated with people, processes and tools that are required to realise ACRE.

The book is also supported by a number of appendices.

- Appendix A, a summary of the modelling notation used in the book.
- Appendix B, some additional information regarding sequence diagrams that may be used by advanced modellers.
- Appendix C, an overview of the process modelling approach that is used in this book with specific examples showing its approach to competency assessment.

Overall, the layout of the book is intended to be as intuitive as possible.

References

1. *ISO 9001:2008. Quality management systems — requirements.* London, UK. International Standards Organisation; 2008
2. Alexander I., Stevens R. *Writing better requirements.* Boston, MA: Addison Wesley; 2002
3. Alexander I., BeusDukic L. *Discovering requirements – how to specify products and services.* London, UK. John Wiley; 2009
4. Holt J. *UML for systems engineering.* London, UK. 2nd edn. IET; 2005
5. Holt J., Perry S. *SysML for systems engineering.* London, UK. IET; 2008
6. OED. Available from http://oxforddictionaries.com/definition/requirement [Accessed 08 August 2011]
7. *INCOSE Systems Engineering Handbook – A Guide for System Life Cycle Processes and Activities. Ver. 3.0.* International Council on Systems Engineering (INCOSE); 2006
8. Runbaugh J., Jacobson I., Booch G. *The unified modelling language reference manual.* 2nd edn. Boston, MA: Addison Wesley; 2004

Part I

Introduction

Chapter 2
Modelling

I can't work without a model.

Vincent Van Gogh (1853–1890)

2.1 Introduction

This is a book concerned with model-based requirements engineering, MBRE, so it makes sense, before we consider the requirements engineering aspects of MBRE in Chapter 4 and subsequent chapters, to first consider the modelling aspects.

This chapter introduces the concept of modelling, discusses why modelling is so important in the context of the 'three evils' of engineering, establishes (somewhat appropriately in a book on requirements modelling) the basic requirements for modelling and concludes with a brief introduction to the systems engineering modelling language (SysML) that is used as the modelling language throughout the rest of the book.

2.2 Why we model

It is vital to understand the reasons why we must model things. To justify the need for models, three simple examples are considered. These examples aim to justify modelling in general terms and not simply with regard to requirements engineering, which demonstrates the need for flexibility when modelling – indeed, the three examples used here are not directly related to requirements engineering. These examples taken from Reference 1 were created by Grady Booch, the master of modelling and co-developer of the Unified Modelling Language (UML) on which the SysML is based. They are based on Booch's doghouse, house and office block examples.

2.2.1 The kennel (doghouse)

For the first example of modelling, consider building a kennel, a small house where dogs can spend time without being exposed to the elements. The basic requirement is to keep the dog happy. Therefore, the kennel should be waterproof and large enough to fit the dog inside. The entrance must be larger than the dog itself. The inside should also be large enough for the dog to be able to turn around in order to leave. Dogs are particularly bad at walking backwards, which makes this last point crucial. Finally, the dog should be comfortable enough to sleep in the kennel and thus some bedding or cushions may be in order (Figure 2.1).

Figure 2.1 The kennel (doghouse)

If you were to build this kennel, it would be wise to have the following basic skills and resources:

- *Basic materials such as timber and nails*. The quality is not really that important because the structure is to be built for a dog. Old pieces of wood around the house or a trip to a hardware store will suffice.
- *Payment*. The money needed to pay for the kennel would be your own, but is unlikely to be a large outlay. In terms of your personal income, it would be a fraction of a week's salary – perhaps the cost of a social evening out.
- *Some basic tools such as a hammer, a saw, and a tape measure*. The tools need not be of very good quality, provided they get the job done.
- *Some basic building skills*. You need not be a skilled craftsman, but basic hand-to-eye co-ordination would be an advantage.

These skills and resources relate directly to the concepts of person, process and tool discussed in Chapter 1. The *person* who builds the kennel must have the necessary skills and basic carpentry knowledge (*competence*) along with materials and equipment (the *tools*) to carry out the job (execute the *process*). If you have these skills and resources, you would probably end up with a kennel that is functional and in which the dog would be happy to shelter from the rain.

If the kennel is somewhat less than functional and the dog is not very happy with its new accommodation, you can always start again (after destroying the one built earlier) and try a bit harder, learning from past mistakes. It would also be possible to destroy the kennel and then deny all knowledge of ever having built one in the first place, thus avoiding embarrassment later. Alternatively, you can get rid of the dog and buy a less demanding pet such as a tortoise, as there is no need to build a kennel for an animal that carries its own accommodation on its back. After all, the dog is in no position to argue or complain.

This is Booch's first example of modelling: the kennel or, for our trans-Atlantic readers, the doghouse.

2.2.2 The house

Consider now, maybe based on the resounding success of your kennel, that you are planning to build a house for your family. This time the requirements would be somewhat different. There should be adequate space for the whole family to sit and relax, and adequate number of bedrooms to sleep. A kitchen, maybe a dining room and one or more bathrooms and toilets would also be required. As there will be more than one room, some thought should be given to the layout of the rooms in terms of their location in the house and to one another (Figure 2.2).

Figure 2.2 The house

If you were to build a house for your family, you would (hopefully) approach the whole exercise differently from that of the kennel:

- You would have to start with some basic materials and tools, but the quality of these resources would no doubt be of a higher concern than those used for the kennel. It would not be good enough to simply drive down to a local hardware store and pick up some materials because the quantity required would be far greater and it would not be possible to guess, with any degree of accuracy, the amount of materials required.
- Your family would also be far more demanding and vocal than the dog. Rather than simply guessing your family's requirements, it would be more appropriate

to ask them their opinions and perhaps get a professional architect in to listen to and discuss their needs.

- Unless you have built many houses before, it would be a good idea to draw up some plans. If you were hiring skilled craftsmen to do the job, you would certainly have to draw up plans in order to communicate your requirements to the builders. These plans may require some input from an architect so that they achieve a standard that may be used effectively by the people who will be building the house.
- The house would also have to meet building regulations and require planning permission. This may involve contacting the local council or government representative and possibly applying for permission to build the house. This, in turn, would almost certainly involve submitting plans for approval before any work could be started.
- The money for the construction of the house would probably be yours, and therefore you would have to monitor the work and ensure that the people involved stick to the plans in order to get the job done in time and within budget and to meet your family's original requirements. The scale of the financial outlay is likely to be in the order of several years' salary, and would probably be borrowed from a bank or building society and would thus have to be paid back, regardless of the outcome of the project.

Again, we can relate these skills and resources to the concepts of person, process and tool discussed in Chapter 1. A number of *people* involved, such as builders, architects and electricians, must have the necessary skills and knowledge (the *competence*) along with materials and equipment (the *tools*) to carry out the job (i.e. to execute the various *processes*). However, their skills don't just relate to building or wiring a house, but must also include the skills needed to work and communicate with other craftsmen and the customer in order to understand the requirements of the job they do. The processes they follow are necessarily more complex, since legislation, such as building and wiring regulations, that governs how they do their job has to be followed.

If the house turns out not to suit the requirements, the consequences would be more serious than in the case of the kennel. The house cannot be knocked down and constructed again as in the case of the kennel, because considerably more time and money are spent in the construction of a house. Moreover, you cannot simply get a less demanding family (in most cases), and living with the consequences of failure is not worth thinking about.

This is Booch's second example: the house.

2.2.3 The office block

Taking the two building projects that have been discussed so far even further, imagine that your ambition knows no bounds and that you decide to build an entire office block (Figure 2.3).

Consider the resources that would be required for the construction of the office block:

Figure 2.3 The office block

- It would be infinitely stupid to attempt to build an office block by yourself.
- The materials required for building an office block would be in significantly larger quantities than that used in the construction of the house. The materials would be bought directly from the source and may even need to be brought in from specialist suppliers, perhaps even from different counties or countries.
- You will probably be using other people's money and thus the requirements for the building will probably be their requirements. Moreover, their requirements may change once you have started building the office block.
- More permissions are required to build an office block than a house and many more regulations must be considered. Consider, e.g. the environmental conditions that the office building would have to meet – the building must not block anyone's light, it may be required to blend in with its surroundings, or it may have to conform to the so-called 'carbon footprint' legislation.
- You will have to carry out extensive planning and be part of a larger group that is responsible for the construction of the office block. Many teams from different areas of work (builders, plumbers, electricians, architects etc.) will be involved, all of whom must intercommunicate.

Unsurprisingly, these points again relate to the concepts of the person, process and tool. On a project of this size, different kinds of persons are involved in carrying out much more complex processes using far more (and more specialised) tools and materials than is the case when building a house. If you get the right teams involved and enjoy a degree of luck, you will produce the desired building.

If the project does not meet the investor's requirements, you would face severe repercussions, including no further work and the loss of reputation.

This is Booch's third example: the office block.

2.2.4 The point

The three examples from Booch may seem a little strange and somewhat trivial at first glance; however, there is a very serious and fundamental point behind all of these.

Nobody in their right mind would attempt to build an office block with basic do-it-yourself (DIY) skills. In addition, there is the question of resources, and not only in terms of the materials needed. To build an office block, you would need the knowledge to access the necessary human resources (including people such as architects, builders and crane operators), plenty of time and plenty of money.

It's quite strange and worrying that many people approach for building a complex system with the skills and resources of a kennel builder, without actually knowing whether it is a kennel, house or office block that is to be built. When contemplating any complex system, you should assume that it would be, or has the potential to turn into, an office block building. Do not approach any project with a 'kennel' mentality. If you approach a project as if it were an office block and it turns out to be a kennel, you will end up with a very well made kennel that is the envy of all canines. If, however, you approach a project as if it were a kennel and it turns out to be an office block, the result will be pure disaster!

One of the reasons why it is so easy to misjudge the size and complexity of a project is that, in many cases, many elements of the system will not be tangible or comprehensible. Consider, e.g., a smartphone. Who can tell, simply by looking at these increasingly ubiquitous devices, what components it is made of? And even if the device is dismantled in order to understand the hardware used in its construction, what about its functionality? Yes, you will expect it to function as a phone and a camera and probably also as a web browser and a music player. But what else? With the tens of thousands of applications available for the more popular models, it is impossible to know what else the smartphone is capable of, without powering up the device and looking at what is installed and trying the applications out to understand what they do. Also, what about the infrastructure that is needed to support the creation and delivery of such applications, let alone that needed to allow the phone to be used for its most fundamental purpose of making phone calls? How complicated is the infrastructure? Who runs it? How is it paid for? The fact is that all projects that involve complex systems will have an intangible element about them, whether it is a control system, a process, or any thing else.

The important term that is used here is 'complexity' and not size, as size is not necessarily a reflection of the complexity of a system. The next section discusses complexity in more detail.

2.3 The three evils

Projects fail and disasters occur for many reasons. However, there are three underlying reasons why things go wrong, the 'three evils' of *complexity*, a *lack of understanding* and *communication* issues.

2.3.1 Complexity

The concept of complexity is illustrated in two ways – one that emphasises the importance of relationships and the one that uses a brontosaurus to visualise the nature of the evolution of complexity.

For the first example, consider five boxes that represent five requirements for a system, as shown in Figure 2.4(a). Each of these requirements may very well be understood by someone who is reading the figure, but this does not necessarily mean that the system they describe and what it does or is capable of doing is understood.

Consider Figure 2.4(b); it is quite clear that this figure is more complex than the previous one, although nothing has changed in the requirements but only the relationships between them.

Consider now Figure 2.4(c); it is, again, obvious that this figure is more complex than its predecessor and far more complex than the first.

In fact, the more the relationships are added between the requirements, the higher the complexity. More and more lines can be drawn onto this figure and the complexity will increase dramatically, despite the fact that the complexity of each of the five requirements has not increased.

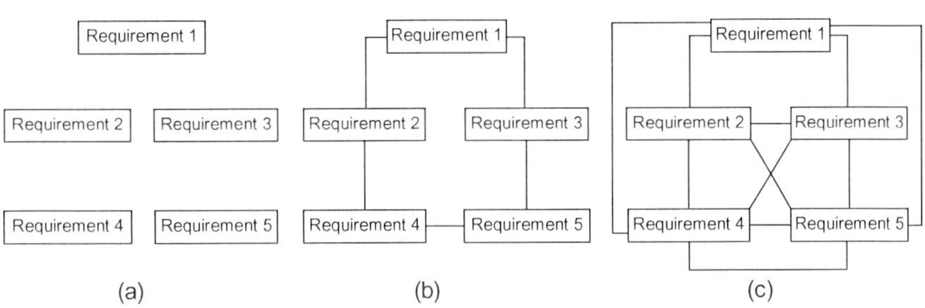

Figure 2.4 Complexity manifesting through relationships

The point here is that just because someone understands each requirement for a system does not mean that the system itself is understood. Undoubtedly, part of the complexity of a system manifests itself simply in the large number of requirements that are needed to describe it. However, much of the complexity resides in the relationships *between* the requirements. Unfortunately, many organisations approach the understanding and management of requirements in the fashion depicted in Figure 2.4(a) or (b), treating requirements either as simple lists or as lists with relatively few links between them and making the mistake of thinking that just because they have requirements stored in requirements tools, they must have a good understanding of the requirements. They never analyse the requirements of the model in a way that reveals the complexity residing in the connections; they never get to Figure 2.4(c).

The second way of illustrating the complexity is through the concept of the 'brontosaurus of complexity'. In this slightly bizarre analogy, complexity is visualised as a brontosaurus, in that the complexity of a system at the outset is represented by the dinosaur's head and, as the project life cycle progresses, this complexity increases (travelling down the neck), increases even further (through the belly) before reducing and finally ending up at the tail of the brontosaurus.

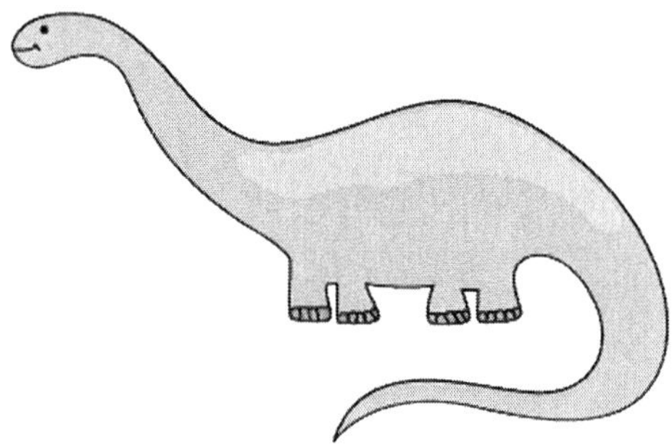

Figure 2.5 A brontosaurus

This analogy fits with the shape of the brontosaurus, i.e. '*thin at one end, much much thicker in the middle, and then thin again at the far end*' [2]. The perceived complexity of a project is almost always low to begin with, but balloons during the analysis of the project, as all the requirements and the constraints are gathered and their relationships begin to be understood; the project is well and truly in the 'belly of the brontosaurus'. As further analysis is carried out, with requirements put into context and scenarios developed to validate the requirements, the requirements become optimised and the project heads towards the 'tail of the brontosaurus'. By applying the brontosaurus of complexity analogy, it is shown that one must go from the head (initial ideas and requirements) to the tail (optimised, analysed requirements that can be used to develop the system), but it is impossible to do this without going through the belly of the brontosaurus.

Consider the situation when a project is at the head of the brontosaurus; this may be visualised as the illustration in Figure 2.4(a). Here we have the requirements for the project, perhaps captured as a simple list in a requirements management tool, or more likely at the start of a project, spread across a number of source documents, emails and the like. These requirements correspond to having the *Source Element View* (and perhaps the *requirements definition view*) defined. These views (and the other views needed when undertaking full MBRE) are described in detail in Chapter 4 onwards.

As the complexity of the project increases and we move down the neck of the brontosaurus, our understanding of the requirements increases, as shown in

Figure 2.4(b). In fact, the more the relationships are added between the requirements (and hence the more interactions between them), the closer to the belly of the brontosaurus we actually get. It is in the belly of the brontosaurus where much of the requirements work is achieved, where we put requirements in context, establishing and understanding the relationships between them and ensuring that they are validated. This would correspond to having the *Context Definition View, Requirement Context View* and some *Validation Views* defined. These views are discussed in detail in Chapter 4 onwards.

Many projects fail because they never leave the belly as their requirements are not addressed properly; in some cases, the projects are left even higher up in the neck. If a project stays in the head or neck, there is a great danger of the system being oversimplified and the complexity inherent in the system is never uncovered until it is too late. If the project remains in the belly, however, the complexity has been realised, but it has not been managed effectively. Our goal is to reach the tail of the brontosaurus with all the views necessary for effective realisation of MBRE, with views such as full *Validation Views* and *Traceability Views* being added to that produced so far.

Unfortunately, when a project is in the belly of the brontosaurus, it may seem to the project personnel that the world is at an end and that there is no understanding of the project as a whole. Successfully developing a system is about being able to see the brontosaurus as a whole and that there is life after the belly.

In a final twist to this analogy, there is a major difference between complexity and the brontosaurus. Complexity is difficult to visualise, but definitely exists in any system, whereas a brontosaurus is easy to visualise (Figure 2.5) but never actually exists in any system (it was demonstrated in 1974 that the brontosaurus was actually the body of the Apatosaurus and the head of the Camarasaurus).

2.3.2 Lack of understanding

Key to the successful delivery of any system is an understanding of the requirements for the system. Failing to arrive at a clear understanding of the requirements will cascade throughout the whole project. It is widely accepted that mistakes made during the early stages of a system's life cycle cost many times more to fix during later stages of the life cycle [3], so it makes sense to get things right as early as possible and this means having the requirements for the system stated in as concise and unambiguous a fashion as possible. Consider the following examples of lack of understanding that can affect requirements:

- When considering requirements, many engineers talk about the *customer* requirements or the *user* requirements, treating each of these terms as being the same, but the reality is that they are very different. Failure to understand the difference between stakeholders or, even worse, failing to consider all the stakeholders involved is one way in which lack of understanding manifests itself when looking at requirements. The customer (or probably more correctly the *sponsor*) and the user represent two of the biggest stakeholder groups, but they are not the only stakeholders involved. Sometimes it is the lack of understanding of the requirements of the smallest stakeholder in a project that can lead to major problems. Consider, e.g., the snail darter, a tiny little fish that is native to the waters of Tennessee. In 1975, the snail darter was placed on the newly created endangered species list. Even though the snail darter is a tiny little mud fish, not much bigger than a paperclip, it created a giant controversy. In *Tennessee Valley Authority (TVA) v. Hill 437* U.S. 153 (1978), the United States Supreme Court affirmed an injunction that stopped the Tennessee Valley Authority from completing the Tellico Dam project because the proposed project would destroy the snail darter's habitat. Even though over $100 million had been spent by 1978, and the dam was substantially finished, this decision led to over 12 months delay and required a change in legislation before the dam could be completed and put into operation [4].

- As discussed earlier in this chapter, when looking at the brontosaurus of complexity, requirements are often managed as relatively simple lists, leaving us in the neck of the brontosaurus. These requirements may be well written and complete in terms of descriptions and sources, but show a lack of understanding of the *context* in which they apply, their *relationships* to other requirements, the *stakeholders* that have an interest in each requirement, and the scenarios that need to be considered in order to *validate* the requirements.
- As discussed earlier, many organisations hold requirements in requirements management tools. They then assume that because the requirements are managed in this way and that because they know how to use all the useful functionality provided by such a tool, they understand the requirements. Being able to use a word processor to produce all your documentation does not mean that the documentation is well written. It just means that you can prepare the documentation easily. Similarly, having your requirements held in a requirements management tool does not mean that you understand them, but many organisations make this mistake, resulting in requirements that are not fully understood. Managing requirements is not the same as understanding them.

Of course, these examples are merely a representation of some ways that a lack of understanding can manifest itself when dealing with requirements, there are many other places where problems may occur.

2.3.3 Communication

The last of the three evils is the problem in communication or, more correctly, *ineffective* communication. The richness and complexity of human communication is what separates humans from other species. One of the earliest recorded examples of project failure is that of the Tower of Babel, as described wonderfully by Fred Brookes [5]. The Tower of Babel was started as a very successful project and the first few stages of the project went off without a hitch and the project was running on schedule, within budget and meeting all the project requirements. However, one of the key stakeholder's requirements was not considered properly, which was to cause the downfall of the project. When the stakeholder intervened, the communication between project personnel was effectively destroyed.

Communication problems may occur at any level of an organisation or a project:

- *Person-to-person level.* If individuals cannot communicate on a personal level, there is little hope for the success of the project. This may be because people speak different languages and have different technical backgrounds personalities.
- *Group-to-group level.* Groups, or organisational units within an organisation, must be able to communicate effectively with one another. These groups may have different technical backgrounds, such as hardware group and software group, or the groups may span boundaries, such as management and technical group, and marketing and engineering group. Such groups often use language specific to themselves, making inter-group communication difficult.
- *Organisation-to-organisation level.* Different organisations speak different languages – each organisation has its own specific terms for different concepts and an industry-specific terminology as well. When two organisations work in a customer–supplier relationship, the onus is often on the supplier to speak the customer's language so that communication can be effective, rather than on the customer to speak the supplier's language. After all, if the supplier won't make an effort to speak the customer's language, it is quite likely that they would not remain customers for very long.
- *System-to-system level.* Even non-human systems must be able to communicate with one another. Technical systems must be able to communicate not only with technical systems but also with financial systems, accountancy systems, environmental systems etc.
- *Any combination of the above.* Just to make matters even more confusing, any combination of the above communication types is also possible.

These problems, therefore, lead to ambiguities in interpreting any sort of communication, whether it is a spoken language or an application-specific or a technical language. When considering requirements, we have the coming together of many of the different organisational and project levels as described earlier in this chapter, and there is, therefore, a strong need for a common language. If we cannot communicate effectively, what hope is there that we will be able to understand the requirements for a project?

2.3.4 The vicious triangle

Having established that these three evils exist, matters become even worse. Each of these three evils does not exist in isolation, but will feed into one another. Therefore, unmanaged complexity will lead to a lack of understanding and communication problems. Communication problems will lead to unidentified complexity and a lack of understanding. Finally, a lack of understanding will lead to communication problems and complexity.

The three evils, therefore, form a triangle of evils that is impossible to eliminate. In fact, the best that one may hope for is to address each of these evils in its entirety and try to minimise each evil. In fact, it is important to always consider these three evils at all stages of a project, and it is, perhaps, even more important to consider them when looking at the requirements for a system. If the complexity of the requirements is not managed, or not communicated properly so that the real requirements for the system are not properly understood, what chance is there of delivering what the customer wants?

2.4 Modelling

Having considered why modelling is so important and having discussed the three evils, it is time to introduce a definition of modelling and some of the principles associated with it that build the foundations for the rest of the book. Modelling allows us to identify complexity, aid understanding and improve communication – all important if requirements are to be understood properly and the core skills needed when adopting a model-based approach to requirements engineering.

2.4.1 Defining modelling

To understand modelling, it is important to define it. We define a model as a *simplification of reality* [1]. It is important to simplify reality in order to understand the system. This is because as human beings we cannot comprehend complexity.

If a model is a simplification of reality, there are many things that may be thought of as a model:

- Mathematical models, such as equations that represent different aspects of a system, which may be used as part of a formal analysis or proof.
- Physical models, such as mock-ups, which may be used to provide a picture of what the final system will look like or may be used as part of a physical simulation or analysis.
- Visual models, such as drawings and plans, which may be used as a template for creation or the basis of analysis.
- Text models, such as written specifications, which are perhaps the most widely used of the tools at our disposal. Regarding text as a model can be initially quite surprising, but the moment we start describing something in words, we simplify it in order to understand it. One of the ACRE views, the Requirement Description View, is often heavily text-based, yet it is still part of the requirements model.

This is by no means an exhaustive list, but it conveys the general message.

It is important to model so that we can identify complexity, increase our understanding and communicate in an unambiguous (or as unambiguous-as-possible) manner.

To model effectively, it is essential to have a common language that may be used to carry out the modelling. Many modelling approaches exist, such as graphical, mathematical and textual, but regardless of the approach, there are a number of requirements for any modelling language:

- The choice of model
- The level of abstraction

- Connection to reality
- Independent views of the same system

Each of these requirements is discussed below.

2.4.2 The choice of model

The choice of model refers to the fact that there are many ways to solve the same problem. Some of these will be totally incorrect, but there is always more than one correct way to solve the problem at hand. Although all these approaches may be correct, some will be more *appropriate* and, hence, more correct for the application. For example, if one wants to know the answer to a mathematical equation, several approaches are open: you may simply ask someone else what the answer is, you may guess the answer, you may apply formal mathematical analysis and formulae or you may enter the equation into a mathematical software application. All may yield a correct answer, but the most appropriate approach will be application dependent. If you were merely curious to know the answer to the equation, guessing or asking someone else may be more appropriate. If, however, the equation were an integral part of the control algorithm for an aeroplane, something more formal would be more appropriate.

It is important to have a number of different tools in order to choose the most appropriate solution to a problem, rather than just relying on the same approach every time.

Therefore, one requirement for any modelling language is that it must be flexible enough to allow different representations of the same information in order to choose the optimum solution. That is, a modelling language should be flexible enough to allow different *views* of the information to be produced. The concept of views is central to the ACRE and is covered in detail in Chapter 5, with example views from a case study system described in Chapter 6.

2.4.3 The level of abstraction

Any system may be considered at many different levels of abstraction. For example, an office block may be viewed as a single entity from an outside point of view. This is known as a high level of abstraction. It is also possible to view a tiny part of the office block, e.g. the circuit diagram associated with a dimmer light switch in one of the offices. This is what is known as a low level of abstraction.

In addition to high and low levels of abstraction, it is also necessary to look at many intermediate levels of abstraction, such as each floor layout on each level, each room layout, the lifts (or elevators), the staircases and so on. Only by looking at something at high, low and in-between levels of abstraction, it is possible to gain a full understanding of a system.

Therefore, the second requirement for any modelling language is that a system must be able to be represented at different levels of abstraction. This is equally true for requirements. We want to capture very high level of requirements, such as 'building office block' to requirements at a very low level of abstraction such as

'The elevator buttons should be ACME XYZs', with consistency and traceability between these different levels.

2.4.4 Connection to reality

It has already been stated that by the very nature of modelling we simplify reality, and there is a very real danger that we may oversimplify reality to such a degree that the model loses all connection to reality and hence all relevant meaning.

One type of modelling in which it is very easy to lose the connection to reality is mathematical modelling. Mathematical modelling is an essential part of any engineering endeavour, but it can often be seen as some sort of dark art as very few people possess a sufficient knowledge as to make it usable and, indeed, many people are petrified of maths. Consider the example of the mathematical operation of differentiation that is used to solve differential equations. As every school child knows, differentiation can be applied in a quite straightforward manner to achieve a result. What this actually means in real life, however, is another matter for discussion. Differentiation allows us to find the slope of a line which when taken at face value, and particularly when at school, can be viewed as being utterly meaningless. To take this example a little further, we are told that integration is the *opposite* of differentiation (what is the opposite of finding the slope of a line?) which turns out to be the measurement of the area underneath a line. Again, when first encountered, this can be viewed as being meaningless. In fact, it is not until later in the educational process when studying subjects like physics or electronics that one realises that finding the slope of a line can be useful for calculating rate of change, velocity, acceleration etc. It is this application, in this example, that provides the connection to reality and hence helps communicate 'why'.

The third requirement for any modelling language, therefore, is that it must be able to have a strong connection to reality and hence meaningful to observers who, in many cases, should require no specialist knowledge, other than an explanation, to understand the meaning of any model.

2.4.5 Independent views of the same system

Different people require different pieces of information depending on who they are and what their role is in the system. It is essential that the right people get the right information at the right time. Also, for the purpose of analysing a system, it is important to observe a system from many different points of view. For example, consider the office block again where there would be all different types of people who require different information. The electricians require wiring diagrams and not colour charts or plumbing data; the decorators require colour charts and not wiring diagrams and so on.

There is a potentially very large problem when considering things from different points of view, and this is consistency. Consistency is the key to creating a correct and consistent model, and without any consistency, it is not possible to have or demonstrate any confidence in the system.

The fourth requirement for any modelling language is that it must allow any system to be looked at from different points of view and that these views *must* be consistent. The model-based systems engineering approach taken by ACRE is based on this concept of a number of different, yet consistent, views of requirements. None of these views taken alone is sufficient to describe the requirements for a system; it is the information captured in the *set* of views that gives a full and consistent model of the requirements.

2.5 The SysML

The previous sections have identified many requirements for modelling and have justified why modelling is so important. To model successfully, a common language is required, and the common language which is used throughout this book is the SysML.

The primary audience for SysML is systems engineers. SysML is defined on the website of the OMG (the Object Management Group – the US-based industry standards body who manages and configures the SysML) as 'a general-purpose graphical modelling language for specifying, analyzing, designing, and verifying complex systems that may include hardware, software, information, personnel, procedures, and facilities' [6].

The SysML originates from an initiative between the OMG and the International Council on Systems Engineering (INCOSE) to adapt UML, the Unified Modelling Language (which was originally intended for use by software engineers), for systems engineering applications.

Its rather tangled history begins in March 2003 when the OMG issued the 'UML for Systems Engineering Request for Proposal'. In response to this RFP, there were several major drafts of SysML, which differed in terms of the content and concepts. At one point, the team developing the SysML split into two groups, each led by a different CASE (Computer-Aided Systems/Software Engineering) tool vendor, both groups produced separate specifications for the standard, one of which was then chosen as the basis for the official SysML by the OMG. This version was made official in September 2007. For a full description of the SysML language, including details of its history, see Reference [7].

2.5.1 The two aspects of a SysML model

Any SysML model has two aspects: 'structural' and 'behavioural'. It is vital that both of these aspects exist for any system, otherwise the system will not be fully defined.

The structural aspect of the model shows the 'things' or entities in a system and the relationships between them. It is crucial to remember that the structural aspect of the model shows 'what' the system looks like and 'what' it does, but not 'how'. The structural aspect of the model may be thought of as a snapshot in time of any system.

The behavioural aspect of the model shows 'how'. It demonstrates how a system behaves over time by showing the order in which things happen, the conditions under which they happen and the interactions between things.

Each of these two aspects of the model must exist, even in models that are developed purely to aid in understanding of requirements where, as will be seen in Chapter 4 onwards, it is important to define both requirements (i.e. structure) and show how they can be validated (behaviour).

Understanding the need for these two aspects is absolutely fundamental to understanding the SysML, which has five diagrams that can be used to realise the structural aspect of the model and four diagrams that can be used to realise the behavioural aspects, although not all nine diagrams are necessary when modelling requirements. Structural and behavioural modelling and the five of the nine SysML diagrams that are used when modelling requirements are the subject of the next chapter.

2.6 Summary

This chapter has introduced, appropriately for a book on model-based requirements engineering, the concept of *modelling*. We define a model as a *simplification of reality* created in order to help us understand the system under consideration because as human beings we cannot comprehend complexity. A model, whether mathematical, physical, textual or visual, allows us to address the so-called *three evils*, giving us techniques to identify *complexity*, increase our *understanding* and *communicate* in an unambiguous (or as unambiguous-as-possible) manner.

To model effectively, it is essential to have a common language that fits for the purpose, allows any system to be represented at different levels of abstraction, is easy to understand without requiring specialist knowledge and can be used to allow the system to be looked at from different points of view. The modelling language that is used to realise the MBRE approach described in this book is the SysML and it is described in more detail in the next chapter.

References

1. Booch G., Rumbaugh J., Jacobson I. *The Unified Modeling Language User Guide*. 2nd edn. Boston, MA: Addison-Wesley; 2005
2. Elk A. *The brontosaurus sketch*. Monty Python's Flying Circus. BBC TV; 1974
3. Pressman R. *Software Engineering: A Practitioner's Approach: European Adaptation*. Maidenhead: McGraw-Hill; 2000
4. http://en.wikipedia.org/wiki/Snail_darter_controversy. *Snail darter controversy*. [Accessed February 2011]
5. Brookes F. P. *The Mythical Man-Month*. Boston, MA: Addison-Wesley; 1995
6. http://www.omgsysml.org. This is the official website for the SysML language. [Accessed February 2011]
7. Holt J., Perry S. *SysML for Systems Engineering*. London: IET; 2008

Chapter 3

The SysML notation

> Concern for man and his fate must always form the chief interest of all technical endeavours. Never forget this in the midst of your diagrams and equations.
>
> *Albert Einstein (1879–1955)*

3.1 Introduction

As discussed in Chapter 2, the modelling notation used throughout this book and recommended for use in the ACRE is the Systems Modelling Language (SysML). Although Chapter 2 has discussed modelling in general, this chapter looks at the SysML notation in some detail. Any SysML model has two aspects, the structural and the behavioural. Both aspects must exist and must be consistent. There are nine SysML diagrams to realise the model, five diagrams that can be used to realise the structural aspect (Figure 3.1) and four diagrams that can be used to realise the behavioural aspect (Figure 3.2).

In brief, the five SysML structural diagrams used are as follows:

- A block definition diagram describes the system hierarchy and system/component classifications. It allows properties and behaviour of system elements to be modelled.
- The package diagram is used to organise the model.
- The internal block diagram describes the internal structure of a system in terms of its parts, ports and connectors.
- The parametric diagram represents constraints on system property values, allowing engineering analysis models to be produced as well as defining complex constraint relationships that can be used in verification and validation of requirements.
- The requirements diagram captures requirements hierarchies and requirements derivation. It allows a requirement to be related to model elements that satisfy or verify the requirement.

For a full description of these diagrams, see References 1 and 2.

The four SysML behavioural diagrams used are as follows:

- The use case diagram provides a high-level description of functionality that is achieved through interaction among systems or system parts. It shows requirements in context.

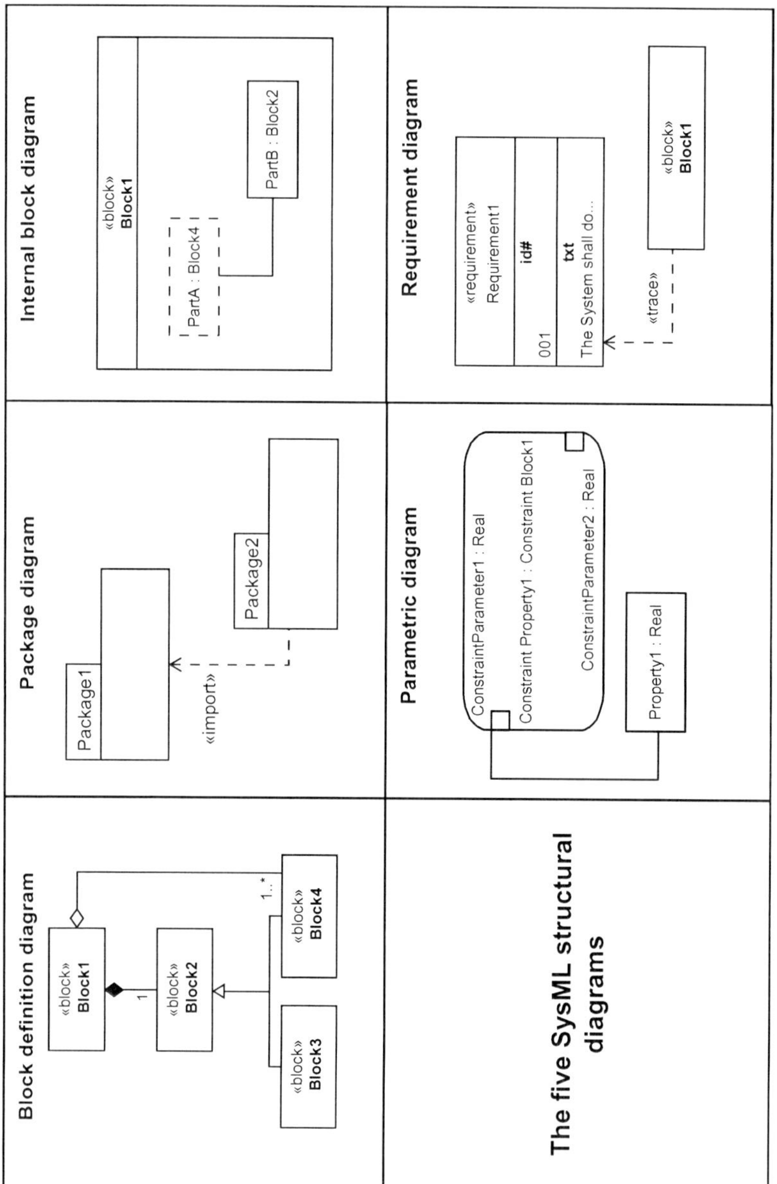

Figure 3.1 SysML structural diagrams

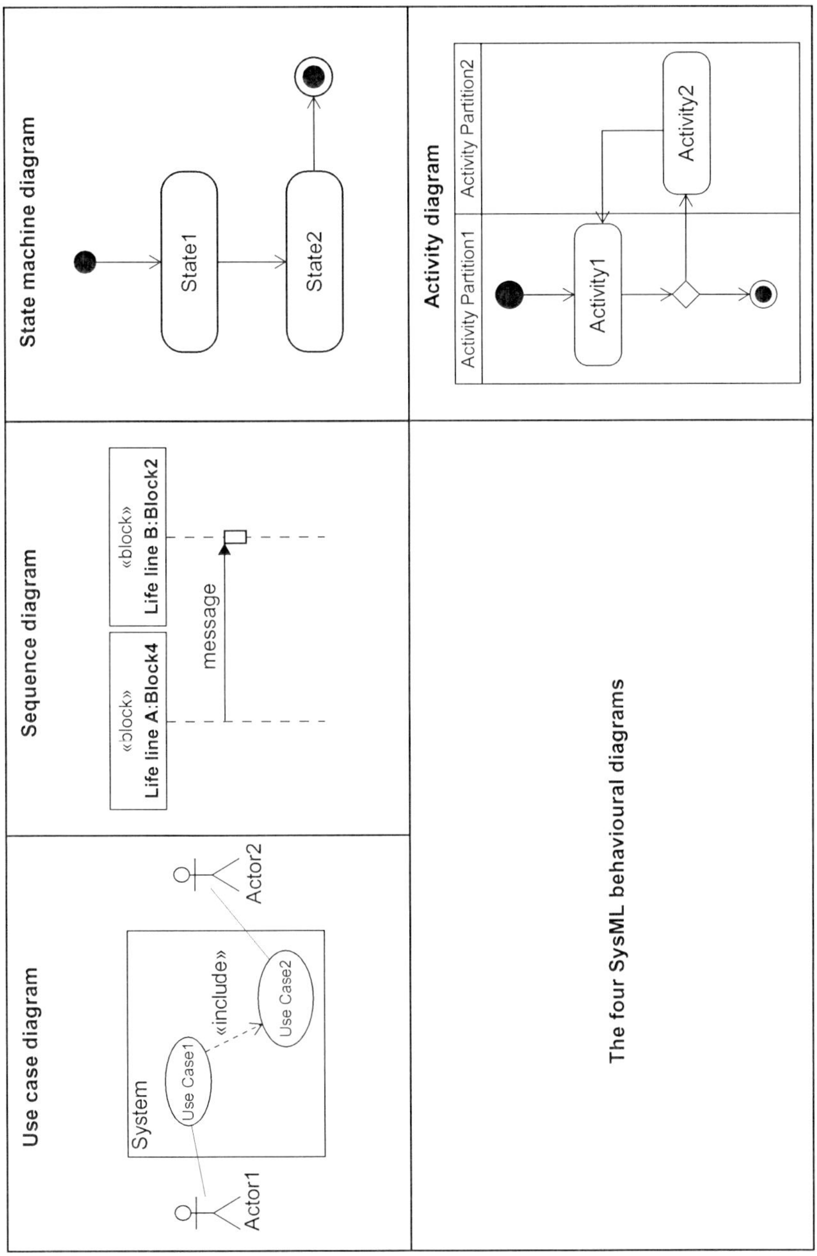

Figure 3.2 SysML behavioural diagrams

- The sequence diagram represents the interaction between collaborating parts of a system, allowing the messages between system elements to be modelled in order to capture behavioural scenarios.
- The state machine diagram describes the state transitions and actions that a system or its parts perform in response to events.
- The activity diagram represents the flow of data and control between activities and is often used to model the internal behaviour of the operations of system elements.

For a full description of these diagrams, see References 1 and 2.

Although SysML contains nine diagrams, only five are used for the ACRE approach to model-based requirements engineering described in this book:

1. Block definition diagram
2. Requirements diagram
3. Parametric constraints diagram
4. Use case diagram
5. Sequence diagram

This introductory section discusses structural and behavioural modelling and concludes with a look at *stereotypes*, a standard way of extending the SysML. The rest of the chapter covers each of the five diagrams used by ACRE and uses an example to illustrate them. For each diagram there are four sub-sections: an *overview* that introduces the diagram, sections on *notation* and *usage* and a *summary* section. The notation for the five diagrams is summarised in Appendix A.

3.1.1 Structural modelling

The structural aspect of a model shows the 'what' of the system. It identifies and defines system *elements*, defines their *properties*, identifies their *behaviours* and identifies the *relationships* between the system elements.

There are five structural diagrams in SysML, as shown in Figure 3.1. Three of these diagrams are used in the ACRE approach to model-based requirements engineering: the block definition diagram, parametric constraints diagram and requirements diagram. Similar concepts apply to all five of these diagrams, and in order to illustrate the concepts behind structural modelling, one of the five structural diagrams is used to show some simple examples. The diagram chosen is the block definition diagram because this forms the backbone of the SysML.

There are two basic elements that make up a block definition diagram: the 'block' and the 'relationship'. A block represents a type of 'thing' that exists in the system being modelled. A relationship relates together one or more blocks. Blocks should be named using nouns or noun phrases, and relationships should have names that form sentences when read together with their associated blocks.

Figure 3.3 is a block definition diagram showing two blocks, 'Block1' and 'Block2'. Blocks are represented by rectangles and each block must have a name, which is written inside the rectangle. The rectangles also contain the *stereotype* «block». Stereotypes are discussed further in section 3.1.3.

Figure 3.3 Example blocks

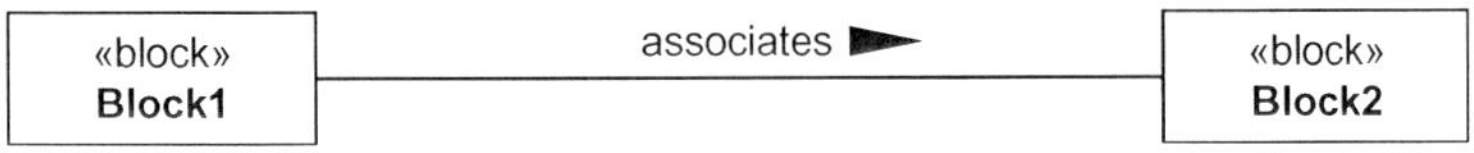

Figure 3.4 Representing relationships

Figure 3.4 shows how to represent a relationship between two blocks. The type of relationship shown is known as an 'association' and is a general type of relationship that relates together one or more blocks. The association is represented by a line that joins two blocks, with the association name written on the line and a direction marker showing which way the relationship should be read. This diagram reads: 'Block1' associates 'Block2'.

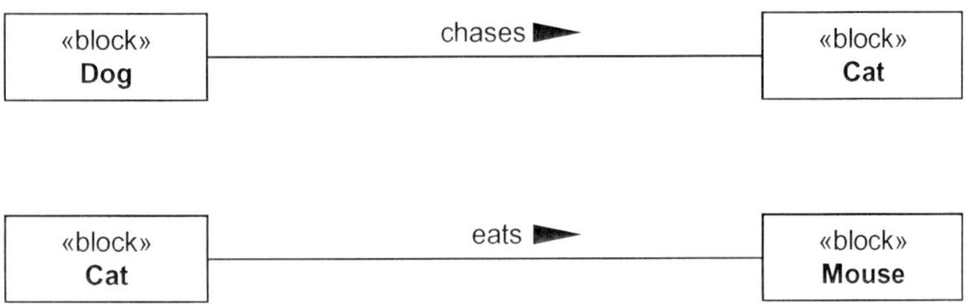

Figure 3.5 Examples of blocks and associations

Figure 3.5 shows two more examples. The top part of the diagram reads: there are two blocks 'Dog' and 'Cat', where 'Dog' chases 'Cat'. Likewise, the lower part of the diagram reads: there are two blocks 'Cat' and 'Mouse', where 'Cat' eats 'Mouse'.

An important point concerning blocks is that blocks are *conceptual* and do not actually exist in the real world. There is no such thing as 'Cat', but there do exist many examples of 'Cat'. A block represents a grouping of things that look and behave in the same way, as, at one level, all examples of 'Cat' will have a common set of features and behaviours that may be represented by the block 'Cat'. What this block really representing is the blueprint of 'Cat', or the essence of 'Cat'.

The direction of an association is shown by a small filled-in triangle, as shown in the examples in Figures 3.4 and 3.5 and in the top part of Figure 3.6. The diagram reads: 'Dog' chases 'Cat' and definitely not 'Cat' chases 'Dog'. If, in the system being modelled, dogs chase cats *and* cats chase dogs, then a second association running in the opposite direction would need to be added to the diagram. This illustrates an important point that there can be *any number* of relationships between two blocks.

The directionality of an association can be augmented through the use of *role names* on each end of the association, as shown in the bottom part of Figure 3.6.

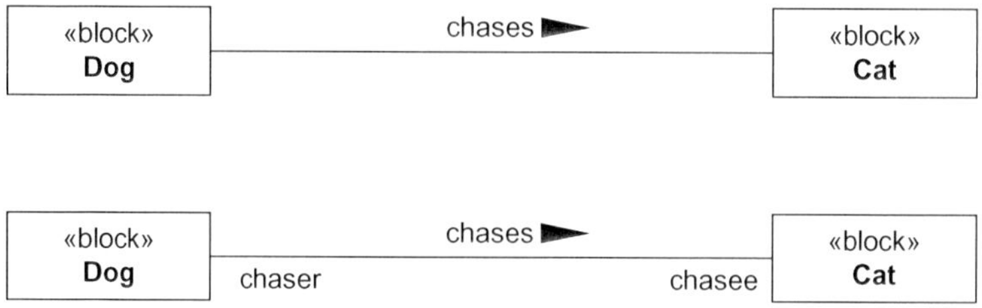

Figure 3.6 Showing direction with role names

In this case, the two roles that have been defined are 'chaser' and 'chasee'. The diagram can now be read as follows: 'Dog', in the role of 'chaser', chases 'Cat', in the role 'chasee'. The association and role names should be chosen so that they make the diagrams as unambiguous as possible.

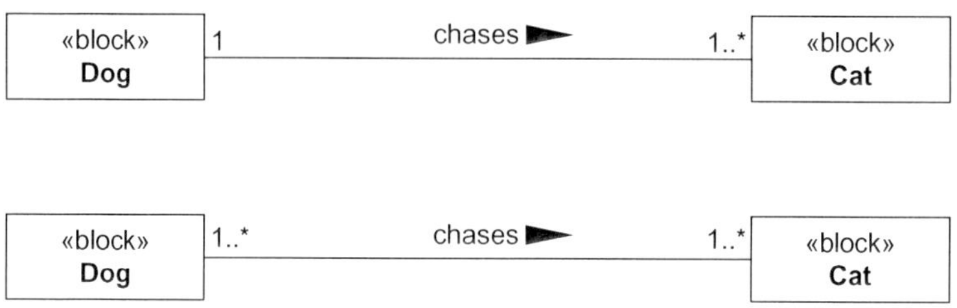

Figure 3.7 Showing multiplicity

In addition to showing the direction of an association between blocks, SysML allows the *multiplicity* of the blocks involved in the association to be shown. The top part of Figure 3.7 shows that each 'Dog' chases one or more 'Cat'. Although the number is one, it does not necessarily indicate that there is only one dog, rather each 'Dog' in the system chases one or more 'Cat'. The multiplicity at the other end of the 'chases' association states '1..*', which means 'one or more' or somewhere between one and many. Therefore, the association shows that each 'Dog' chases one or more 'Cat', and that each 'Cat' is chased by only one 'Dog'.

The lower part of the diagram shows a case where the multiplicity has been changed, which changes the entire meaning of the model. In this case, the diagram is read as follows: one or more 'Dog' chase one or more 'Cat'. This could mean that a single dog chases a single cat, a single dog chases any number or a herd of cats, or an entire pack of dogs chases a herd of cats.

The appropriate multiplicity depends, of course, on the system being modelled. As the model evolves from requirements to detailed design, the multiplicities may require change. Common multiplicities are the following:

0..1	indicates an optional value
1	indicates exactly 1
0..*	indicates any number, including zero
*	same as 0..*
1..*	indicates 1 or more

In fact, any subset of the non-negative integers can be used to specify a multiplicity. If, e.g., a dog can chase between one and three cats, the multiplicity at the 'Cat' end of the association in Figure 3.7 would be changed to '1..3'.

Modelling blocks and the relationships between them is an essential part of structural modelling, but the amount of detailed information for each block is very low. If the block 'Cat' represents *all* cats that look and behave in the same way, it is important to show how a cat looks or behaves. This information is added to a block by using *properties* and *operations*.

The common properties of a 'Cat' are represented using SysML *properties*. It is very important to limit the number of general properties that are identified to only those that are relevant, because it is very easy to get carried away and over-define the amount of detail for a block. For this example, suppose that we wish to represent the features 'age', 'colour', 'favourite food' and 'weight' on the block 'Cat'. These features are represented on the block as properties, one for each feature, as shown in Figure 3.8.

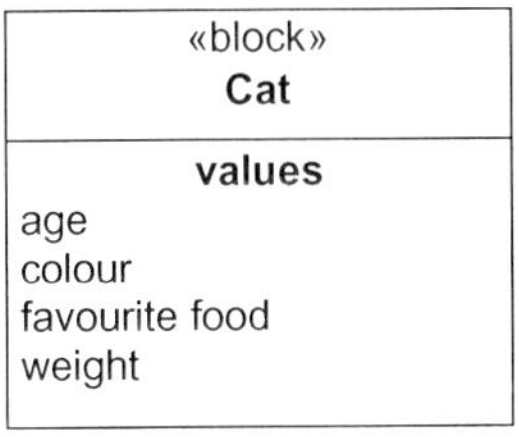

Figure 3.8 Properties of the 'Cat' block

When modelling, it is possible to add more detail, such as the type, default value and so on. As properties represent features of a block, they are usually represented by nouns and they must also be able to take on different values. For example, 'colour' is a valid property, whereas 'red' would not be, as 'red' would represent an actual value of a property rather than a property itself. It is possible for 'red' to be a property, but this would mean that the property would have a Boolean type (true or false) to describe a situation where we would only be interested in red cats and not any other type. SysML allows three different kinds of property to be defined: *value, part* and *reference* properties. As can be seen from the *values* heading in the compartment in Figure 3.8, the four properties shown for the 'Cat' block are all value properties. The differences between these three kinds of properties are discussed in detail in section 3.2.2.

Properties provide a mechanism to represent features of a block – to show what it looks like – but they do not describe what the block *does*. This is shown using *operations*. Operations show what a block does, rather than what it looks like, and are thus named using verbs or verb phrases. An example is shown in Figure 3.9.

In the case of the block 'Cat', we have identified three things that the cat does, which are 'eat', 'sleep' and 'run'. Operations are represented in SysML by adding an *operations* rectangle to the block and writing the operation names within it.

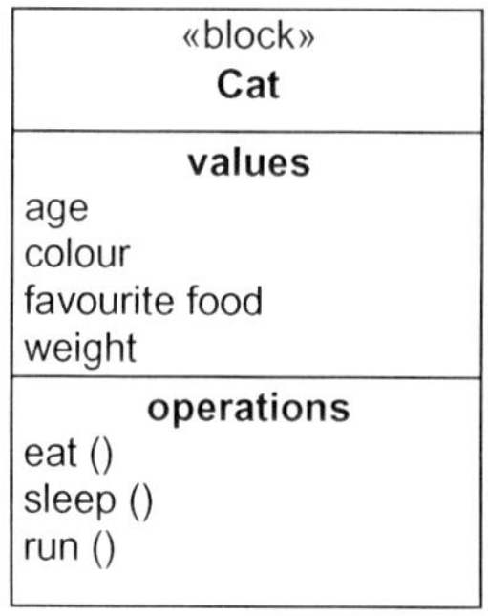

Figure 3.9 Operations of the 'Cat' block

The name of an operation is followed by two brackets. These are not optional and form part of the name of the operation. Thus, in Figure 3.9 the operation name is 'eat()' and *not* 'eat'. Extra detail can be added to operations as necessary, such as parameters and return values. For example, an operation that takes two integers, adds them and returns the result as an integer as follows:

```
add( a : Integer, b : Integer) : Integer
```

This illustrates the full form of an operation in SysML, namely:

```
operation name( parameter name : parameter type, ... ) : return type
```

An operation can have any number of parameters, as shown by the ', ...' in the general form earlier.

Although a number of properties and operations have been defined for the 'Cat' block in Figure 3.9, this does not mean that these properties and operations have to be shown on every diagram. Most SysML tools will allow the property and operation compartments to be turned on or off for a block, allowing a block to be shown at different levels of detail on different diagrams as has been done on Figures 3.7 and 3.9. As with everything in the SysML, only use as much detail as is necessary, rather than as much as is possible.

3.1.1.1 Adding more detail to relationships

Although properties and operations allow more detail to be added to blocks, SysML allows the modeller to add more detail to relationships by defining some special types that are commonly encountered in modelling. One of these types of relationship, the *association*, has already been discussed. This section covers the other two main relationship types that exist, namely the *aggregation* and the *specialisation/generalisation* relationships. For details of other types of SysML relationship, see References 1 and 2.

Aggregation allows emphasis to be placed on the 'whole/part' relationships between system elements. An example of aggregation is shown in Figure 3.10.

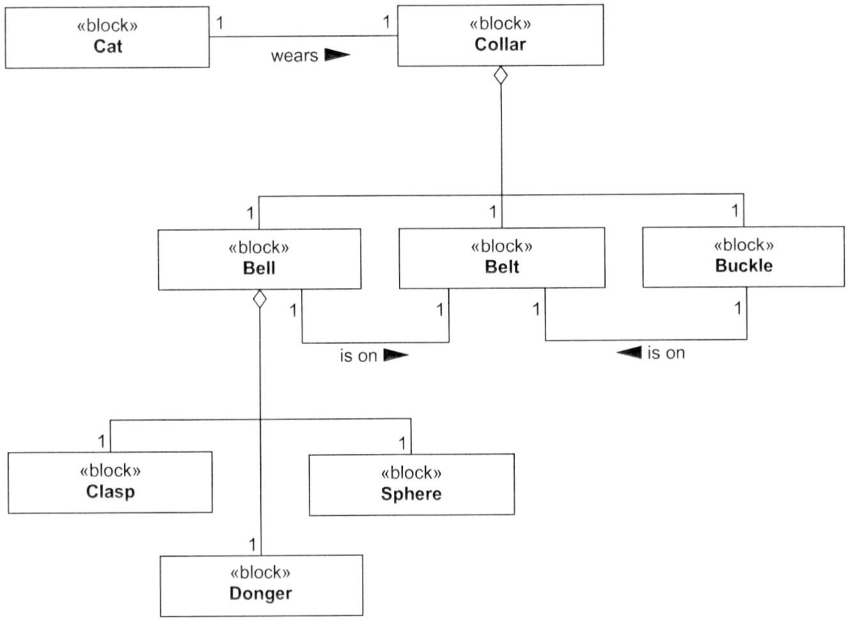

Figure 3.10 An example of aggregation

The diagram in Figure 3.10 makes use of aggregation to show the structure of the 'Collar' worn by a 'Cat'. The aggregation is shown by the use of a diamond at the 'whole' end of the relationship. The diagram could be read as follows: A 'Cat' wears a 'Collar' that is made up of a 'Bell', a 'Belt' and a 'Buckle'. The 'Bell' is on the 'Belt' and the 'Buckle' is on the 'Belt'. The 'Bell' is made up of a 'Clasp', a 'Donger' and a 'Sphere'. Like associations, the aggregation can also take multiplicities and role names. Aggregations can also be named like an association, although this is rare. When carrying out requirements engineering following ACRE, Figure 3.10 would be an example of a *System Context Definition View* (discussed in detail in Chapter 5).

A variant on aggregation also exists in SysML. This is known as *composition* and looks the same but with a black filled-in diamond. The difference between them is to do with uniqueness of ownership and is discussed in section 3.2.2. Figure 3.10 would probably be better drawn using composition rather than aggregation.

In addition to showing structural hierarchies using aggregation or composition, it is often necessary to model type hierarchies or taxonomies. SysML allows this through the use of the specialisation/generalisation relationship. An example of its use is shown in Figure 3.11.

Specialisation/generalisation is shown by a line with an unfilled triangular arrowhead at one end. 'Specialisation' refers to the case when a block being made is more special or is being refined in some way. Specialisation may be read as 'has types'. In the somewhat science fiction inspired types of lifeform shown in Figure 3.11, 'Lifeform' has types 'Animal', 'Vegetable' and 'Mineral'.

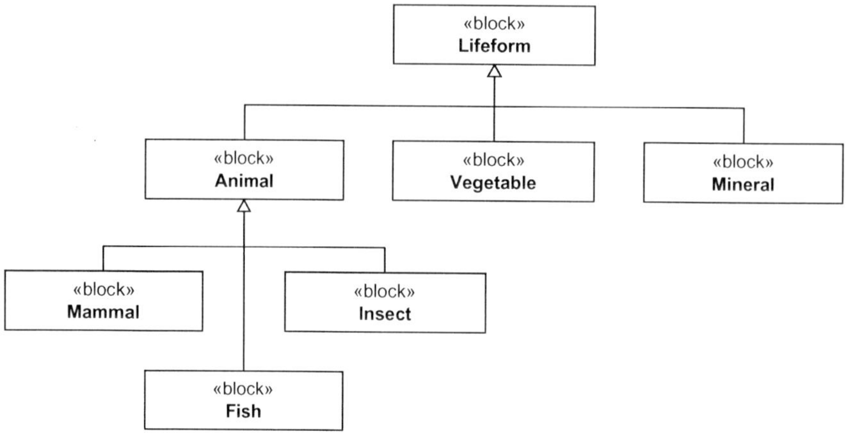

Figure 3.11 Example of specialisation/generalisation

If the relationship is read the other way around, then the triangle symbol is read as 'is a type of', which is a generalisation. Thus, in Figure 3.11, 'Mammal', 'Fish' and 'Insect' are types of 'Animal'. Therefore, if the relationship is read one way, the block becomes more special (specialisation), and if read the other way, the block becomes more general (generalisation). When carrying out requirements engineering following ACRE, Figure 3.11 would be an example of a *Stakeholder Context Definition View* (discussed in detail in Chapter 5).

Specialisation is used to show *child* blocks of a *parent* block. An important property of the specialisation/generalisation relationship is *inheritance*: child blocks inherit their appearance and behaviour from their parent blocks, but are different in some way in order to make them special. In SysML terms, this means that a child block will inherit any properties and operations that its parent block has, but may have additional properties or operations that make the child block special.

Consider the left-hand part of Figure 3.12. This shows that 'Cat' is a type of 'Mammal'. Taken together with Figure 3.11, a 'Cat' is therefore also a type of 'Animal' and a type of 'Lifeform'. A 'Cat' therefore inherits all the properties and operations of all its parent blocks. Given the information in Figures 3.11 and 3.12, a 'Cat' also has two operations, namely 'suckle young()' and 'breathe air()'.

A block can have additional properties and operations that are not found in any of its ancestor blocks (parent block, parent block of parent block etc.), and on the right-hand side of Figure 3.12 the 'Cat' block is shown with some additional properties that have been defined for 'cat'. The inherited operations from 'Mammal' are also shown for completeness. Remember that inheritance only runs, as is generally the case in life, from parent blocks to child blocks. Thus, defining the 'age' property in 'Cat' does *not* mean that it is inherited by its parent 'Mammal'. If, when modelling this lifeform-type hierarchy, it is found that 'age' is

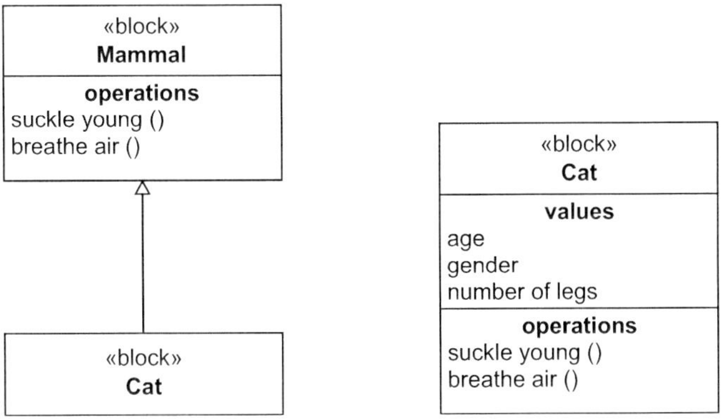

Figure 3.12 Inheritance

being added into types of 'Mammal', types of 'Fish' and so on, this is a sign that the property should be moved up the hierarchy to 'Animal' or perhaps even 'Lifeform' so that it is defined in one place and inherited by all the descendant blocks.

Further information on structural modelling and the notation used on block definition diagrams can be found in section 3.2 and in References 1 and 2.

3.1.2 Behavioural modelling

The behavioural aspect of a model shows the *how* of the system. It identifies the behaviour of the system at the *system* level, *between* system elements, *within* system elements and within *operations* of system elements.

There are four behavioural diagrams in SysML, as shown in Figure 3.13. Two of these diagrams are used in the ACRE approach to model-based requirements engineering: the use case diagram and the sequence diagram.

Although the SysML structural diagrams can be used to model any level of a system's structure, from high-level systems to low-level components, different SysML behavioural diagrams are typically used to model the behaviour of the system at the levels identified earlier. Rather than picking one of the diagrams for discussion, as has been done in the structural modelling section, this section discusses behavioural modelling more generally by relating it to system structure, considering consistency and looking how the various SysML behavioural models relate to typical engineering activities.

Figure 3.14 shows a typical and generic structural hierarchy for a system. There are many types of 'System'; 'Natural System' and 'Man-made System' are just two examples. A 'System' is typically made up of one or more 'Subsystems'. These can themselves be further broken down into one or more 'Assemblies' and then down into one or more 'Components'. Typically, there is interaction between elements at each of these levels: systems interact with systems, subsystems with subsystems and so on.

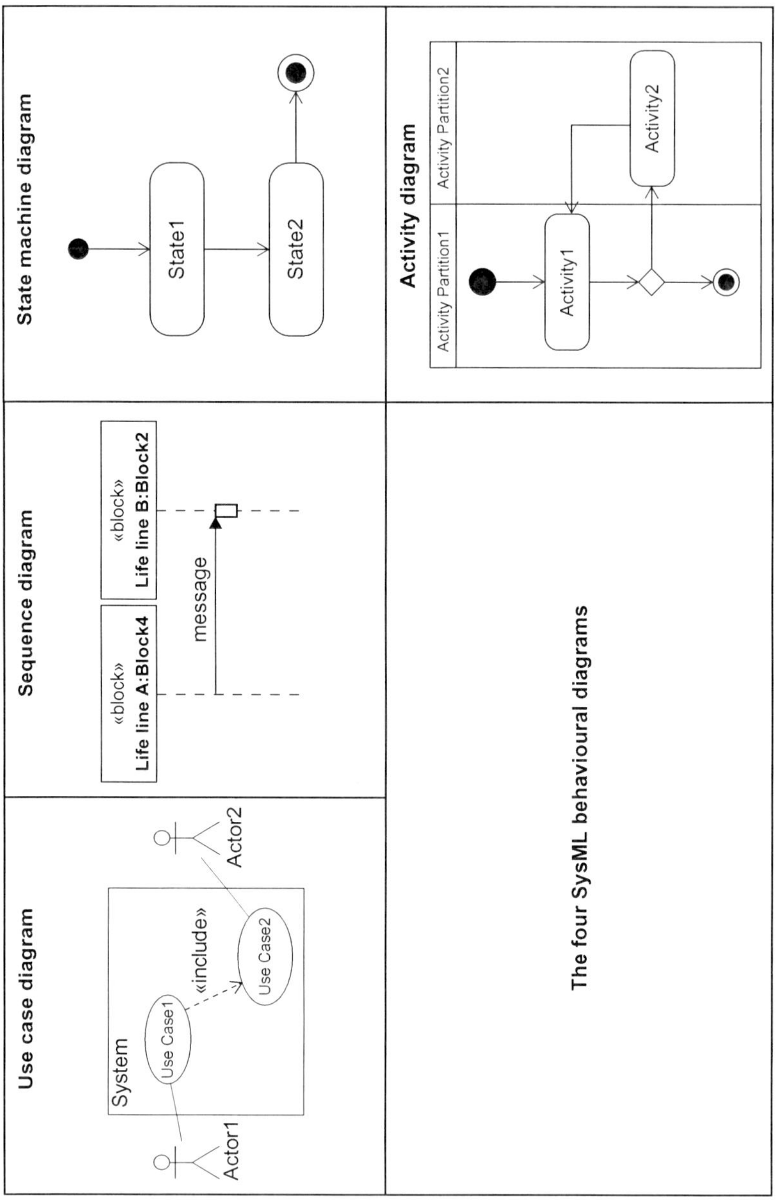

Figure 3.13 SysML behavioural diagrams

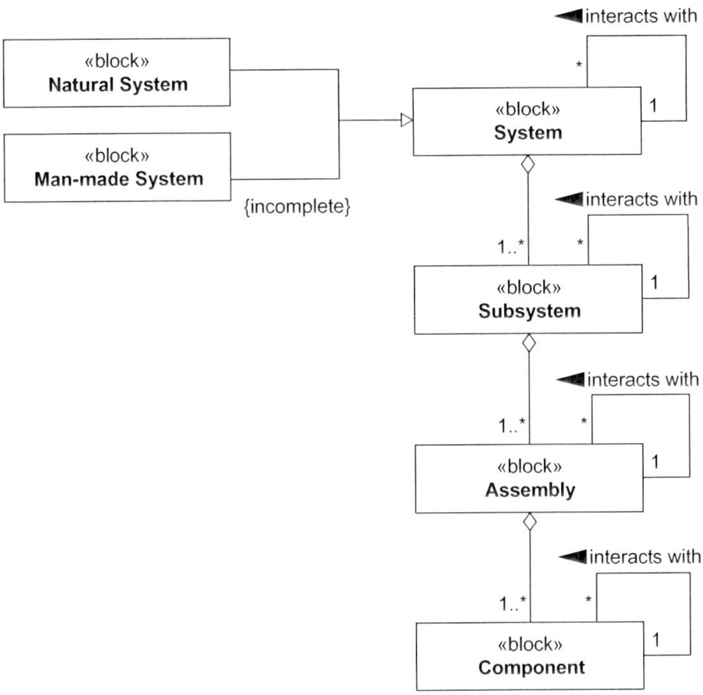

Figure 3.14 Typical structural hierarchy

The system hierarchy from Figure 3.14 is reproduced in Figure 3.15, which also shows the typical SysML behavioural diagrams that are used at each level as discussed below:

- At the system level, it is the highest level of behaviour that needs to be modelled, that of the *system itself* and the interactions between the system and stakeholders (which may, of course, themselves be other systems). Such behaviour is usually modelled using use case diagrams and sequence diagrams. Use case diagrams allow system requirements in context to be modelled and sequence diagrams allow scenarios for the use cases to be developed.
- At the subsystem level, it is the behaviour *between* system elements that is modelled, allowing the interfaces between subsystems to be explored from a behavioural perspective via typical scenarios showing how subsystems work together to achieve some goal. The sequence diagram is most often used to model behaviour at this level.
- At the assembly level the behaviour *within* an assembly is often of most interest, allowing exploration of how assemblies respond to events and under what circumstances they carry out their operations to be made. The state machine diagram is typically used at this level.
- At the component level, it is the low-level internal behaviour of a component that is of interest. This usually equates to modelling the behaviour of the *operations* of a component for which an activity diagram is used.

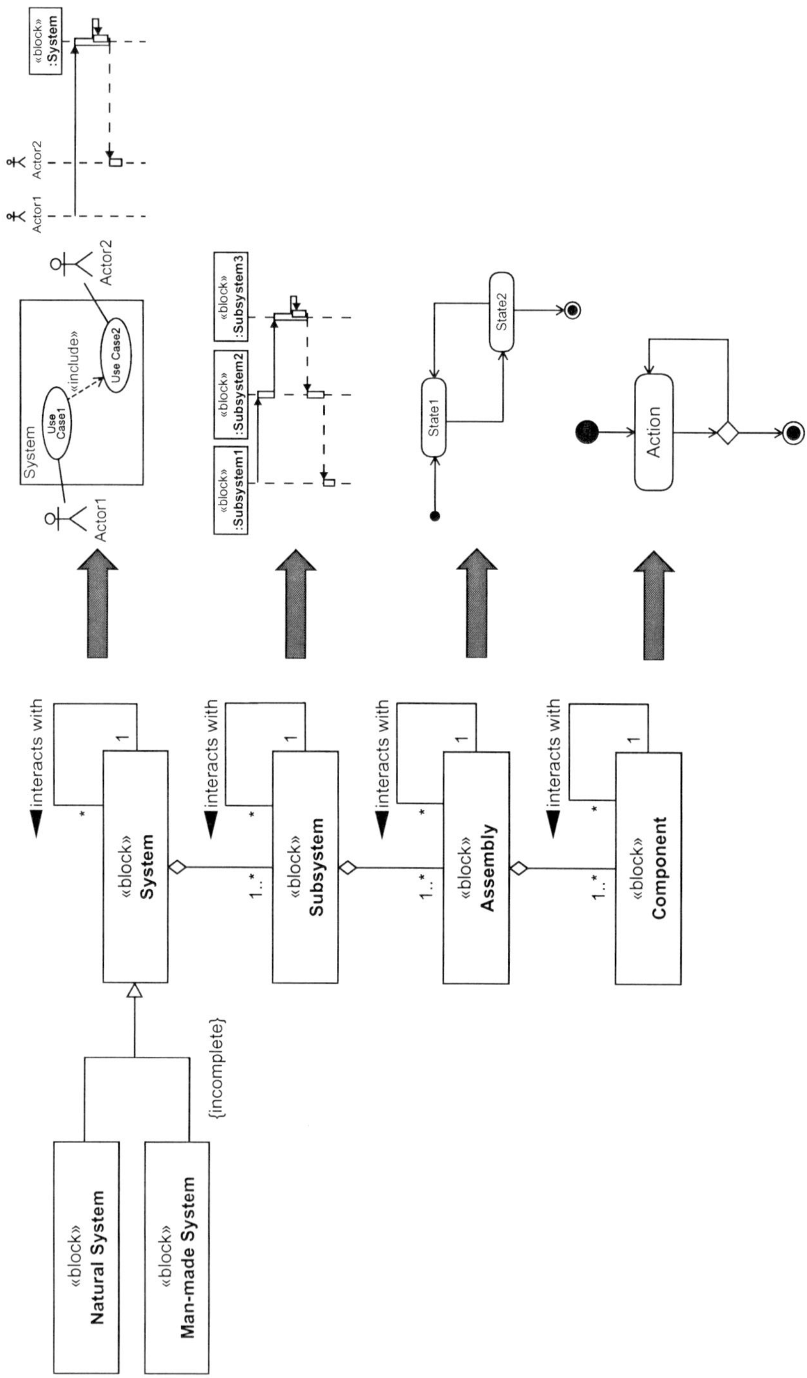

Figure 3.15 Relating behavioural diagrams to structural level

It is essential to remember while thinking about system hierarchies and behavioural modelling that one person's subsystem is another's system (and similarly for the other levels). This means that behavioural modelling is never as simple and clear cut as described earlier. Although the system developer may model subsystem behaviour using sequence diagrams, the subsystems may also be developed by other suppliers. For them, they will be considered as systems and therefore modelled using use case and sequence diagrams. Similar considerations apply across the whole hierarchy, which means that the behavioural diagrams used depend entirely on context.

The various structural and behavioural diagrams produced as part of a system model must, as discussed in Chapter 2, be consistent with one another. Consider Figure 3.16 which contains the same structural and behavioural diagrams as shown in Figure 3.15, but which has double-headed arrows at each level between the structural and behavioural diagrams and between the behavioural diagrams across each level. The double-headed arrows in Figure 3.16 show the kinds of consistency that must exist if the various SysML diagrams of a system are to give a *model* of the system rather than simply being a collection of *pictures*.

There has to be consistency between the structural and behavioural aspects of a system at the same level. For example, the systems appearing as life lines on a sequence diagram must exist as elements on a structural diagram, such as a block definition. At the component level, each operation of a component, modelled as a block, should have its behaviour modelled using an activity diagram; conversely, each activity diagram should correspond to the operation on a block representing a component.

Similarly, consistency between behavioural diagrams across system levels is also essential. For example, the messages between subsystems on a sequence diagram may correspond to events received by or the signals sent out by the state machine diagram for an assembly. These same messages between subsystems may correspond to the self-messages sent from a system's life line to itself on a sequence diagram at the system level.

In addition to considering behavioural diagrams with respect to the typical system levels at which they are used and in terms of the consistency that is essential to ensure the diagrams give a model of the system, it is also possible to consider them in terms of engineering activity (Figure 3.17).

The structural and behavioural diagrams of Figures 3.15 and 3.16 are shown in Figure 3.17, but this time with boxes indicating the type of engineering activity in which they are often used. Four types of activity are shown on the diagram:

- *Requirements analysis*, in which the main system elements are identified along with their requirements and validating scenarios. To fully understand the system-level requirements, it may be necessary to consider the structural aspects of the system at the next level of decomposition, namely that of the subsystem. The behavioural aspects at the subsystem level would also be investigated.

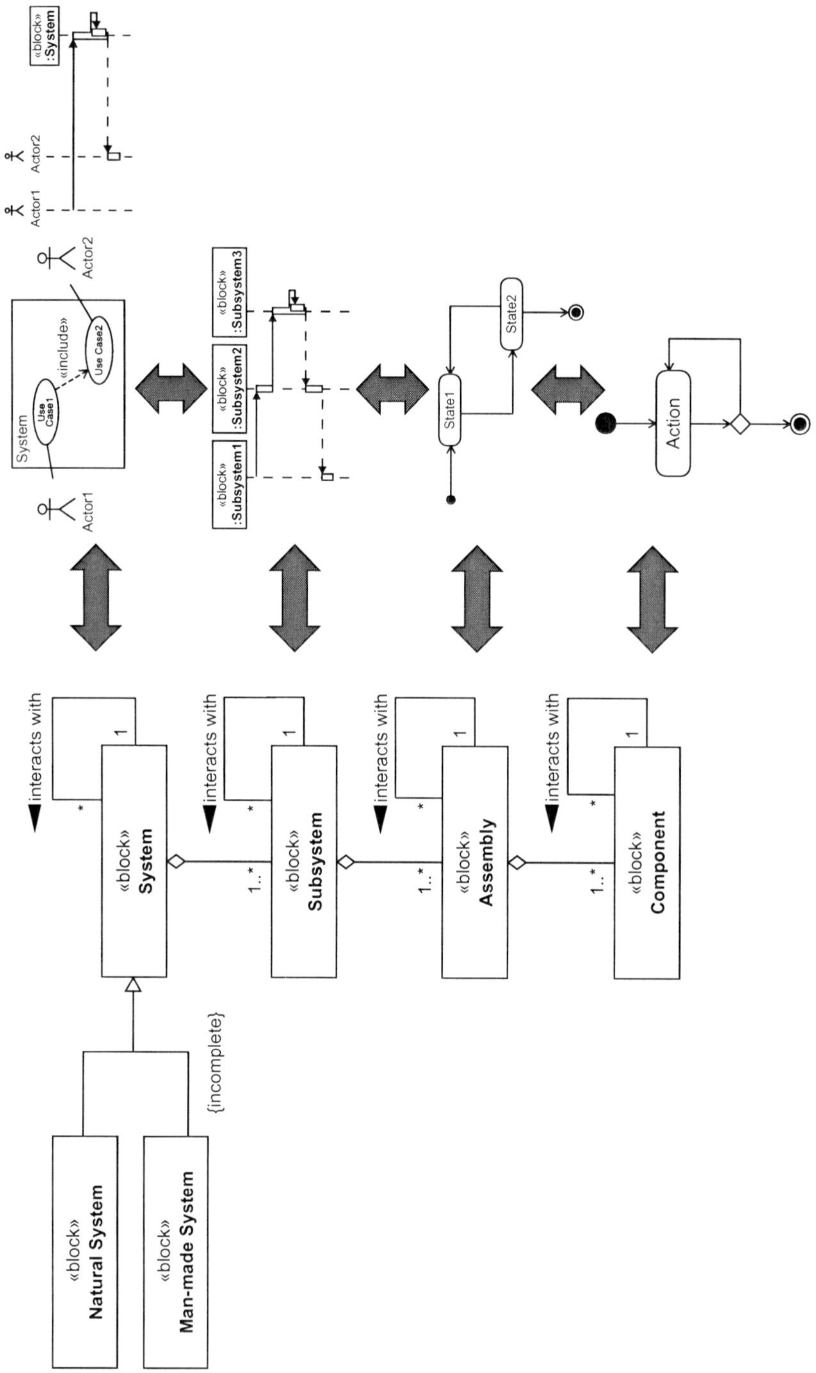

Figure 3.16 Structural and behavioural consistency

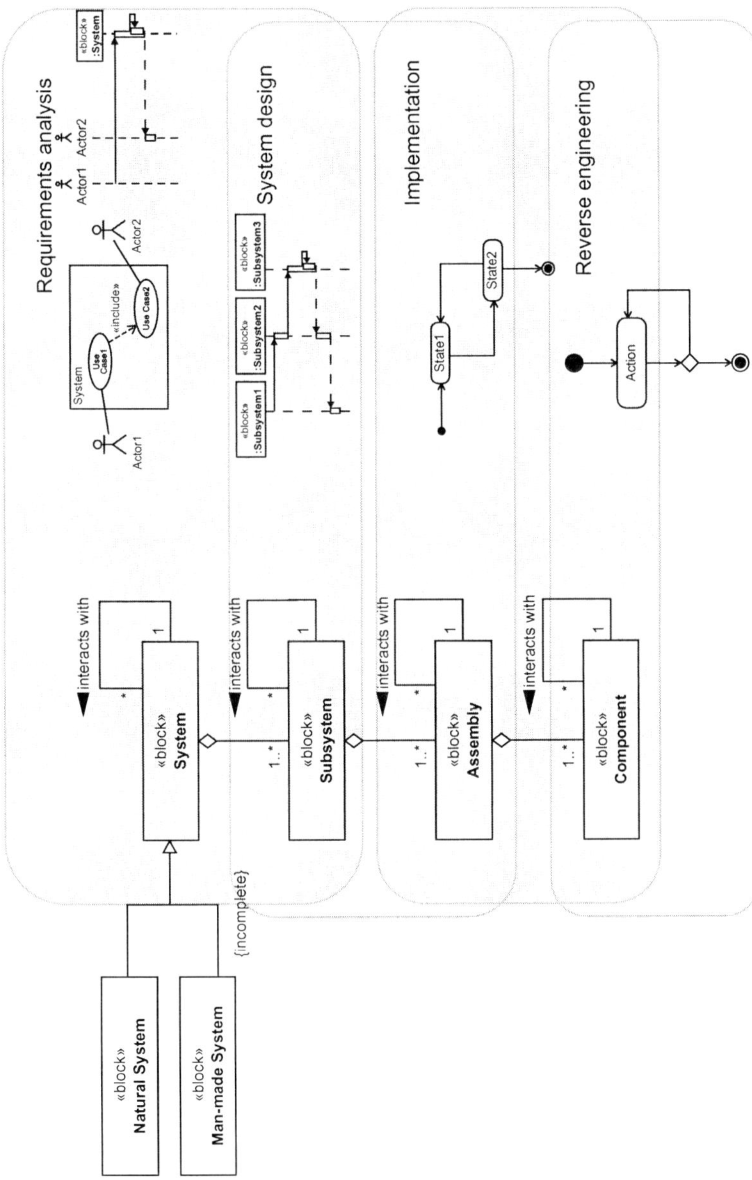

Figure 3.17 Typical behavioural diagram usage by engineering activity

- *System design*, in which the main subsystems that make up the system are modelled structurally and behaviourally. The subsystems may be broken down into assemblies, with the behaviour within each assembly modelled so that the events that govern an assembly's behaviour and its response to such events can be explored.
- *Implementation*, in which the low-level assemblies and components are modelled. The internal behaviour of assemblies is captured as the low-level behaviour of component operations, allowing the flow and manipulation of information and data within operations to be modelled.
- *Reverse engineering*, in which the system is often investigated from the bottom up, starting with the lowest level components of the system and modelling their structure and low-level behaviour. This can be used to begin abstracting up through the higher levels of the system hierarchy to allow assemblies, subsystems and entire systems to be reverse engineered.

This book deals, unsurprisingly given its title, with the 'Requirements analysis' engineering activity of Figure 3.17. Having considered structural and behavioural modelling, each of the five SysML diagrams used in ACRE are discussed in detail in the following sections. However, before looking in detail at these diagrams, this section concludes with a discussion of *stereotypes*, one way in which the SysML can be extended and tailored.

3.1.3 Stereotypes

Stereotypes are the most widely used of all the SysML extension mechanisms and represent a powerful way to define new SysML elements by tailoring the SysML to our needs.

To use stereotypes effectively, it is necessary to first spot one within a model. Visually, this is very simple, as stereotypes are indicated by enclosing the name of the stereotype within a set of double chevrons. The SysML block itself contains the «block» stereotype.

Figure 3.18 Example stereotypes

Figure 3.18 shows two example stereotypes: «testCase» applied to a block and «validate» applied to a dependency. A dependency, represented by a dashed line with an open arrowhead, can be considered to be the weakest of the SysML relationships since it simply shows that there is some kind of (usually) unspecified relationship between the connected diagram elements. Dependencies are not named and cannot have any multiplicities associated with them. SysML makes use of a number of stereotyped dependencies, particularly in the requirement and use case diagrams described in sections 3.3 and 3.5. In Figure 3.18, a *new* stereotype is used, the one *not* found in the standard SysML, in order to show that a test case validates a use case.

Stereotypes can be defined for any of the standard SysML elements. Unfortunately, the method by which stereotypes are defined varies from SysML tool to SysML tool. However, a common diagrammatic method of defining a stereotype, found in many tools, is shown in Figure 3.19.

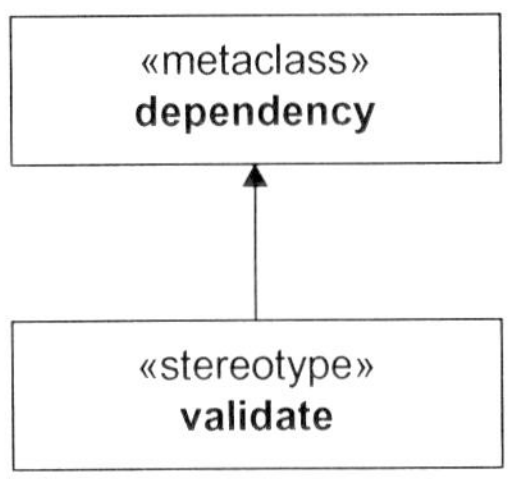

Figure 3.19 Defining a stereotype

The diagram in Figure 3.19 shows the definition of the «validate» stereotype. The diagram shows two blocks 'dependency' and 'validate' that are related together by a special type of specialisation/generalisation known as *extension*. An extension is used specifically when defining stereotypes. An extension is represented graphically by a filled-in triangle very similar to the specialisation/ generalisation symbol.

The new stereotype to be defined, in this case 'validate', is shown in a block which is itself stereotyped «stereotype». The element that is being stereotyped, in this case 'dependency', is shown in a block containing the «metaclass» stereotype. The most important point to understand here is that 'dependency' is a SysML element; it is part of the standard definition of SysML as indicated by the «metaclass» stereotype. The two blocks are then connected with an extension relationship. This shows that the «validate» stereotype can be applied to *only* a dependency, as defined in Figure 3.19. In addition to the graphical definition, it is considered good modelling practice to provide a textual description of the stereotype that describes its intended use.

3.2 Block definition diagrams

The remainder of this chapter looks at each of the SysML diagrams used in ACRE starting with the block definition diagram. These diagrams are discussed using an example system taken from the world of escapology. The system consists of an escapologist who is placed in a rectangular coffin which is then placed into a hole. Concrete is pumped into the hole, under computer control, until the hole is full. The escapologist has to escape from the coffin and the concrete-filled hole before his breath runs out. Figure 3.20 shows the set up for the escape.

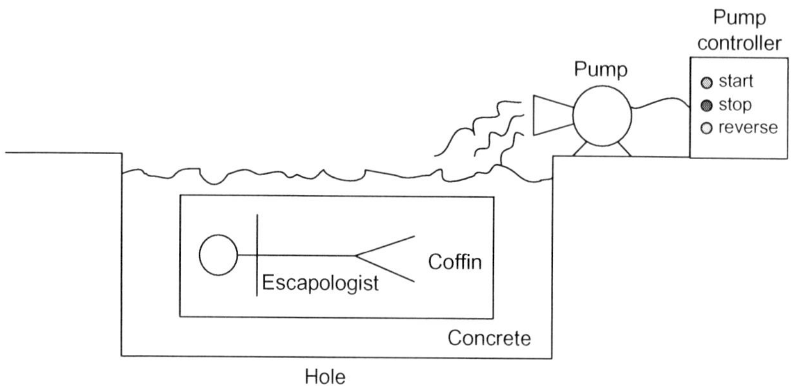

Figure 3.20 The coffin escape stunt

This is a classic escapology stunt that has been performed by many people. It is also a dangerous one, and escapologists have lost their lives performing it because the system requirements and constraints were not properly understood or evaluated. One such performer was Joe Burrus who died on 30 October 1990, when the weight of the concrete crushed the coffin he was in. This example is chosen for two reasons:

- It is an example of a system that includes hardware, software, people and process. There are multiple roles involved, meaning that there are multiple *contexts* to be considered, and it is an example of a safety-critical system – get the requirements wrong and there is the strong possibility that the escapologist would die.
- It is similar in nature to the system described in the case study in Chapter 6, an escapology stunt modelled and then carried out by the authors. This will give a continuity of example across this chapter and Chapter 6.

3.2.1 Overview

Perhaps the most widely used of the SysML diagrams is the block definition diagram. Block definition diagrams realise a structural aspect of the model of a system, showing what conceptual 'things' exist in a system and what relationships exist between them. They have relationships with all the other SysML diagrams and can be used to model just about anything, e.g. from materials, physical equipment and software to people, documents and processes.

The block definition diagram has been introduced earlier in the discussion on structural modelling, and this section expands upon that information, covering more of the syntax and showing further examples that are all taken from the escapology example that is used throughout this chapter to illustrate each of the diagrams.

3.2.2 Notation

The notation for block definition diagrams is shown in Figure 3.21. Readers should be aware that this shows only *part* of the notation and concepts relating to SysML block definition diagrams, covering the main concepts and notation used throughout this

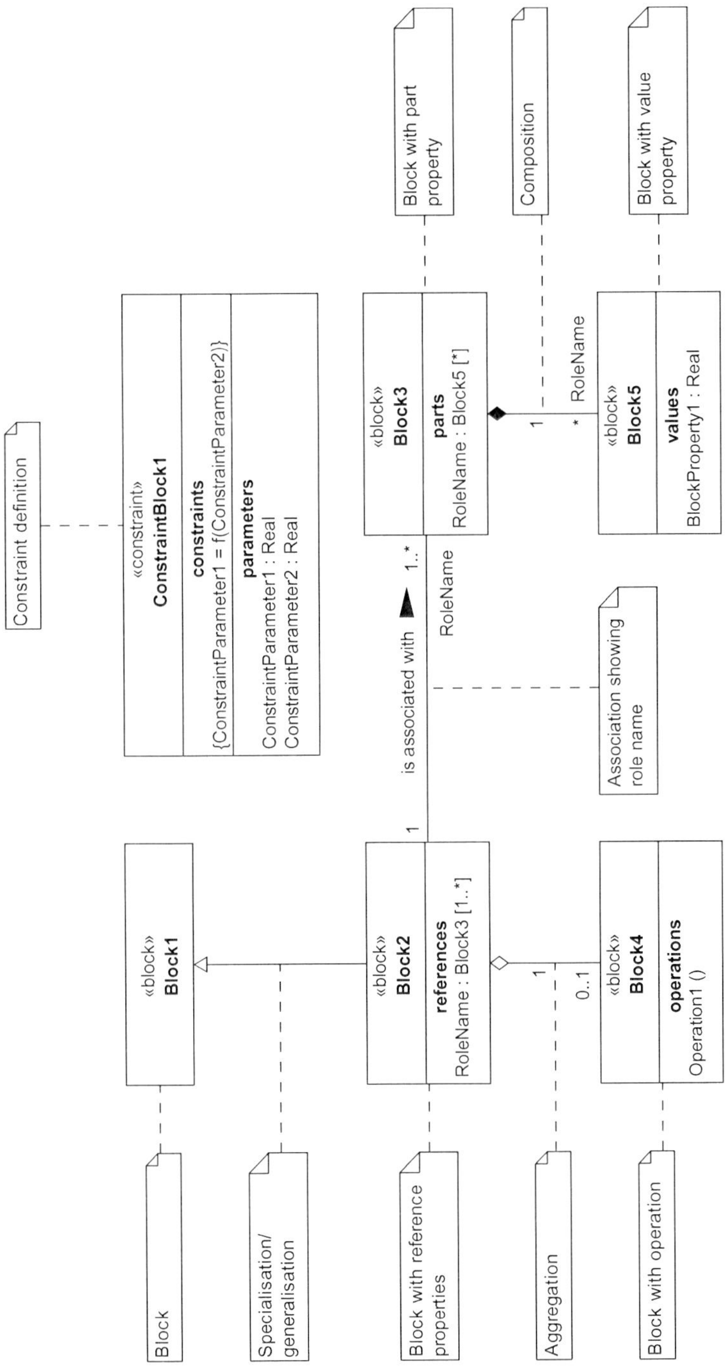

Figure 3.21 Block definition diagram notation

book. For a full discussion of the SysML block definition diagram notation, see Reference 1. In addition, given the coverage of block definition diagrams in section 3.1.1, this section does not describe in detail notation and concepts covered previously.

The main element on the block definition diagram is the *block*, five examples of which are shown in Figure 3.21. 'ConstraintBlock1' is a special case of a *constraint definition* block used to define a *parametric constraint* and is discussed further in this section.

Each block may have a number of *properties* and a number of *operations*. 'Block4' is an example of a block with a single operation defined, 'Operation1()'. Recall from section 3.1.1 that operations can also have *parameters* and a *return value*. Most SysML tools allow parameters to be omitted from diagrams if desired. If this is done, then adding a note to the diagram informing readers that the parameters have been omitted is good modelling practice. Otherwise a reader may assume that the operations have no parameters.

Four main types of relationship are shown between the blocks:

- *Association*, which defines a simple relationship between one or more blocks. An example is shown between 'Block2' and 'Block3'. In SysML, this is also known as a *reference association*.
- *Specialisation/generalisation*, which shows a 'has types' relationship used to show parent and child blocks and which carries with it the notion of *inheritance*. An example is shown between 'Block1' and 'Block2'.
- *Aggregation*, which shows that a block 'is made up of' other blocks. An example is shown between 'Block2' and 'Block4'. In SysML, this is also known as a *shared* or *shared part association*.
- *Composition*, which shows that a block 'is composed of' other blocks. An example is shown between 'Block3' and 'Block5'. In SysML, this is also known as a *part association*.

Associations, aggregations and compositions can all be adorned with *multiplicities* and *role names* as shown in Figure 3.21. Aggregations and compositions are related, as indicated by the similarity in notation, with composition being a stronger form of aggregation. This difference is largely to do with that of the *uniqueness of ownership* of the blocks that form the 'parts' of the whole ('Block4' and 'Block5' in Figure 3.21). If the part can be a part of more than one owning block *at the same time*, it is a *shared part* and aggregation should be used. If the part can only be part of one owning block, composition should be used. Readers who may be familiar with UML should note that the semantics of aggregation and composition in SysML are somewhat different from their semantics in UML where composition also brings with it the idea that the life of the parts are bound up with that of the whole. This is not true of composition in SysML.

The types of relationship used are also closely related to the three types of properties that a block can have:

- *Value* properties, which represent properties that cannot be identified except by the value itself, e.g. numbers or colours. An example of this is 'BlockProperty1' of type 'Real' on 'Block5'.

- *Part* properties, which are owned by the block. That is, properties that are intrinsic to the block but which may have their own identity. These are properties that can also be shown through the use of aggregation or composition. 'Block4' is a part property of 'Block2' and 'Block5' is a part property of 'Block3'.
- *Reference* properties, which are referenced by the block, but *not* owned by it. These are properties that can also be shown through the use of an association. 'Block3' is a reference property of 'Block2' in Figure 3.21.

Each of these three types of properties can be shown in a block in their own compartment labelled 'values', 'parts' or 'references' as appropriate. In fact, this is the only way to show 'value' properties. 'Part' and 'reference' properties can be shown either in a compartment or through the appropriate relationship. For example, in Figure 3.21, 'Block3' is shown as being a reference property of 'Block2' both through the explicit 'references' compartment in 'Block2' and through the association between 'Block2' and 'Block3'. A further example is given in Figure 3.22.

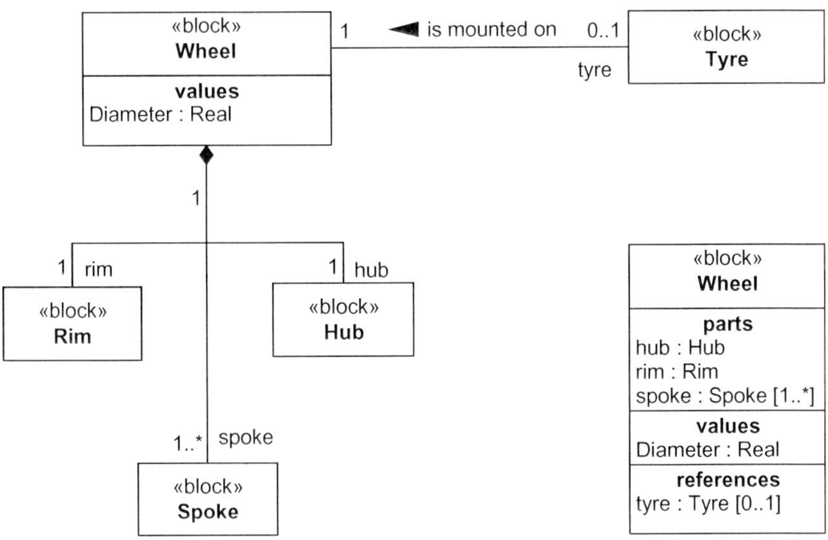

Figure 3.22 Types of property – alternative representations

The left-hand part of Figure 3.22 shows that a 'Wheel' is composed of a 'Rim', a 'Hub' and one or more 'Spokes'. It has a 'Diameter' that will be represented as a 'Real'. Zero or one 'Tyre' is mounted on the 'Wheel'.

Now consider the 'Wheel' block on the right. This shows exactly the same information but in a different format which uses named property compartments rather than separate blocks and relationships. Since the 'Rim', 'Hub' and one or more 'Spokes' are all intrinsic parts of the 'Wheel' – that is, they can be thought of as having their own identity but form indivisible elements of the 'Wheel' – they are modelled as *part* properties. The 'Tyre' is not owned by the 'Wheel'. It can be removed and attached to another 'Wheel'. For this reason, it is modelled as a *reference* property. Finally, the 'Diameter' is simply a number – it does not have any individual identity – and is therefore modelled as a *value* property.

The diagram on the right is more compact than that on the left although perhaps not as clear, useful perhaps when producing summary diagrams. The authors, generally, prefer the more visual approach, but ultimately it is a question of utility, diagramming style and taste. Most SysML tools will allow any or all of these compartments to be omitted from a diagram, allowing a simple version of a block to be shown as in Figure 3.23.

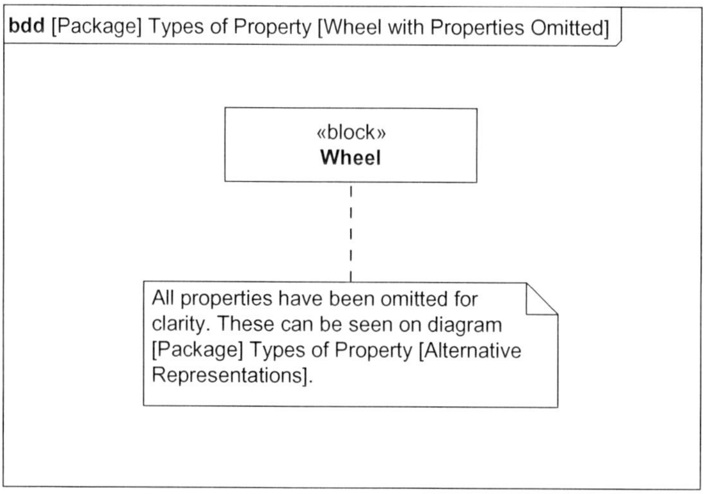

Figure 3.23 Block with properties omitted

Note that in Figure 3.23, a note has been added to the diagram informing a reader of the diagram that properties have been omitted and directing them to a diagram that shows these properties. Remember, modelling is intended to aid communication and understanding, not hinder it.

Block definition diagrams are also used for the *definition* (but not the usage) of *parametric constraints*. An example of such a parametric constraint definition is 'ConstraintBlock1' shown in Figure 3.21. This notation is described in section 3.4, but it should be noted here that 'normal' blocks and 'constraint definition' blocks would not normally be mixed on the same diagram.

3.2.3 Usage

This section presents some examples of block definition diagrams. Further, more detailed, examples taken from a real project undertaken by the authors can be found in the case study in Chapter 6.

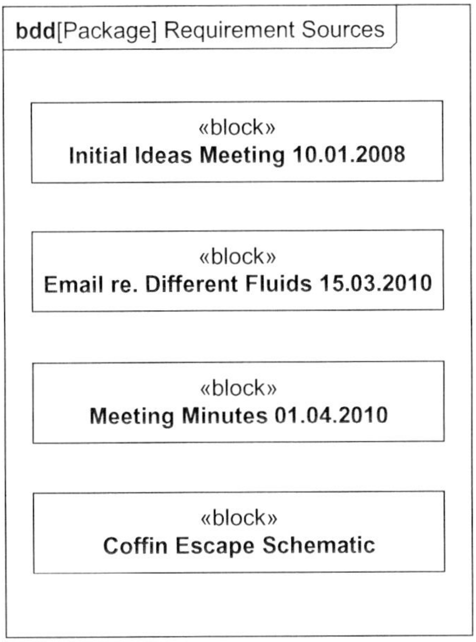

Figure 3.24 Example block definition diagram

Figure 3.24 is an example of a simple block definition diagram that is using blocks to represent requirement sources. In ACRE terms, this is an example of a *Source Element View*, which is described in Chapter 5. Here, blocks are being used to represent things like meeting minutes, emails and sketches. Each block can have the appropriate email text stored as part of the description of the block, something which all good SysML tools will allow. However, displaying this text on the diagram is more problematic because different tools support the display of descriptive text in different ways. One possible solution is shown in Figure 3.25.

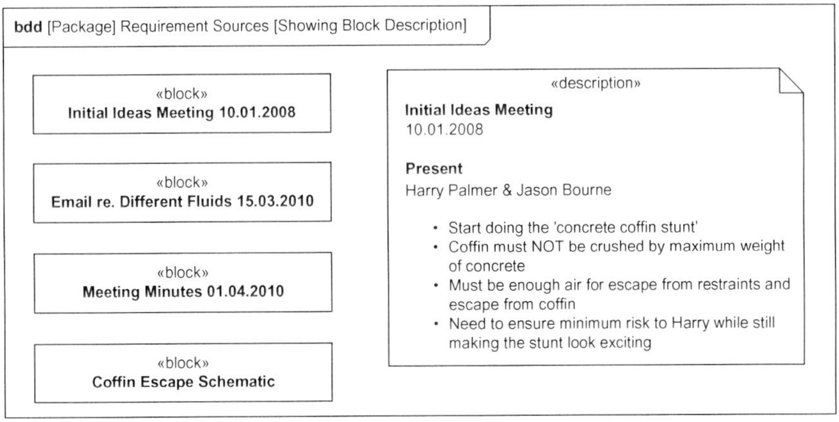

Figure 3.25 Block definition diagram showing use of stereotyped note

The diagram in Figure 3.25 is the same as that in Figure 3.24 except that a SysML note has been added to the diagram. This is the symbol that looks like a piece of paper with the corner folded over. This note has been stereotyped «description» to show that it is not a standard note but one containing descriptive text for another diagram element, in this case for the 'Initial Ideas Meeting 10.01.2008' block to which it is connected. Such 'description' notes are *not* part of the standard SysML but a suggested way that such descriptive information can be displayed in a SysML-compliant way. Of course, this raises other issues regarding consistency; the text in the 'description' note must be kept consistent with that in the description of its connected block. Again, unfortunately, different SysML tools handle this in different ways from not at all, to the need for users to write scripts within a tool to do this, to full automatic support.

Figure 3.25 is a simple block definition diagram, showing only blocks and a note, but none with relationships between any of the blocks. More commonly, block definition diagrams also show how blocks are related to each other. An example of this is shown in Figure 3.26.

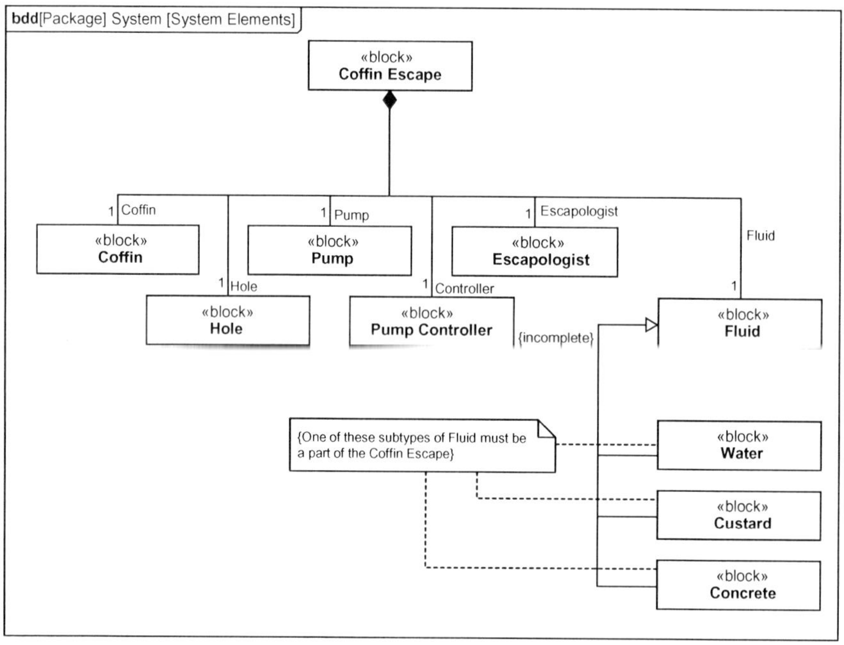

Figure 3.26 Block definition diagram showing compositions and generalisations

The diagram in Figure 3.26 is used to show the high-level structure of a 'Coffin Escape' stunt. In ACRE terms, this is an example of a *System Context Definition View*, described in Chapter 5. Composition is used to show that the 'Coffin Escape' is composed of a 'Coffin', 'Hole', 'Pump', 'Pump Controller', 'Escapologist' and 'Fluid'. Again, the flexibility of the block definition diagram is shown, with blocks representing a stunt, a person, physical equipment and material. The compositions also show the use of multiplicities and role names.

In addition, the diagram shows different types of 'Fluid' through the use of generalisation relationships. Although somewhat redundant given the multiplicity on 'Fluid', the note on the diagram makes it clear that one of the types of 'Fluid' ('Water', 'Custard' or 'Concrete') must be used in the stunt. The text in the note is enclosed in curly braces, {thus}. This means that the text is a formal constraint on the model (not to be confused with parametric constraints, which are something completely different and which are discussed in section 3.4) rather than a simple note. The contents of simple notes are *not* considered to be a formal part of the model and can be ignored, whereas constraint text *cannot* be ignored.

Given that the 'Coffin Escape' is composed of a number of parts (as indicated by the composition relationships), these parts could also be shown in a parts compartment of the 'Coffin Escape' block as shown in Figure 3.27.

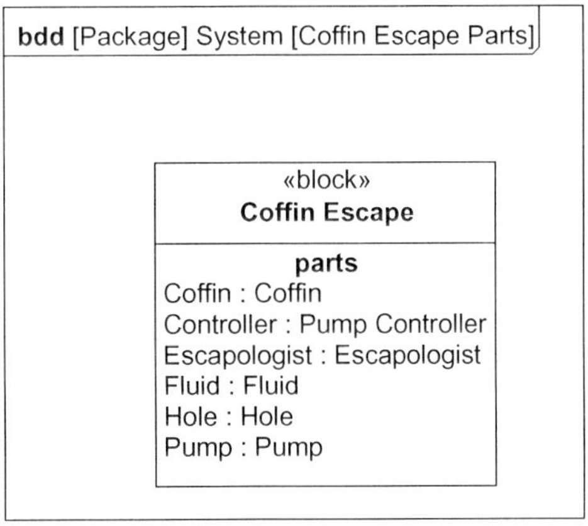

Figure 3.27 Block definition diagram showing parts

In Figure 3.27, each of the component parts of the 'Coffin Escape' from Figure 3.26, i.e. those blocks related to 'Coffin Escape' through a composition relationship, is shown as part properties in the parts compartment for the 'Coffin Escape'. Each property is named using the role name on the corresponding composition relationship. This use of role names in naming part and reference properties is widespread in SysML.

Sometimes the initial values of a block's value properties are known. If so, the SysML allows these initial values to be defined and shown on a block definition diagram, as shown in Figure 3.28. This diagram defines a 'Fluid' that has 'Density' as a value property. Three types of 'fluid' are also defined through the use of generalisation: 'Water', 'Custard' and 'Concrete'. Since these are types of 'Fluid', they inherit all the properties and operations defined for 'Fluid'. This means that they all have the 'Density' value property. The initial value of 'Density' is known for each of these fluids and is shown following the 'Density' value property, so

'Water' has an initial 'Density' of 1000 kg/m^3 and 'Custard' a 'Density' of 1070 kg/m^3. These are *initial* values taken by instances of 'Water', 'Custard' or 'Concrete', there is nothing to stop them changing throughout the lifetime of such instances.

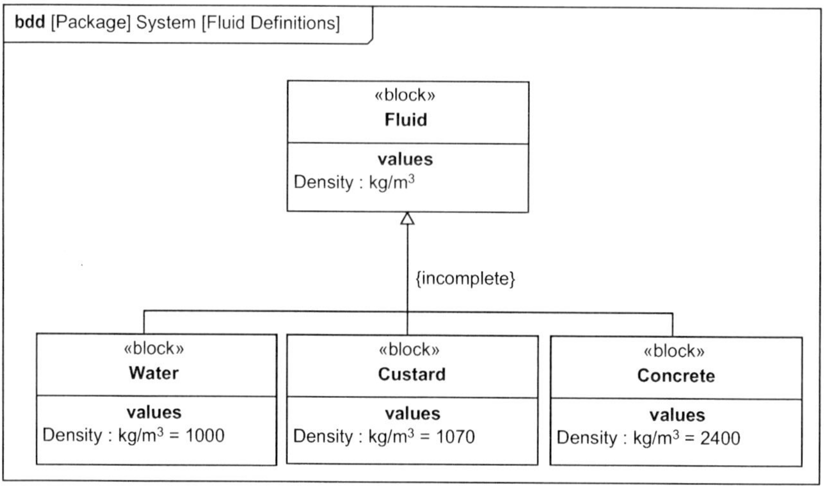

Figure 3.28 Block definition diagram showing default values for value properties

Another piece of notation shown in Figure 3.28 is the '{incomplete}' text next to the generalisations. As discussed earlier, any text in a SysML model that is contained in curly braces is considered to be a constraint. {incomplete} is a standard piece of notation that is used to indicate that not all the subtypes of a block, as indicated by generalisation, are shown. Unfortunately, this is often used in two different ways, either to indicate that other subtypes *are* known but not shown or to indicate that other subtypes *may* exist but haven't been thought about yet. Again, the use of a note to indicate which of these usages is meant can help to make the diagram clearer.

The final block definition diagram, shown in Figure 3.29, is an example of how blocks can be used to model the process aspects of a system. These can be processes carried out by the people, as here, or completely automated processes. Five different types of 'Stunt Process' are shown that are central to the execution of the coffin stunt. That is, each process is part of the escapology 'system' that is being modelled. The activities that are carried out in each process are model using the operations of a block, an approach that is described in detail in Reference 3.

3.2.4 Summary

SysML block definition diagrams are used to show abstractions of 'things' and their relationships to each other. Blocks have properties that further describe them and operations that show *what* behaviour a block exhibits but not *how* such behaviour is carried out. Relationships can be used to show that a block is part of

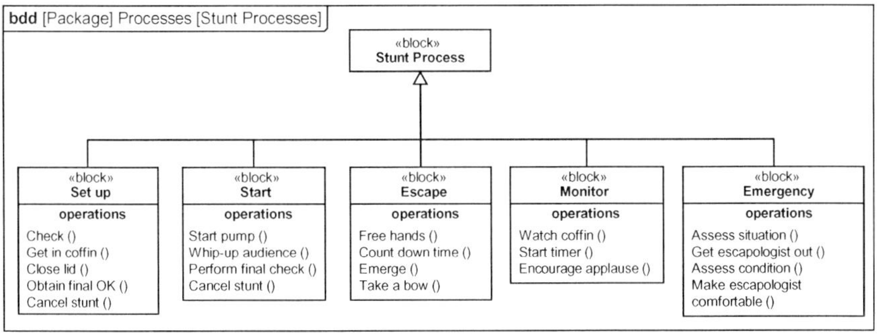

Figure 3.29 Block definition diagram used to show processes that form part of a system

another block (using aggregation or composition), is a type of another block (using specialisation/generalisation) or has a simple relationship with another block (using association).

In ACRE, the block definition diagram is widely used. It is used to show the following:

- Stakeholders on a *Stakeholder Context Definition View*
- System structure on a *System Context Definition View*
- Source Elements on a *Source Element View*
- The definitions of rules on a *Definition Rule Set View*
- The definition of constraints on a *Constraint Definition View*

All these views are described further in Chapter 5, and examples of all these views can be found in Chapter 6.

3.3 Requirement diagrams

3.3.1 Overview

The SysML has a dedicated 'requirement diagram' that is used to represent requirements and their relationships. This diagram is, in essence, a tailored block definition diagram consisting of a stereotyped block with predefined properties and a number of stereotyped dependencies and fixed-format notes. As discussed in Chapters 4 and 5, this diagram is used within ACRE to realise the *descriptions* of requirements in the Requirements Description View. The various relationships provided by the requirement diagram also form an essential and central part of the Traceability Views that are a fundamental aspect of a model-based approach to requirements engineering.

3.3.2 Notation

The notation used in SysML requirement diagrams is shown in Figure 3.30.

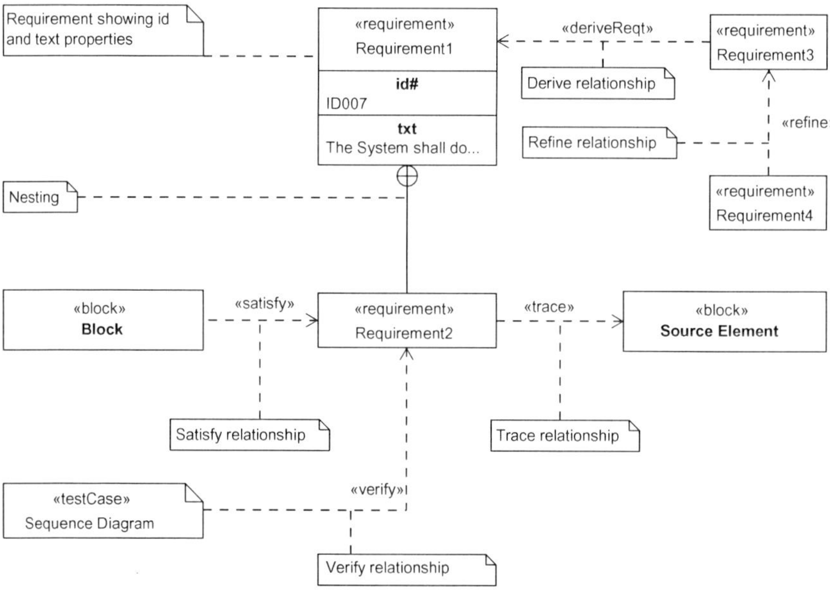

Figure 3.30 Requirements diagram notation

Central to the requirement diagram is the *requirement*. This is shown in SysML as a rectangle with the stereotype «requirement». The rectangle also contains a human-readable name for the requirement. In addition, all requirements have two properties predefined by SysML: the *id#* and *txt* properties. The 'id#' property is to hold a unique identifier for the requirement. The 'txt' property holds descriptive text for the requirement. The display of 'id#' and 'txt' is optional and Figure 3.30 shows these compartments for 'Requirement1' and omits them for 'Requirement2', 'Requirement3' and 'Requirement4'. Other properties may be defined for requirements, such as 'priority' and 'owner', and Appendix C of the SysML specification gives examples. See Reference 2.

A requirement may be decomposed into one or more *sub-requirements*, e.g. when the requirement is not atomic in nature and it is desired to decompose it into a number of related atomic sub-requirements. In SysML, this decomposition relationship is known as *nesting* and is indicated with a 'nesting' relationship such as that shown between 'Requirement1' and 'Requirement2'.

When carrying out requirements analysis, it is often necessary to *derive* additional requirements. A derived requirement is one that is not explicitly stated by a stakeholder but has been derived by systems engineers from an explicit statement of requirement as part of the requirements analysis process. Such derived requirements can be linked back to their source requirements in SysML by using a 'derive' relationship, an example of which is given in Figure 3.30 showing that 'Requirement3' is derived from 'Requirement1'.

The SysML requirement diagram also supports four other types of relationships that are used in the following ways:

- *Satisfy relationship*. This is used to show that a model element satisfies a requirement. It is used to relate elements of a design or implementation model to the requirements that elements are intended to satisfy. Although Figure 3.30 shows a 'satisfy' relationship between a block and a requirement, it can be used between *any* SysML model element and a requirement.
- *Trace relationship*. This is used to show that a model element can be traced to a requirement or vice versa. This provides a general purpose relationship that allows model elements and requirements to be related to each other. One common use of the 'trace' relationship within ACRE is to show how Requirement Descriptions trace to Source Elements. See Chapters 4 and 5 for a description of these ACRE concepts. An example of this is shown by the 'trace' relationship between 'Requirement2' and 'Source Element' in Figure 3.30.
- *Refine relationship*. This is used to show how model elements and requirements can be used to further refine other model elements or requirements. This could be, e.g., one requirement refining another as shown in Figure 3.30, where 'Requirement4' refines 'Requirement3'. In ACRE, it is used where a use case on a Requirement Context View is developed to refine a requirement on a Requirement Description View. The Requirement Context View and Requirement Description View are discussed in Chapter 5.
- *Verify relationship*. This is used to show that a particular *test case* verifies a given requirement and so can only be used to relate a 'test case' and a SysML requirement. However, a 'test case' is *not* a specific type of SysML element. Rather it is a stereotype, «testCase», which can be applied to any SysML operation or behavioural diagram to show that the stereotyped element is a test case intended to verify a requirement. This stereotyped element – the 'test case' – can then be related to the requirement it is verifying via the 'verify' relationship. The test case is shown on a requirement diagram as a SysML note containing the name of the SysML element or diagram that is acting as a test case along with the stereotype «testCase». This is shown in Figure 3.30 by the 'verify' relationship between the test case called 'Sequence Diagram' and 'Requirement2'.

In the context of ACRE, the stereotype can be applied to a sequence diagram to show that the Validation View that it represents is a test case for a use case on a Requirement Context View. However, since SysML allows only a verify relationship to end on a requirement rather than a use case, this verification of use cases will have to be done via the requirement that the use case refines, i.e. the test case verifies the requirement and the use case refines the requirement. Validation Views are discussed in Chapter 5.

Unfortunately, the definition of the «testCase» stereotype in the SysML specification (see Reference 2) prevents the stereotype being applied to SysML parametric diagrams. This is a missed opportunity since parametric diagrams, discussed later from a SysML point of view and in Chapter 5 from an ACRE point of view, are an ideal mechanism by which more formal scenarios (test cases) can be modelled than is possible using sequence diagrams. Readers who are adopting the techniques and approaches described in this book are urged to use the SysML's stereotyping mechanisms to define their own test case stereotype that *can* be applied to parametric diagrams. Similarly, a verify stereotype could be defined that can take a use case as a target given the issues with the verify relationship discussed earlier.

These types of relationship allow the modeller to explicitly relate different parts of a model to the requirements as a way of ensuring the consistency of the model. However, where possible one of the specific types of relationship, such as 'satisfy', should be used in preference to the more generic 'trace' relationship, which has weakly defined semantics since it says nothing about the *nature* of the relationship other than that the two elements can be traced in some general and unspecified manner.

It should also be noted that, although shown in Figure 3.30 using stereotyped dependencies, these relationships can also be shown in SysML using special versions of the note. These 'callout' notes can be useful when relating elements in widely different parts of a model since it avoids the need to produce additional diagrams specifically to show the relationships. However, they can lead to inconsistency, particularly when modelling is not being done using a tool (or is being done using a tool that does not enforce consistency). Using the stereotyped dependencies gives an immediate and direct indication of the relationship since the two elements are explicitly connected by the dependency. Using the notes method hides the immediacy of the relationship inside the text of the note and also requires that two notes are added to the model: one to the source of the relationship and one to the target. If one of these notes is omitted, the model will be inconsistent. For full details of the callout notation see Chapter 4 of Reference 1.

3.3.3 Usage

This section presents some examples of requirement diagrams and related diagramming elements. Further, more detailed, examples of requirement diagrams for a real project undertaken by the authors can be found in the case study in Chapter 6.

Figure 3.31 shows a number of SysML requirements, each of which has its *id#* and *txt* property shown. Some of these requirements are broken down further into sub-requirements via *nesting*. At least two of these requirements, ES004 and ES005, have descriptive text in their *txt* property that could considered to be untestable. In the case of ES005, the sub-requirements further describe what is meant by '... the risk to the escapologist is minimised'. However, in the case of ES004 further analysis is required. This might result in a number of *derived* requirements being created, as shown in Figure 3.32.

The three requirements ES004-D001, ES004-D002 and ES004-D003 shown in Figure 3.32 are each derived from ES004 and show how the vague and untestable

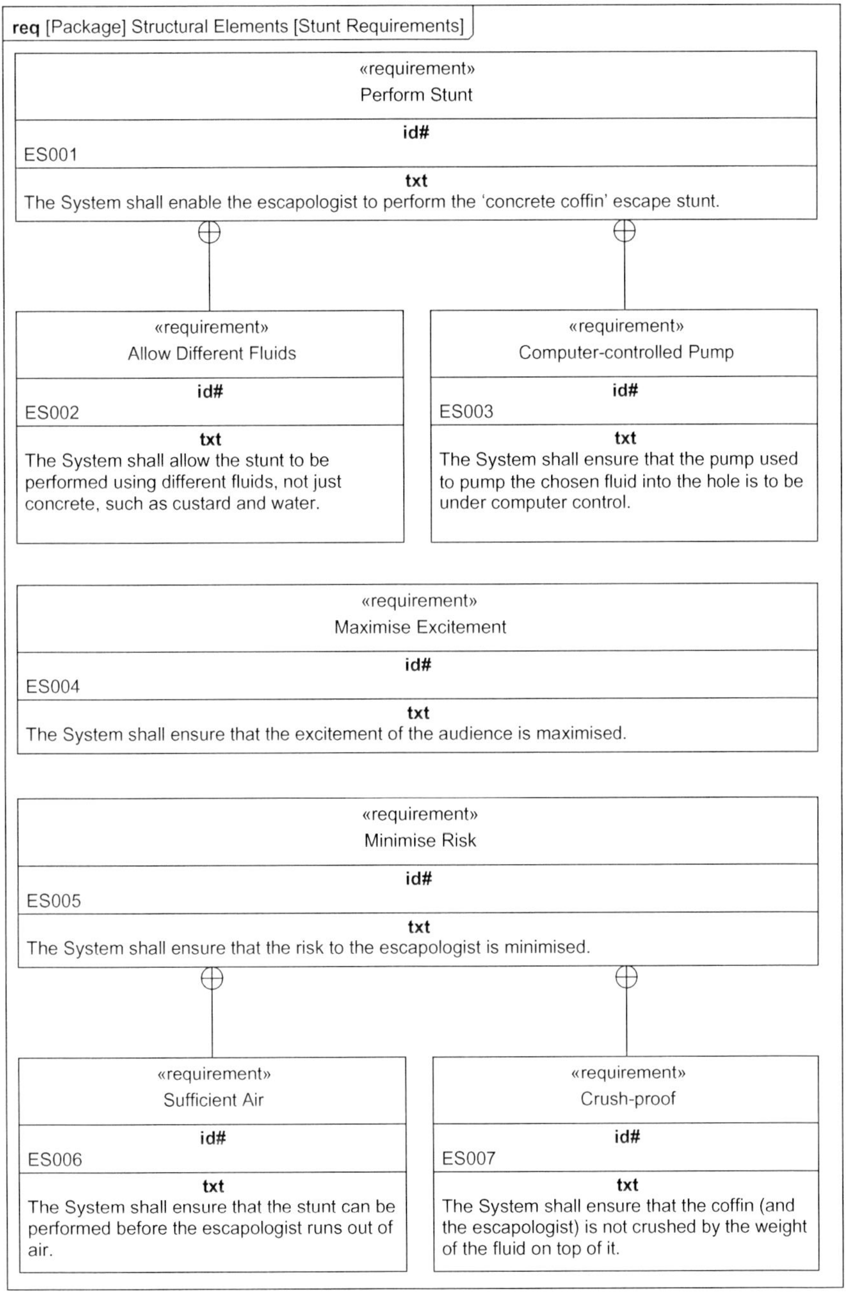

Figure 3.31 Example requirement diagram

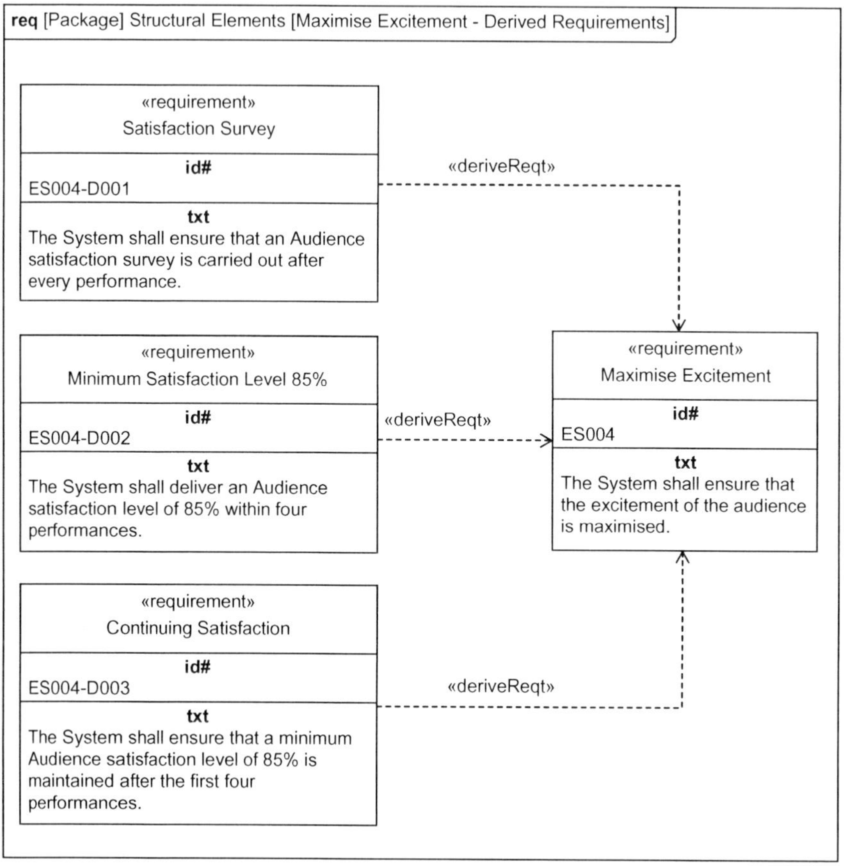

Figure 3.32 Requirement diagram showing derived requirements

requirement 'The System shall ensure that the excitement of the audience is maximised' may be further specified in a way that is testable.

Sometimes turning off the id# and txt properties of a requirement can make a diagram easier to read, particularly when additional information such as «trace» relationships are shown. This has been done in Figure 3.33, which shows the same requirements as are shown in Figure 3.31, but with the id# and txt compartments hidden and «trace» relationships added linking the requirements to SysML blocks representing the *source* of the requirements. There is no significance in the sizing of the various requirement blocks; it has been done simply to ease the layout of the diagram.

A similar diagram is given in Figure 3.34, which concentrates on a single requirement, showing how it traces to Source Elements and in addition showing a use case that refines the requirement. A seemingly obvious, but often overlooked, aspect of modelling is highlighted by Figure 3.34, namely that of keeping diagrams as simple as possible. There is often a temptation to overload diagrams with too

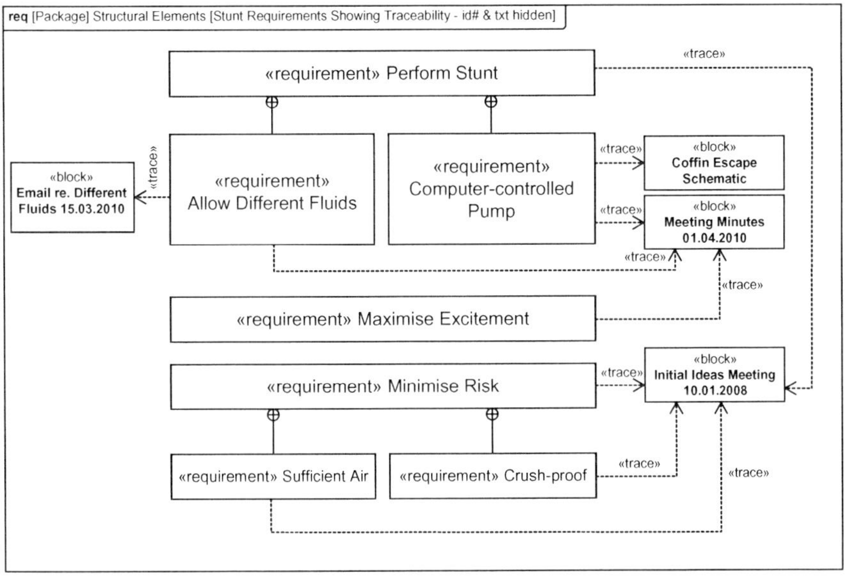

Figure 3.33 Requirement diagram showing «trace» relationships

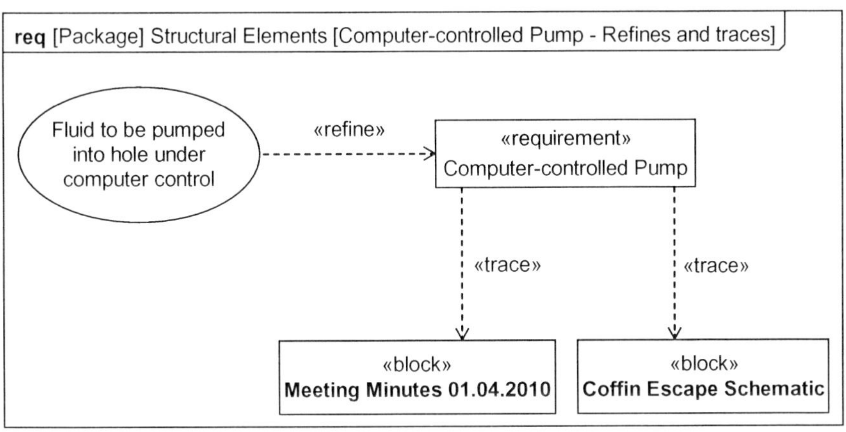

Figure 3.34 Requirement diagram showing «refine» and «trace» relationships

many elements, so they add to the complexity and lack of understanding of the system rather than helping. The information given in the four example diagrams earlier *could* have been shown on a single diagram, but this would have made the communication of the understanding of the requirements and their relationships to other model element harder to achieve. Any sensible modelling tool will allow model elements to be reused on a number of different diagrams, and this is to be encouraged not only for requirements diagrams but also for *any* of the SysML diagrams. If you find a diagram becoming too complex (more than around 9 or 10 elements, as a crude heuristic), break it down into a number of simpler diagrams. Miller's comments on the limits on our capacity to process information are as valid today as when they were first written and apply just as much to SysML models. See Reference 4.

3.3.4 Summary

SysML requirement diagrams are used to show *requirements* and their relationships to each other and how they trace to and are satisfied by, refined by and verified by other model elements such as *Source Elements* and *use cases*. Wherever possible, use of the more specific types of relationship (such as 'satisfy') is preferred over the more generic 'trace'. Each requirement has a name, unique identifier and a description. Most SysML tools allow the identifier and description to be hidden, if desired, in order to simplify diagrams. Additional requirement properties such as 'priority' may be defined if needed and examples are given in the SysML specification (see Reference 2).

In the ACRE, the SysML requirement diagram is used to show descriptions of requirements on a *Requirement Description View*. The Requirement Description View is described further in Chapter 5.

3.4 Parametric diagrams

3.4.1 Overview

In addition to the somewhat informal constraints that can be added to SysML diagrams by enclosing note text in curly braces, {thus}, an example of which can be seen in Figure 3.26, the SysML has the concept of the *constraint block* and the associated *parametric diagram*. Constraint blocks allow for the definition and use of networks of constraints that represent rules that constrain the properties of a system or that define rules that the system must conform to. As discussed in Chapters 4 and 5, in ACRE, constraint blocks and parametric diagrams are used to define rules constraining Requirement Descriptions in the Definition Rule Set View and can also be used to validate use cases in the Constraint Validation View.

3.4.2 Notation

There are two aspects to parametric constraints in SysML: their *definition* and their *usage*. The notation for both aspects is shown in Figures 3.35 and 3.37, respectively.

A *constraint definition* block is defined using a block stereotyped «constraint» and given a name by which the constraint can be identified. The constraint block has two compartments labelled 'constraints' and 'parameters'. The constraints compartment contains an equation, an expression or a rule that relates together the parameters given in the parameters compartment. Figure 3.35 defines a constraint called 'ConstraintBlock1' with two parameters 'ConstraintParameter1' and

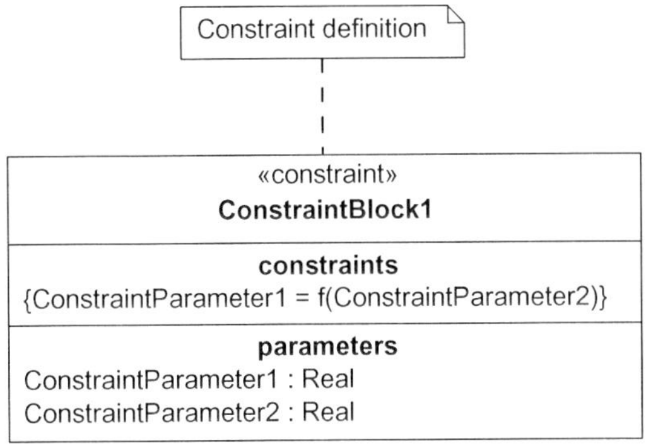

Figure 3.35 Parametric diagram notation – definition of parametric constraint

'ConstraintParameter2', both of which are defined to be of type 'Real'. These parameters are related together by the expression 'ConstraintParameter1 = *f*(ConstraintParameter2)', with *f* representing a function taking 'ConstraintParameter2' as a parameter.

Such constraint blocks are defined on a block definition diagram. A concrete example of a constraint block is shown in Figure 3.36.

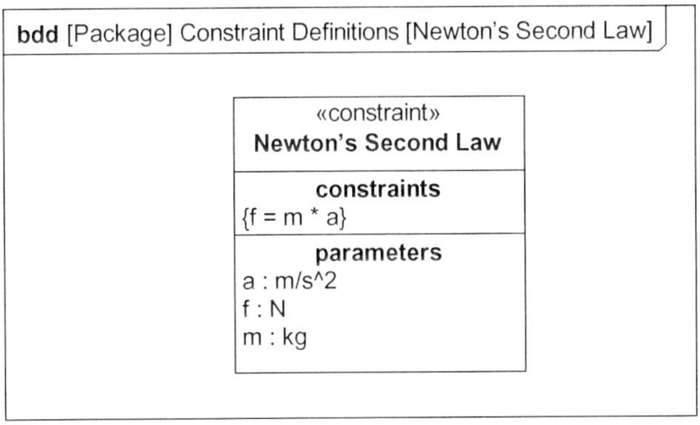

Figure 3.36 Example constraint block definition

The example in Figure 3.36 defines a constraint called 'Newton's second law' that relates the three parameters *f*, *m* and *a* given in the parameters compartment by the equation $f = m \times a$, as shown in the constraints compartment. Although constraint blocks are defined on block definition diagrams, it is convention that such definitions are not mixed with regular blocks on the same diagram.

Once constraint blocks have been defined, they can be used any number of times on one or more parametric diagrams, the notation for which is shown in Figure 3.37.

Each constraint block can be used multiple times on a parametric diagram. The use of a constraint block is shown as a round-cornered rectangle known as a

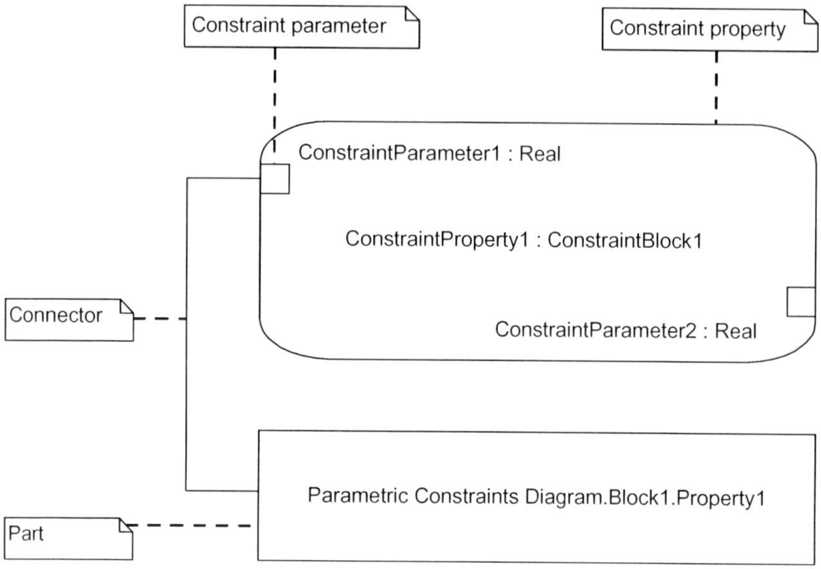

Figure 3.37 Parametric diagram notation – use of parametric constraint

constraint property. Each constraint property is to be named *name:Constraint name*. This allows each use of a constraint to be distinguished from other uses of the same constraint. In Figure 3.37, a single constraint block, 'ConstraintBlock1', is being used and it has been given the name 'ConstraintProperty1'.

Small rectangles attached to the inside edge of the constraint property represent each *constraint parameter*. These are named and their names correspond to the parameters defined for the constraint block in its definition.

These constraint parameters provide connection points that can be connected, via *connectors*, to other constraint parameters on the same or other constraint properties or to block properties. When connecting a constraint parameter to a block property, the block property is represented on the diagram by a rectangle known as a *part*. In Figure 3.37, a single part is shown, with the name 'Parametric Constraints Diagram.Block1.Property1'. This shows that this is the 'Property1' property of the block 'Block1' in the package 'Parametric Constraints Diagram'. Packages are used to structure SysML models and are outside the scope of this book. See References 1 and 2 for details.

In Figure 3.37, the part 'Parametric Constraints Diagram.Block1.Property1' is connected to 'ConstraintParameter1'. There is nothing connected to 'ConstraintParameter2' and therefore the diagram is incomplete.

3.4.3 Usage

This section presents some examples of parametric diagrams and related diagramming elements. Further examples of parametric diagrams for a real project undertaken by the authors can be found in the case study in Chapter 6.

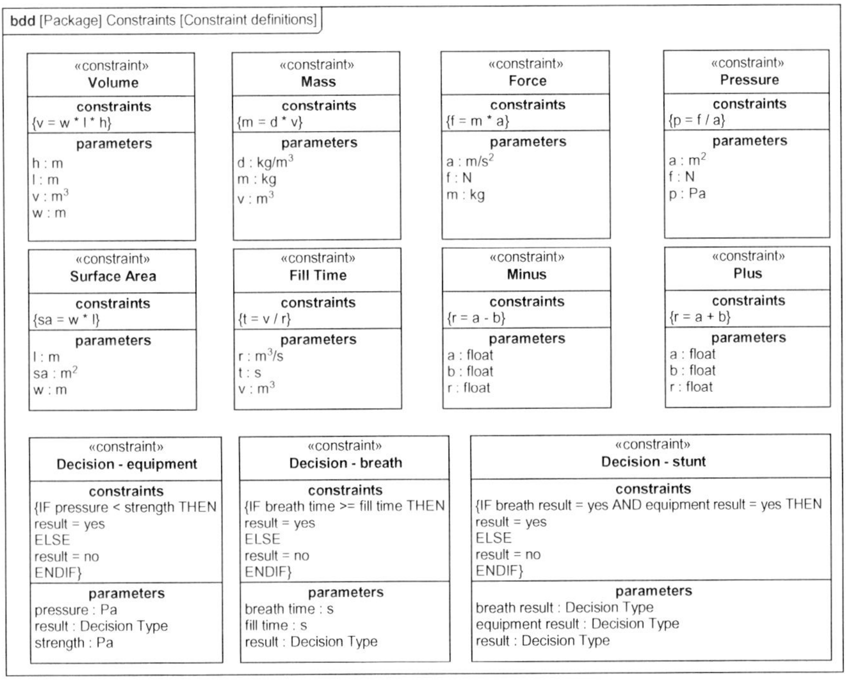

Figure 3.38 Block definition diagram showing definition of parametric constraints

Figure 3.38 shows a number of definitions of constraint for the coffin escape, which are used as the source of examples for this chapter. As noted previously, such constraint blocks are actually defined on a block definition diagram, and good modelling practice has been followed with constraint blocks being kept separate from normal SysML blocks.

It can also be observed that the eight constraint blocks on the top two rows of the diagram are all general constraints that can be used on a number of projects, whereas the three constraints on the bottom row are all specific to the particular system being considered (in this case the concrete coffin escape). For this reason, a better way to organise them would be to split them out into two separate diagrams and perhaps even two separate packages within the model in order to maximise reuse and decouple generic constraints from solution-specific constraints.

Another observation that can be made is that there are three types of constraints defined:

- Constraints representing physical laws or other formulae, such as the definitions of 'Force' or 'Pressure'.
- Constraints representing mathematical and logical operators that make it easier for other constraints to be connected together in a constraint usage network, such as the definitions of 'Plus' and 'Minus'.

- Constraints representing decisions (heuristics) rather than calculation-type constraints, evaluating input parameters against some criteria and returning a result that could be, e.g., a 'yes/no', 'true/false' or 'go/no-go'. The three 'Decision' constraints in Figure 3.38 are examples.

If so desired, the SysML stereotyping mechanism can be used to explicitly mark the constraint blocks as one of these three types, as shown in Figure 3.39. This can be done in order to convey extra information about the constraints, and is discussed in more detail in Reference 1.

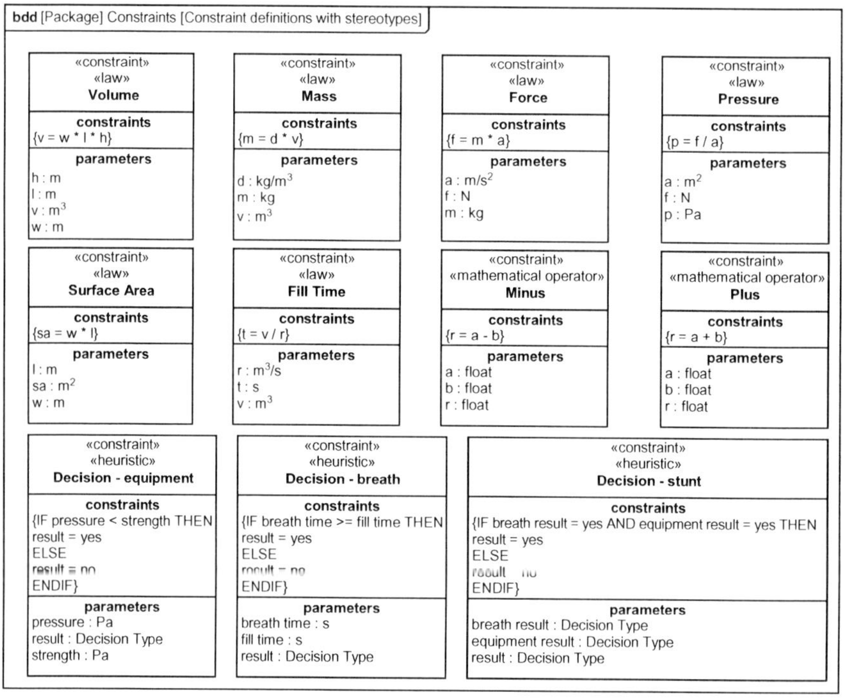

Figure 3.39 Parametric constraints with stereotypes showing type

An example parametric diagram showing the constraints defined in Figure 3.38 is presented in Figure 3.40. This diagram shows the constraints being used to determine a go/no-go decision for the escapologist based on various system properties. That is, the parametric diagram is being used to help validate a use case, namely 'Minimise risk to escapologist'. This can be seen in the call-out note showing that the diagram traces to that use case.

It is worth noting here the failing of the SysML discussed in section 3.3.2. A better relationship from this diagram to the use case would be a «verify» relationship, with the parametric diagram marked as a «testCase», since that is essentially the role that it is playing here. The parametric diagram determines a go/no-go

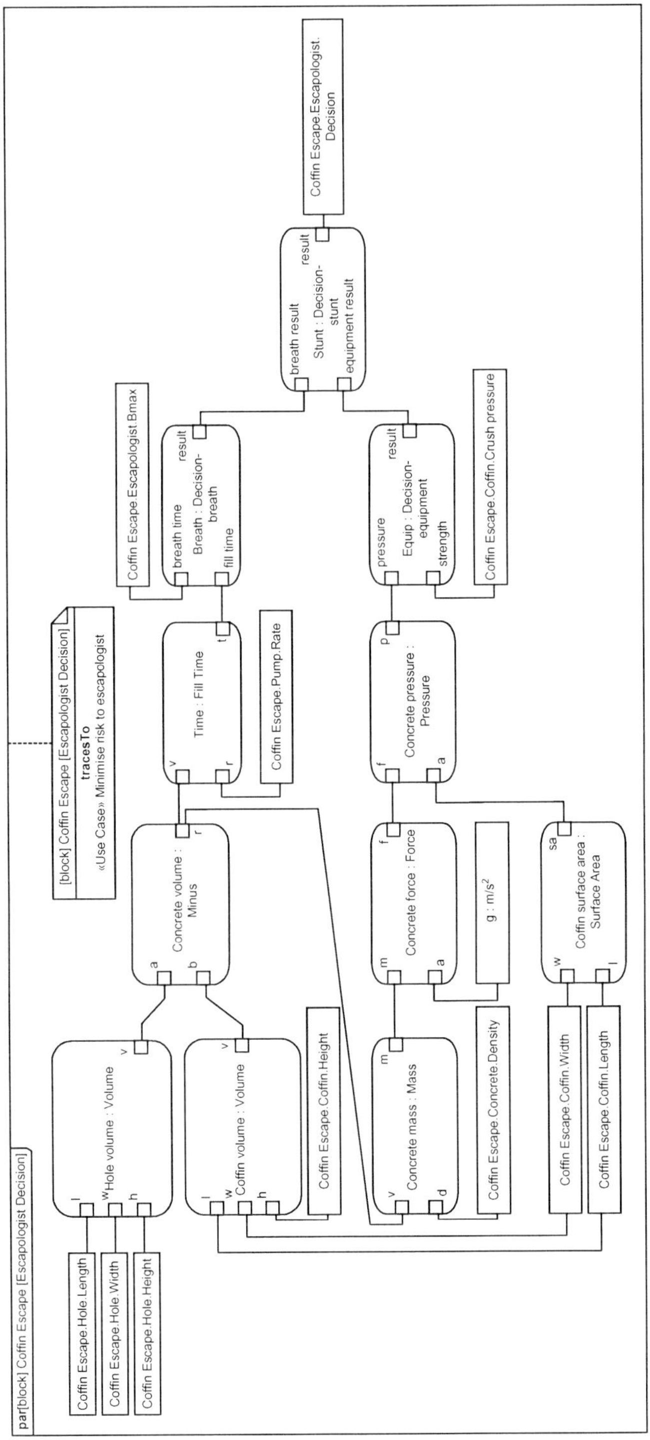

Figure 3.40 Parametric diagram for determining go/no-go decision

decision based on the other system parameters that test whether the use case can be met or not. However, because of the limitations of SysML a simple «trace» has been used.

A convention adopted by the authors, but not part of SysML, is to draw such parametric diagrams with an *implied* left to right direction. In Figure 3.40, the parametric diagram is drawn as though the 'result' constraint parameter, connected to the 'Decision' property of the 'Escapologist' block, is the *output* of the diagram. Similarly, the constraint parameters are arranged around each constraint property with 'inputs' on the left and 'outputs' on the right. This is done as an aid to thinking about and constructing the diagram, and reflects the purpose of the diagram.

However, one could think about going 'backwards' through Figure 3.40: we could use 'Escapologist.Bmax' and 'Pump.Rate' to determine the maximum volume of concrete that can be pumped before the escapologist runs out of breath, and hence the maximum volume of the hole. If the hole is just a little longer and wider than coffin (i.e. we can set values on 'Hole.Length' and 'Hole.Width'), then knowing the maximum volume of the hole would allow the height of the hole to be determined. Perhaps this usage would be considered by the Safety Officer in calculating the hole size. If so, it could be redrawn and linked to the appropriate use case as shown in Figure 3.41.

For a full discussion of the use of parametric diagrams, including how to nest constraints and constraining system behaviour at different levels of accuracy, see Reference 1.

3.4.4 Summary

SysML parametric diagrams show how constraints are related to each other and to properties of system elements modelled using blocks. They use constraint blocks, defined on block definition diagrams, which contain a constraint expression that relates together a number of constraint parameters. In ACRE, they can be used to do the following:

- Validate use cases in a formal way on a *Constraint Validation View*
- Define rules for constraining Requirement Descriptions on a *Definition Rule Set View*

Using the SysML «trace» relationship allows parametric diagrams to be traced to the use cases that they validate. The ACRE Constraint Validation and Definition Rule Set Views are described in Chapter 5.

3.5 Use case diagrams

3.5.1 Overview

The SysML use case diagram realises a behavioural aspect of a model, with an emphasis on functionality rather than the control and logical timing of the system. The use case diagram represents the highest level of abstraction that is available in

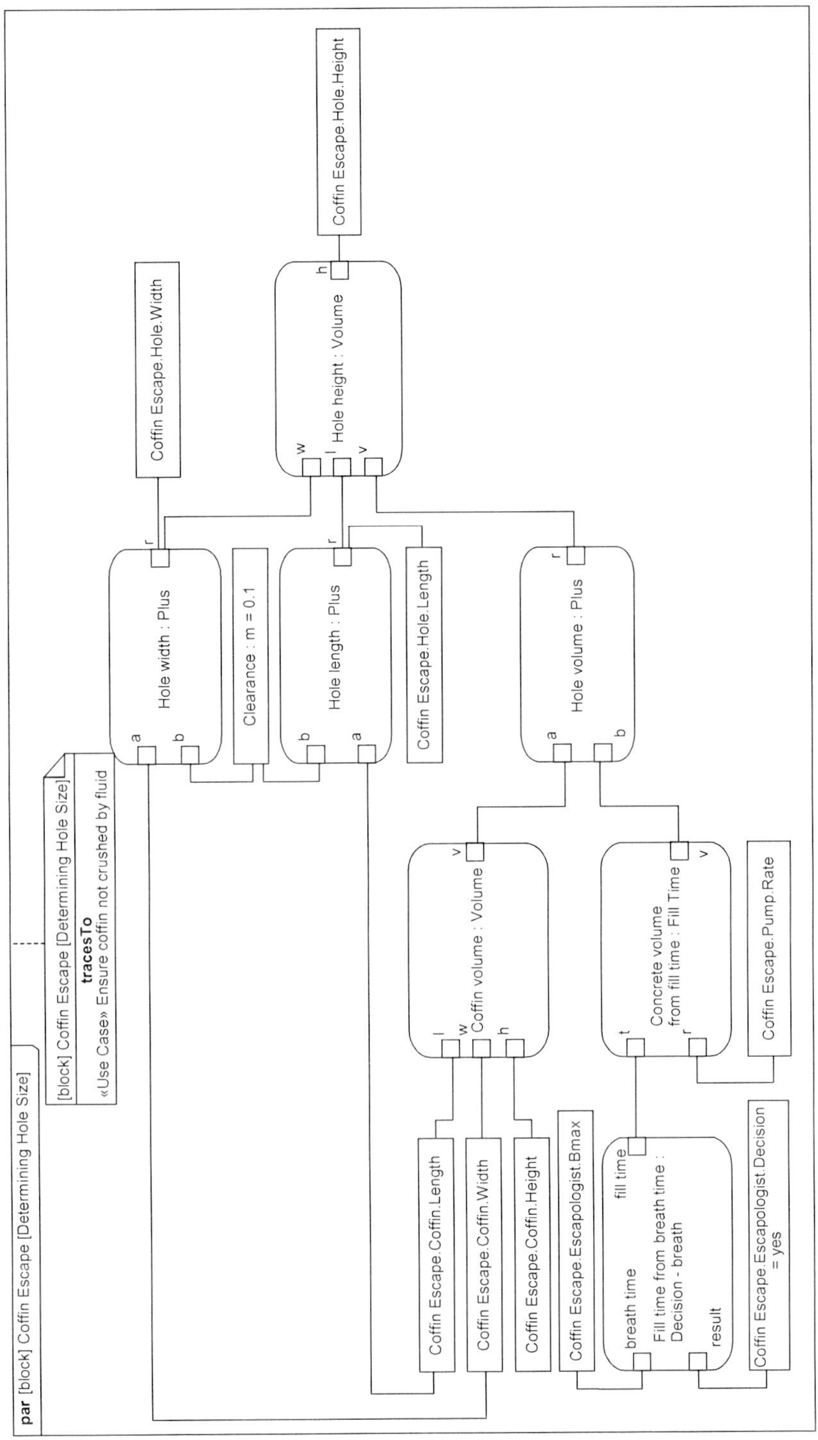

Figure 3.41 Parametric diagram to determine hole size

the SysML. However, use case diagrams are arguably the easiest diagram to get wrong in the SysML. There are a number of reasons for this as given below:

- The diagrams themselves look very simple, so simple in fact that they are often viewed as being a waste of time.
- It is very easy to go into too much detail on a use case model and to accidentally start analysis or design, rather than conducting context modelling.
- Use case diagrams are very easy to confuse with data flow diagrams as they are often perceived as being similar. This is because the symbols look the same as both use cases (in use case diagrams) and processes (in a data flow diagram) are represented by ellipses. In addition, both use cases and processes can be decomposed into lower level elements.
- Use case diagrams make use of perhaps the worst symbol in SysML, the stick person notation used to represent actors. This is discussed further in section 3.6.2.

Nevertheless, use case diagrams are central to the ACRE, being used to model *contexts* of a system via ACRE *Requirement Context Views* described in Chapter 5.

3.5.2 Notation

The notation that is used on a use case diagram is shown in Figure 3.42.

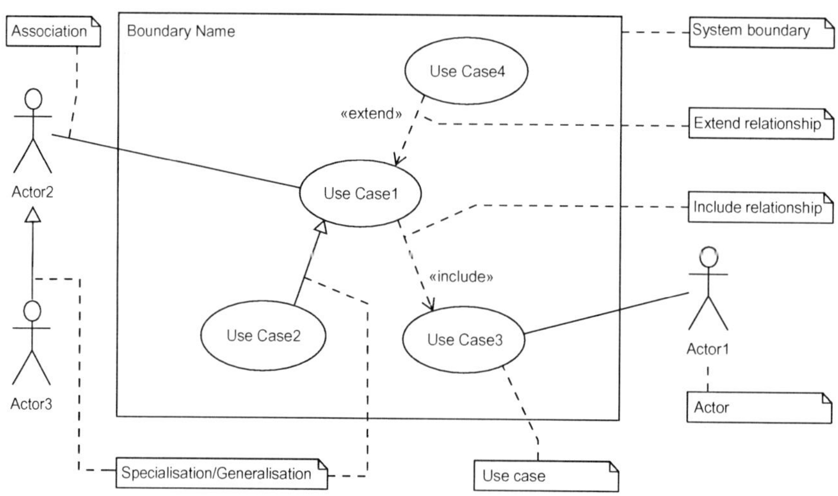

Figure 3.42 Use case diagram notation

Use case diagrams are composed of four basic elements: *use cases, actors, relationships* and a *system boundary*. As a minimum a use case diagram must contain at least one use case; all other elements are optional.

Each use case describes a behaviour of the system that yields an observable result to an actor. It is with the actor that the SysML notation is at its weakest, both in terms of the symbol and the name. The 'stick man' symbol and the name 'actor' suggest that this concept represents that of a *person*. This is not the case. An actor represents the *role* of a person, place *or thing* that interacts with, is affected by or has an interest in the system. So, while an actor can represent a person, it can also be used to represent an organisation, other system or even a piece of legislation or a standard. Furthermore, it is essential to understand that it is the *role* that is represented. This means that you should never see the names of people or organisations or standards on a use case diagram, but the role that they are playing. An actor named 'ISO 15288' would be wrong, but one named 'Systems Engineering Standard' would be correct. It is also worth noting that a given role may be taken by more than one person, place or thing and that a given person, place or thing may take on more than one role.

Use cases are related to actors and to other use cases using a number of different types of relationship as given below:

- *Association relationship*. This is used to relate use cases to actors and, unlike when used on a block definition diagram, is a simple unadorned line with neither name nor multiplicity as can be seen in the association between 'Actor2' and 'Use Case1' in Figure 3.42.
- *Include relationship*. This is used when a piece of functionality may be split from the main use case, e.g., to be used by another use case. A simple way to think about this is to consider the included use case as *always* being part of the parent use case. This is used to try to spot common functionality within a use case. It is highly possible that one or more of the decomposed use cases may be used by another part of the system. It is shown using a dashed line with an open arrowhead, the line bearing the stereotype «include». The direction of the arrow should make sense when the model is read aloud. In Figure 3.42, 'Use Case1' includes 'Use Case3'.
- *Extend relationship*. This is used when the functionality of the base use case is being extended in some way. This means that sometimes the functionality of a use case may change, depending on what happens when the system is running. A simple way to think about this is to consider the extending use case as *sometimes* being part of the parent use case. Extending use cases are often used to capture special, usually error-handling, behaviour. The extend relationship is also shown using a dashed line with an open arrowhead, the line bearing the stereotype «extend». It is important to get the direction of the relationship correct, as it is different from the '«include»' direction. The direction of the arrow should make sense when the diagram is read aloud. In Figure 3.42, 'Use Case4' extends 'Use Case1'.
- *Specialisation/generalisation relationship*. This is exactly the same relationship as found in block definition diagrams and is used when one use case is a specialisation of another. Just like when used with blocks, generalisation between use cases allows for inheritance of behaviour and relationships. For example, consider the use case diagram shown in Figure 3.43 that shows use cases for a system that governs the delivery of training courses. The general use case 'Organise course' is specialised by the two use cases 'Organise in-house course' and 'Organise public course' which inherit the behaviour described in 'Organise course' as well as include the three uses cases 'Setup course', 'Publicise course' and 'Support course' that are included by 'Organise course'.

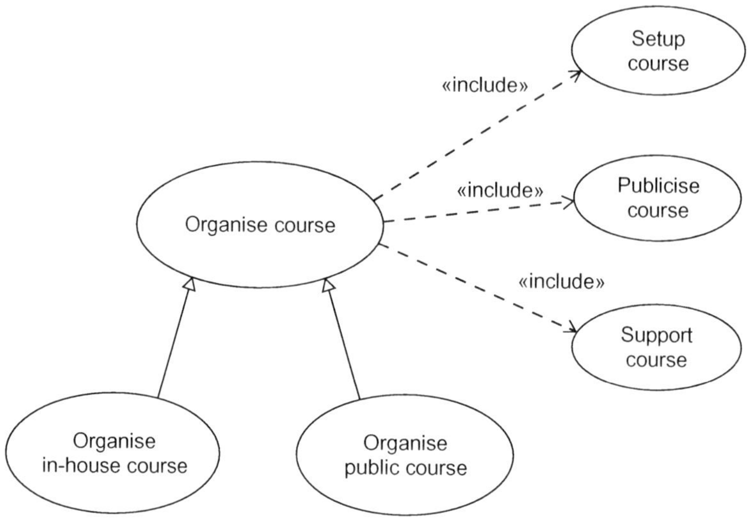

Figure 3.43 Example use case diagram showing generalisation

In a similar way, generalisation can be used between actors, as is shown in Figure 3.42, when one actor is a specialisation of another.

The final element that can appear on a use case diagram is the *system boundary* used when describing the *context* of a system. As its name suggests, the system boundary defines the boundary of the system *from a particular point of view, i.e. context*. Everything inside the system boundary is part of the system; everything outside the boundary is external to the system. Actors are always outside the system, and an association between an actor and a use case that crosses a system boundary indicates that there is an interface between the actor and the system (which may be a sophisticated software and hardware interface but equally can be an interface in which a person passes a note on a piece of paper to another person).

System boundaries are not mandatory on a use case diagram. They are used when use cases are being shown in context. Where a use case diagram is being drawn simply to expand on a use case, as shown in Figure 3.43, then no system boundary is needed.

3.5.3 Usage

This section presents some examples of use case diagrams and related diagramming elements. Further examples of use case diagrams for a real project undertaken by the authors can be found in the case study in Chapter 6. In addition, this section concludes with some guidance notes on common patterns that are often seen in use case diagram that can guide the modeller in refinement of the use case diagrams.

Figure 3.44 shows a use case diagram identifying the high-level use cases for the coffin escape stunt. The context, as indicated by the presence and title of the system boundary, is for the stunt 'system' rather than from the point of view of an individual stakeholder. The relevant high-level stakeholders are shown as actors,

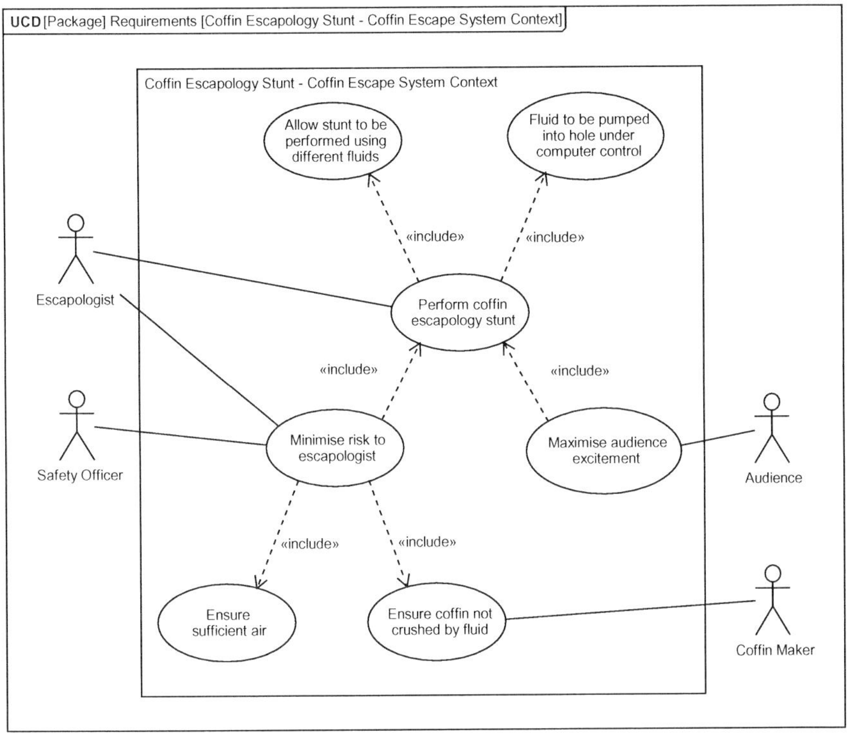

Figure 3.44 Example use case diagram – system context

with associations connecting them to the use cases in which they have an interest, and the relationships between the use cases are also shown. There are two points worth highlighting about this diagram: the number of use cases shown and the use of the «constrain» dependency.

The diagram shows only seven use cases, yet this is the top-level use case diagram showing the use cases for the whole coffin stunt 'system'. Surely, there must be more use cases than this? The answer to this is, of course, yes there are. However, this does not mean that all these use cases have to be shown on a single diagram. Other use case diagrams can be drawn that break these use cases down further and put them into the correct context. Don't forget that these diagrams are produced to aid understanding and communication. A complicated diagram with tens of use cases on it may *look* impressive but is rarely of any practical use (other than for illustrating just how complicated the system is). Consider a system such as an aeroplane. There will be thousands of use cases for the complete system, but how many high-level use cases are there? Probably not many more than 'Take off safely', 'Land safely', 'Have a fully-laden range of *X* km', 'Have a carrying capacity of *X* kg' etc.

The second point to discuss is that of the «constrain» dependency, such as the one between 'Minimise risk to escapologist' and 'Perform coffin escapology stunt'.

The «constrain» dependency is not part of standard SysML, but is an extension used by the authors to show that one use case constrains another in some way. It is created using the SysML stereotyping mechanisms built into the language that allows existing language elements to be extended, which is discussed in detail in Reference 1. Other extensions that are sometimes applied to use case diagrams are the use of the stereotypes «goal» and «capability» on use cases to show those that are modelling organisational or system goals and those that are modelling the capabilities needed to realise those goals.

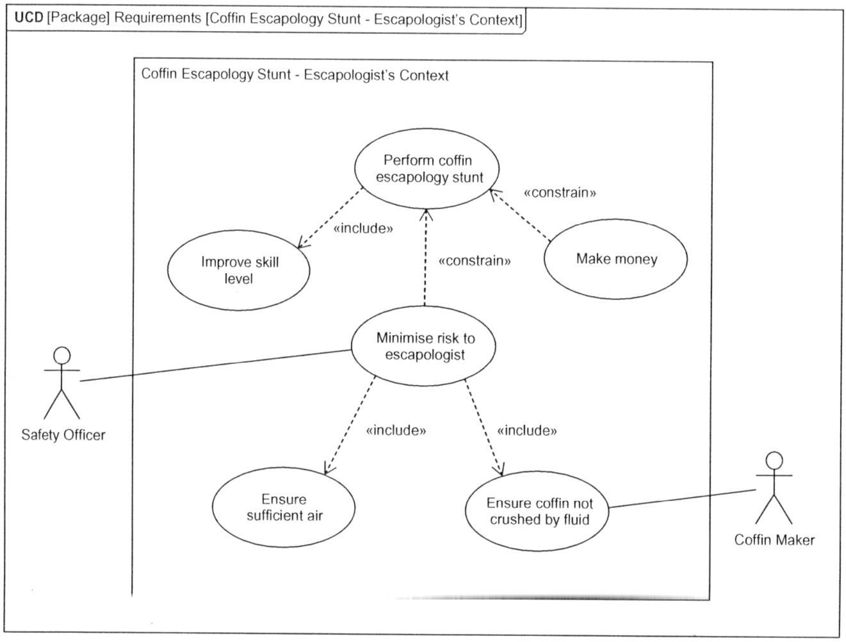

Figure 3.45 Example use case diagram – stakeholder context

Figure 3.45 shows another use case diagram showing use cases in context. However, rather than showing the use cases for the entire system, this diagram shows them from point of view of a single stakeholder, namely the escapologist. Unsurprisingly, some of the use cases shown in Figure 3.44 are also shown in Figure 3.45, since the escapologist is one of (if not *the*) main stakeholders in any escapology stunt. However, some of the use cases in Figure 3.44 (such as 'Maximise audience excitement') are not of direct interest to the escapologist and are therefore not shown in Figure 3.45. Conversely, there are use cases that are relevant only to the escapologist (such as 'Improve skill level'), which are shown in Figure 3.45 but are not relevant from the system context and are therefore not shown in Figure 3.44. This whole idea of *context* is central to ACRE and is discussed in much more detail in Chapter 5. Note also the use of the «constrain» dependency in Figure 3.45.

As discussed in section 3.6.2, a use case diagram does not have to show any actors or contain system boundary. An example of such a use case diagram is shown in Figure 3.46. This diagram is focusing on use case related to the use of different fluids in the stunt and to the computer control of the pump used in the stunt. Two specific types of fluids are identified and are shown via the use of the generalisation relationship between 'Allow stunt to be performed using different fluids' and 'Perform stunt using concrete' and 'Perform stunt using custard'. A use case representing special case behaviour 'Provide computer-controlled emergency fluid removal' extends the standard 'Fluid to be pumped into hole under computer control' use case.

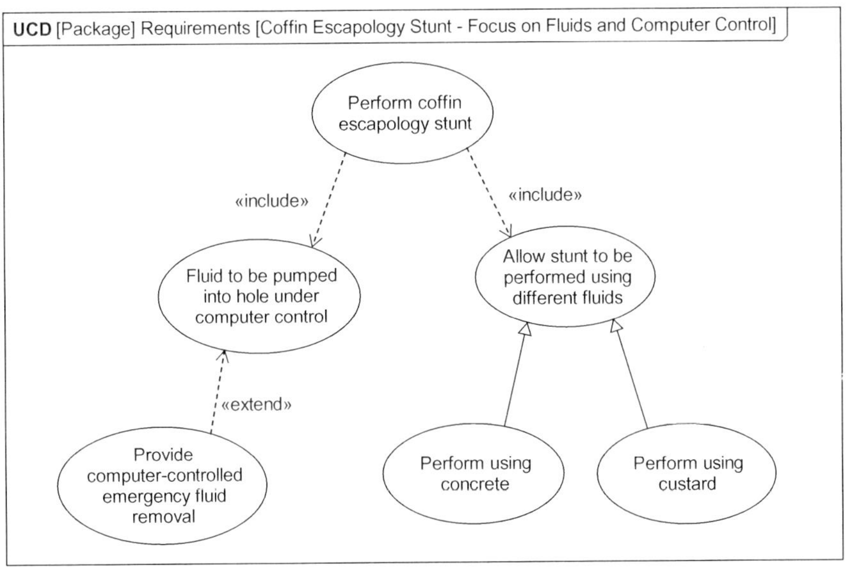

Figure 3.46 Use case diagram without boundary or actors

When developing use case diagrams, there are a number of common patterns that should be looked for as an aid towards the production of good use case diagram. This section concludes with a look at these patterns which cover the following possible situations:

- Use case at too high a level
- Actor at too high a level
- Repeated actors
- Something missing – use cases without actors and actors without use cases

The above four patterns are discussed in sections 3.6.3.1 to 3.6.3.4.

3.5.3.1 Use case at too high a level

One common mistake is to model use cases at too high a level. Consider the case in Figure 3.47.

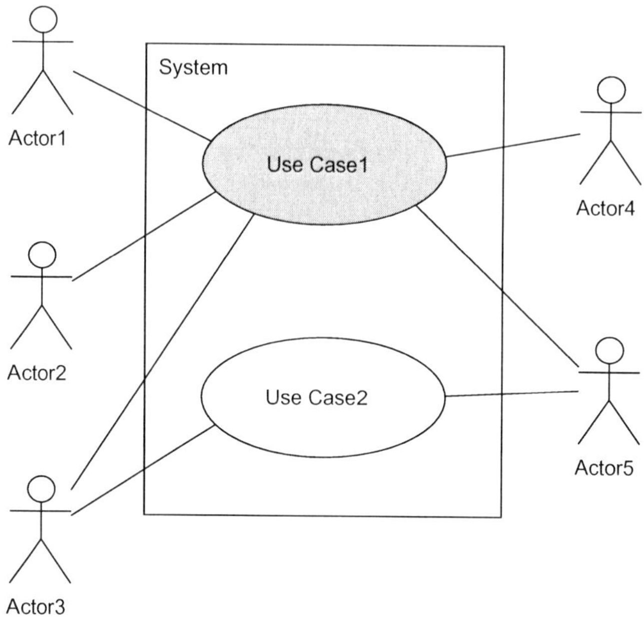

Figure 3.47 Use case at too high a level?

Figure 3.47 shows a use case, 'Use Case1', that is linked to *all* actors. Such a pattern may indicate that the use case is at too high a level and that it should be decomposed further, making use of the «include» and «extend» dependencies to link it to more detailed use cases. The actors would then be associated with the more detailed use cases rather than all being connected to the top-level use case.

3.5.3.2 Actor at too high a level

Another common error is to model actors at too high a level. Consider the case in Figure 3.48.

Figure 3.48 shows an actor, 'Actor2' (drawn in a different style for emphasis), that is connected to *every* use case. Such a pattern may indicate the following:

- The actor is at too high a level and that it should be decomposed further.
- The diagram has been drawn from the point of view of that actor.

If the actor is at too high a level, it should be decomposed further and replaced on the diagram with the new actors. These actors will then be associated with the relevant use cases rather than being associated with all the use cases.

If the diagram has been drawn from the point of view of that actor, i.e. the use case diagram is drawn for that actor's *context*, the actor should be removed from the diagram. The system boundary should indicate that the diagram is drawn for that actor's context.

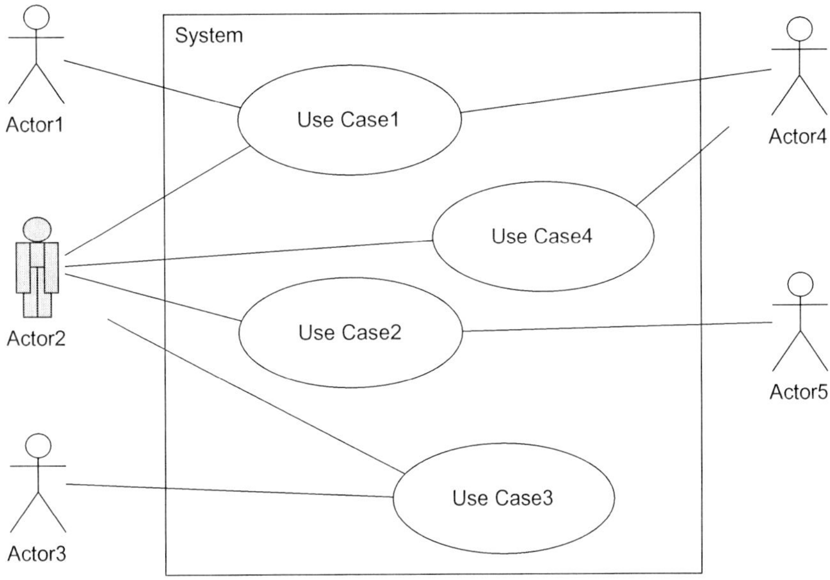

Figure 3.48 Actor at too high a level?

3.5.3.3 Repeated actors

Sometimes a pattern is seen in which two or more actors are connected to the *same* use cases. Figure 3.49 shows this.

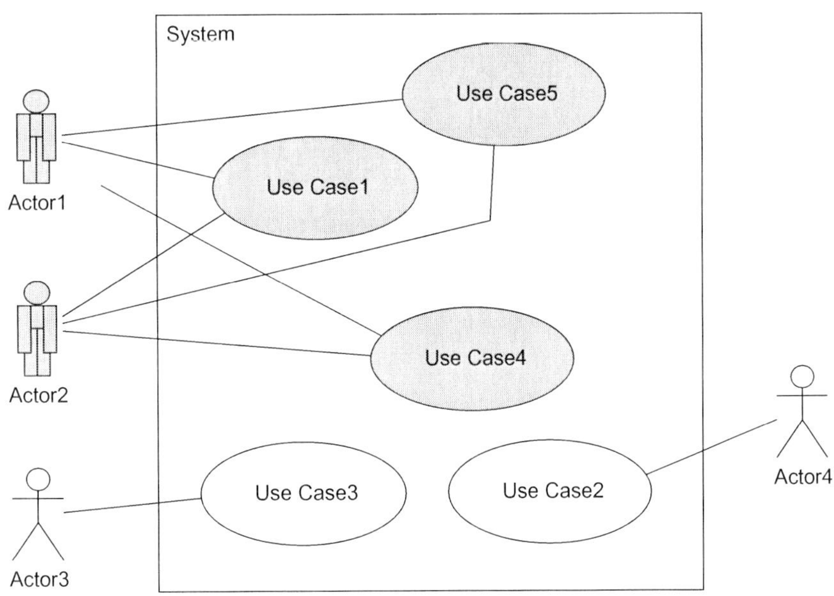

Figure 3.49 Repeated actors?

Here we see two actors 'Actor1' and 'Actor 2', both of which are connected to the same three use cases. This pattern may indicate that the actors are representing the same stakeholder. Alternatively, it may indicate that *instances* of stakeholders have been used (check for names of specific people, organisations, standards etc.). Instances should *never* be used. Remember that a stakeholder represents the *role* of something that has an interest in the project, not an actual instance involved. Any duplicate actors should be removed from the diagram.

3.5.3.4 Something missing – Use cases without actors and actors without use cases

What does it mean if we have use cases or actors that are not related to anything? Consider Figure 3.50.

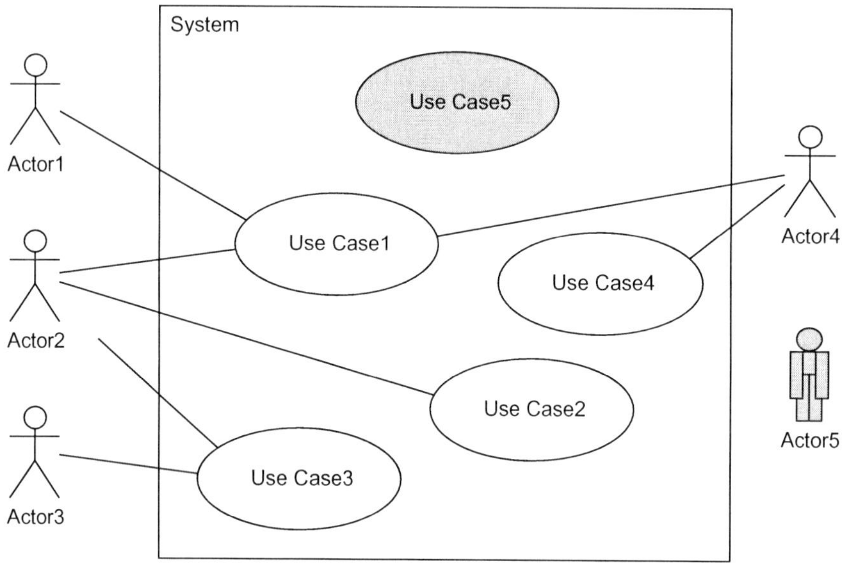

Figure 3.50 Something missing? Basic use case diagram checks

Figure 3.50 has a use case, 'Use Case5', and an actor, 'Actor5', that are not connected to anything else on the diagram.

'Use Case5' has no actors associated with it. There are four possible reasons for this:

1. The use case is not needed and should be removed from the diagram.
2. There is an actor (or actors) missing that should be added to the diagram and linked to the use case.
3. There is an *internal* relationship missing; the use case should be linked to another use case.
4. There is an *external* relationship missing; the use case should be linked to an existing actor.

'Actor5' has no use cases associated with it. There are three possible reasons for this:

1. The actor is not needed and should be removed from the diagram.
2. There is a use case (or use cases) missing that should be added to the diagram and linked to the actor.
3. There is a relationship missing; the actor should be linked to an existing use case.

These two errors are very common, particularly when creating initial use case diagrams, and should be checked for on all use case diagrams.

3.5.4 Summary

Use case diagrams show the highest level behaviour of a system and are used to show requirements in context, along with the stakeholders involved and the relationships between them. This is the central theme of the ACRE, realised in its *Requirement Context View* discussed in detail in Chapter 5.

Care is needed when producing use case diagrams. They should not be over-decomposed so that they start to look like data-flow diagrams and become diagrams detailing the design of the system; they exist to show high-level behaviour as requirements in context. There are a number of common patterns that should be looked for when producing use case diagrams that can help you to spot when use cases or actors are at too high a level, where an actor has been repeated or where there is something missing from a use case diagram.

3.6 Sequence diagrams

3.6.1 Overview

The SysML sequence diagram is a behavioural diagram that is used to show a particular example of the operation of a system by showing how system elements interact with each other over time. The emphasis is on the logical timing of the sequence of interactions (messages) between system elements, with each sequence diagram showing a *scenario* for the system, highlighting pertinent aspects of a particular situation and ignoring all others.

Sequence diagrams relate back to use cases as they allow the ways in which use cases can be satisfied to be explored through different scenarios. In the ACRE, sequence diagrams are used to validate use cases via the ACRE *Stakeholder Scenario View* and *System Scenario View* described in Chapter 5.

3.6.2 Notation

The notation for the sequence diagram is shown in Figure 3.51.

The main element of a sequence diagram is the *life line*, representing a participant in a scenario over a period of time. It is represented by a rectangle with a dashed line hanging below it, as shown in Figure 3.51. The dashed line represents

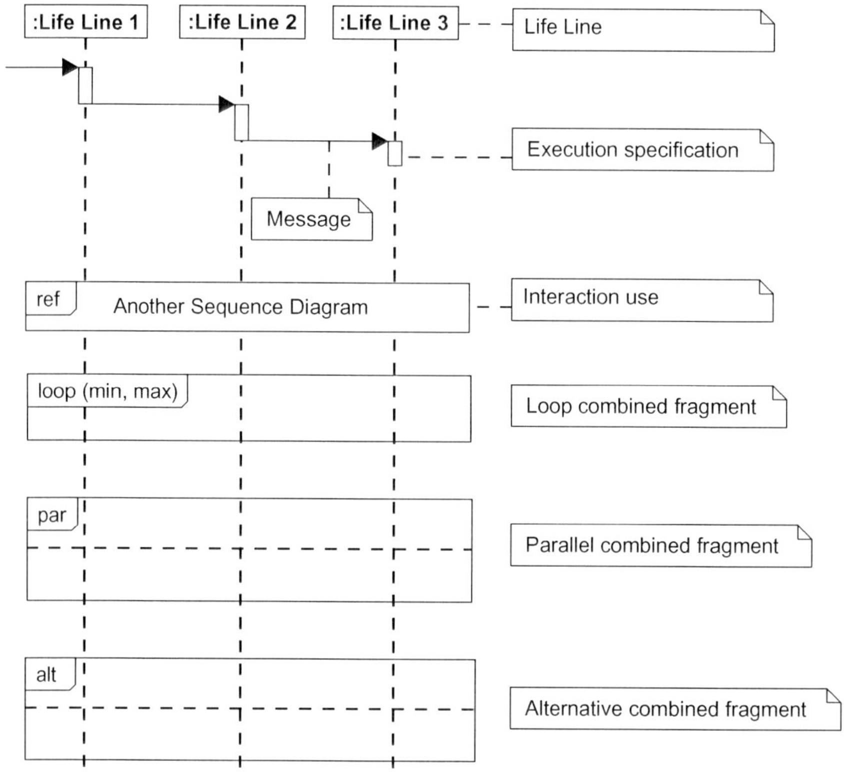

Figure 3.51 Sequence diagram notation

logical time extending down the diagram, with earlier times at the top and later times at the bottom. The sequence diagram is the only SysML diagram in which layout is important, as indicated by this time dimension. A life line will refer to an element from another aspect of the model, such as a block or an actor; it can be thought of as an instance of that element that is taking part in the scenario. This is reflected in the labelling of the life line, placed inside the rectangle, which takes the following form:

```
name : type
```

The *name* part of the label is optional and is used to give the life line a unique identifier in the case where multiple life lines of the same type are used on the same diagram. The *type* indicates the block or actor that the life line is an instance of, and the rectangle can be adorned with the stereotype «block» or the stick man symbol to emphasise that the life line is typed by a block or an actor (see, e.g., Figure 3.52).

The sequence of interaction between life lines is shown by *messages* drawn between the sending and receiving life lines. These messages can be annotated with text describing the nature of the interaction and show the sequence (hence the name of the diagram) of interactions through time. The portion of time during which a

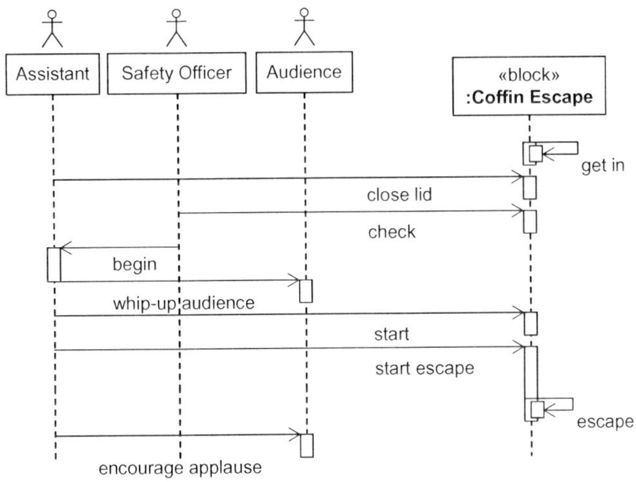

Figure 3.52 Example sequence diagram showing actors as life lines and system as a single block

life line is active is shown by the small rectangles on the dashed line, known as *execution specifications*. A life line can send a message to itself to show that some internal behaviour is taking place. Examples of this can be seen in section 3.7.3.

Complex scenarios can be represented containing looping, parallel and alternative behaviour, shown using various types of *combined fragment*. In addition, a sequence diagram can refer to another via the *interaction use* notation, allowing more and more complicated scenarios to be developed. Examples of the combined fragment and interaction use notation are shown in Figure 3.51, but a more in depth discussion of these notations along with examples are left to Appendix B. However, it is worth sounding a note of caution here. The various combined fragment notations can be nested, allowing very complicated scenarios to be modelled. In particular, the use of the *alternative combined fragment* notation allows alternative paths through a scenario to be shown. What this means is that the sequence diagram is showing more than one scenario. From a SysML perspective, there is nothing wrong with doing this. However, from a modelling perspective such an approach can, in all but the simplest of cases, lead to confusing diagrams. Apart from showing very simple alternatives on a single diagram, the authors would recommend a 'one diagram, one scenario' approach.

There is much more notation available for use on sequence diagrams, including the modelling of timing constraints between messages and the distinction between synchronous and asynchronous messages. See References 1 and 2 for further information.

3.6.3 Usage

This section presents some examples of sequence diagrams. Further examples of sequence diagrams for a real project undertaken by the authors can be found in the case study in Chapter 6.

Figure 3.52 is an example of a sequence diagram that treats the system (in this case the 'Coffin Escape') as a black box; that is, it concentrates on the interactions between stakeholders and the system, modelling the system as a single life line. In addition to showing these interactions, it also shows some interactions that are *internal* to the system, namely the 'get in' and 'escape' messages.

Three other interactions are also worthy of comment, namely the 'begin', 'whip-up audience' and 'encourage applause' messages. These are of interest because they are between *stakeholders* rather than between stakeholders and the system. Some people (and indeed some SysML tools) would consider such interactions as illegal.

Nevertheless, these are essential interactions that are needed to fully describe the scenario (in this case, that of a successful stunt); it is impossible to model this scenario fully without showing them. When considering the system to be the 'Coffin Escape' consisting of equipment, processes and the escapologist (perhaps as modelled in Figure 3.26), the stakeholders shown on Figure 3.52 as actor life lines are outside the system. But this is a question of context. In the wider context of the stunt being performed that includes all the necessary supporting roles and the audience, these stakeholders are *part* of the system and therefore these interactions become interactions between system elements.

Figure 3.53 shows a simple scenario, that of the Assistant starting and stopping the pump used in the stunt. However, unlike in Figure 3.52, the system is no longer treated as a black box. In this figure, the individual elements of the system are shown along with the relevant stakeholder who is shown interacting with one of the system elements (the 'Pump Controller'). The internal interactions between the 'Pump Controller' and the 'Pump' are also shown, as is the behaviour that takes place inside the 'Pump'. Thus, it can be seen that when the 'Pump' receives a 'start' message, it primes itself and then begins pumping. Similarly, on receipt of a 'stop'

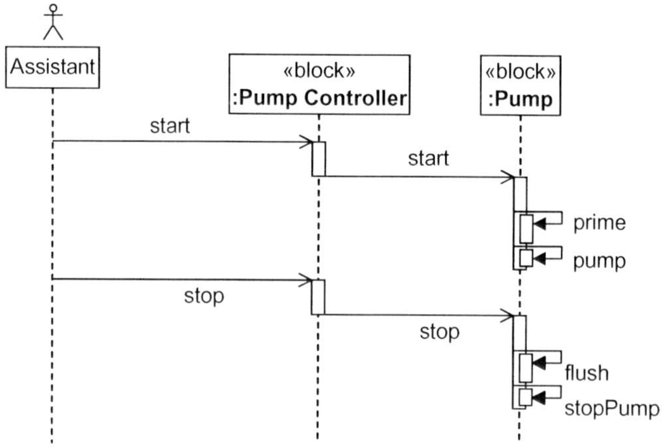

Figure 3.53 Example sequence diagram showing interactions between system elements

message it first flushes itself before stopping. Such white box scenarios are typically developed from black box scenarios, which may have been developed earlier during the requirements engineering process. An equivalent black box scenario for Figure 3.53 is shown in Figure 3.54.

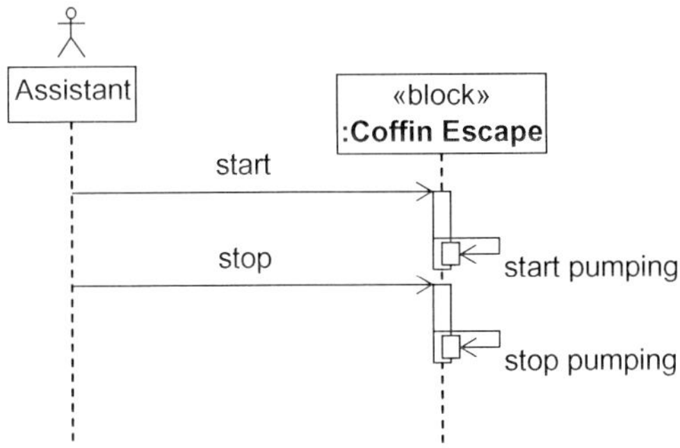

Figure 3.54 Sequence diagram showing equivalent black box scenario

As Figure 3.54 is intended to be the black box scenario from which Figure 3.53 is developed, the diagrams should be consistent. One would expect the interactions between the 'Assistant' and the 'Coffin Escape' system in Figure 3.54 to be the same as those between the 'Assistant' and the relevant system element (in this case the 'Pump Controller') in Figure 3.53, as indeed they are. Similarly, the interactions of the system with itself in Figure 3.54 should be consistent with that between system elements in Figure 3.53. In this case, although the messages are not labelled the same, they are consistent with one another. The difference here is due to the differing levels of abstraction shown on the two diagrams. A single message at the black box system level is refined into a number of messages between and within system elements when the scenario is modelled in more detail.

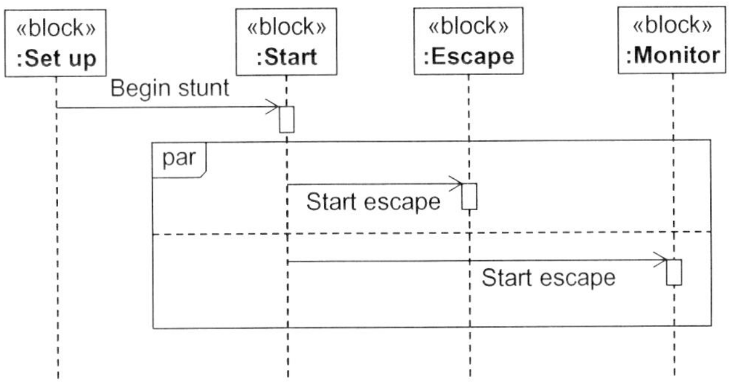

Figure 3.55 Example sequence diagram showing use of parallel combined fragment

The final example in this section, Figure 3.55, presents a scenario where the system elements are not pieces of equipment, rather they represent processes that are carried out as part of the system. The messages between the processes show how one process initiates another, in this case for the scenario showing the successful execution of the stunt. In this scenario the 'Start' process, on completion, has to trigger the 'Escape' and 'Monitor' processes that have to run in parallel. This is shown by the use of the parallel combined fragment, containing two parallel regions, surrounding the two 'Start Escape' messages sent by the 'Start' process.

3.6.4 Summary

Sequence diagrams are used to model *scenarios* that are used to validate use cases. Such scenarios are often modelled as black box scenarios, modelling the system as a single life line or as white box scenarios that show system elements:

- Black box scenarios are often generated when the scenario places the emphasis on the interactions from the point of view of one or more stakeholders, known as the *Stakeholder Scenario View* in ACRE.
- White box scenarios are often generated when the emphasis is on the interactions between system elements, known as the *System Scenario View* in ACRE.

In practice, stakeholders often have to be shown interacting with system elements so that the distinction is often blurred.

3.7 Conclusions

This chapter looks at the two aspects of modelling that are necessary in SysML – the structural and behavioural aspects of a system. Block definition diagrams are used to discuss structural diagrams, and behavioural modelling is discussed by considering how the various diagrams relate to system hierarchy and engineering activity.

A key consideration when modelling is consistency, leading to a good correct, concise and consistent model, which leads directly to confidence in the system. Confidence means that the system is understood and can be communicated to other people.

Remember:

SysML + consistency = model
SysML – consistency = pictures

The nine SysML diagrams are briefly introduced, with the five diagrams that are used by ACRE covered in more detail. For each diagram, its notation is introduced along with examples of its usage. These examples are taken from the world of escapology, namely the so-called coffin stunt, which is chosen since it ties in with the example system used in the case study described in Chapter 6.

References

1. Holt J. & Perry S. *SysML for Systems Engineering.* London: IET; 2008
2. Object Management Group. *SysML* website. Available from http://www.omgsysml.org [Accessed April 2011]
3. Holt J. *A Pragmatic Guide to Business Process Modelling*. 2nd edn. BCS, the Chartered Institute for IT: Swindon; 2009
4. Miller G.A. 'The magical number seven, plus or minus two: Some limits on our capacity for processing information'. *Psychological Review.* 1956; 63:81–97

Part II

Concepts

Chapter 4
Requirements engineering concepts

A consensus means that everyone agrees to say collectively what no one believes individually.

Abba Eban (1915–2002)

4.1 Introduction

This chapter discusses the key concepts associated with the world of model-based requirements engineering. The concepts and terminology introduced here are used throughout the rest of the book and form the basis of the context-based approach to model-based requirements engineering.

To capture and describe the concepts and terminology, an 'ontology' is introduced. An ontology, in the context of this book, provides a visualisation of all the key concepts, the terminology used to describe them and the inter-relationships between said concepts. The ontology, however, is much more than just a data dictionary and plays a pivotal role in the definition and use of any rigorous framework.

The use of ontologies for defining frameworks for architectures, such as enterprise architectures, process architectures, system architectures and so on, is one that is well established and used extensively throughout industry. For examples of the use of ontologies, see References 1–3. Whenever any framework is defined in terms of a set of views, an ontology is essential. It is the ontology that enforces the consistency and rigour demanded by such frameworks.

The ontology that is introduced here covers all of the concepts pertinent to model-based requirements engineering and then, in the next chapter, a number of views are defined based on this ontology. Each view will focus on, and expand upon, a subset of the ontology and instantiate, or realise, specific concepts in the context of a real system or project. The use of the ontology for specifying views is shown explicitly in the next chapter, where the framework is introduced and described in some detail.

4.2 The ontology

Figure 4.1 shows the ontology using a SysML block definition diagram. The diagram is described at a high level in this section, and each concept is fully described and expanded upon in subsequent sections.

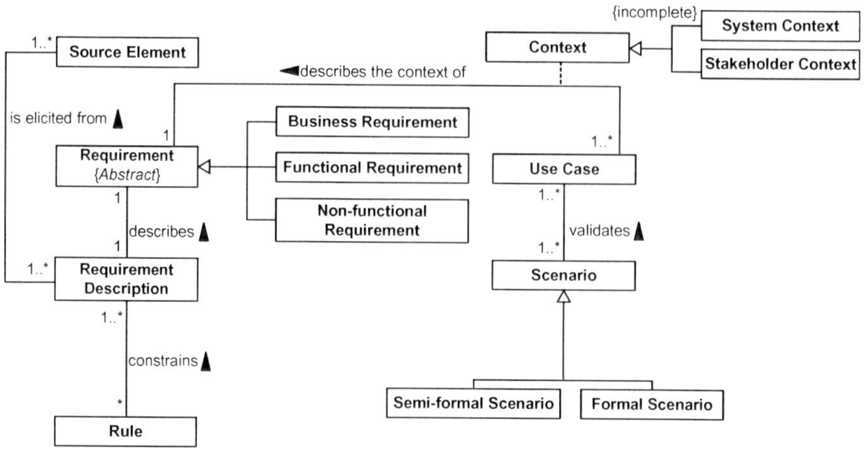

Figure 4.1 The model-based requirements engineering ontology

Figure 4.1 shows that there is an abstract concept of a 'Requirement' that has three types: 'Business Requirement', 'Functional Requirement' and 'Non-functional Requirement'. A 'Requirement Description' describes each 'Requirement', and one or more 'Requirement' is abstracted from one or more 'Source Element'. One or more 'Rule' constrains one or more 'Requirement'.

One or more 'Use Case' describes the context of each 'Requirement' via the 'Context'. Two types of context are shown here: the 'System Context' and the 'Stakeholder Context', although this list is incomplete.

One or more 'Scenario' validates one or more 'Use Case', and there are two types of 'Scenario' – the 'Semi-formal Scenario' and the 'Formal Scenario'.

Each of the elements shown on the ontology is now be described in full in its own section, before the ontology is revisited at the end of the chapter.

4.2.1 The 'Requirement' concept

The prime concept in any requirements engineering approach is that of the 'Requirement'. The requirement is an abstract concept that describes a need or capability of a system or project. Every system has a set of requirements, whether they are formally defined or just exist tacitly inside someone's head. The artefacts that we see when we capture, analyse and document requirements are *not* the requirements themselves, but are representations of each requirement. Of course, a requirement may be represented in any number of different ways, and it is these different interpretations that form the basis of the views that are used in the framework in the next chapter.

A requirement is an abstract and non-tangible concept. Requirements exist as ideas or thoughts, and it is essential, therefore, that we formalise them in some way so that we can capture and reason about them. There is a need, therefore, for requirements to be 'good', but what does this mean and what makes a 'good' requirement? The following points should be addressed when considering a requirement:

- The requirement must be identifiable. Requirements have a habit of evolving over the life of project, and it is important that the original requirement can still be found, even if it has changed beyond all recognition.
- The requirement must be clear. The description of the requirement should be as unambiguous as possible. There are several ways to achieve this, such as stating that when describing a requirement, a verb construct should be used to ensure that the true meaning of the requirement is clearly understood.
- Requirements should also be defined so that they are not solution specific. This is explored later in this section.
- The requirement must be owned. One of the stakeholders must take responsibility for satisfying the requirement.
- The requirement must have an origin. Who or what needs this requirement? This needs to be made explicit.
- The requirement must be able to be verified. It is important that each requirement can be demonstrated to function correctly.
- The requirement must be able to be validated. It is essential that each requirement can be demonstrated that it has been satisfied.
- The requirement must be prioritised. In a world where sometimes not all requirements are met, it is important to know which ones should be satisfied first and in what order. This leads to the need for assigning a priority to each requirement.

Many types of requirement may exist which are often classified into a number of groups using a requirement 'taxonomy', which is simply a classification hierarchy. There is no definitive view on exactly what these types of requirements are, and it may vary depending on the project, the industry or the company. Examples of different types of classification are discussed later in this chapter. Any approach should be flexible enough to allow any number of different types of requirement to be defined. For the purposes of this book, a simple taxonomy is used, which is shown in Figure 4.2.

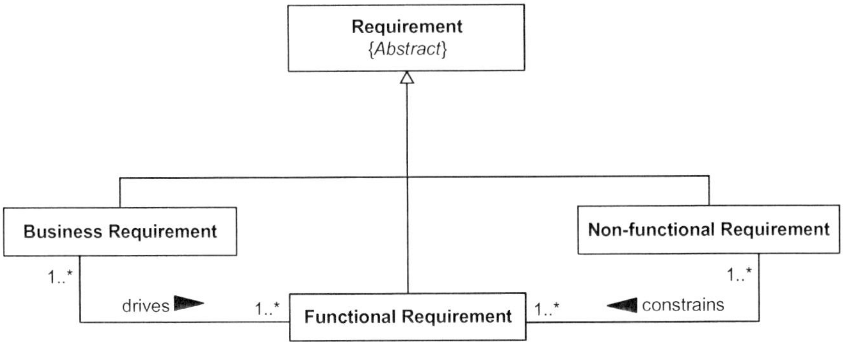

Figure 4.2 Simple taxonomy for types of requirement

Figure 4.2 shows three types of 'Requirement', which are 'Business Requirement', 'Functional Requirement' and 'Non-functional Requirement'.

4.2.1.1 The 'Business Requirement'

A 'Business Requirement' is used to state the needs or capabilities of a business. This includes business drivers that impact the entire organisation and all the projects within it. These requirements will be, by necessity, described at a very high level. Examples of business requirements include the following:

- 'Make money'. This often is an essential business requirement for commercial businesses. Depending on how this requirement is defined within its business context, it may be that all projects in the business must contribute directly to this high-level requirement, or it may be that some projects may contribute indirectly (such as research projects). For not-for-profit businesses and charities, this requirement will, of course, be defined differently.
- 'Be a world leader', or variants thereof. Many businesses define a 'mission statement' that is intended to inspire all employees and give them a common goal to aim for. In many cases these mission statements are not very accurate, comprehensible or feasible and hence mission statements often have a very bad press. When used correctly, these mission statements can be powerful business drivers, and one way to ensure this is to make sure that the mission is correctly reflected in the business requirements for the organisation.
- 'Keep customer happy', and its variants. Many organisations purport that they are committed to keeping their customers satisfied at all costs. If this is indeed the case, this should be reflected in the business requirements.
- 'Provide service *X*' or 'Provide product *X*'. Again, the core nature of the business, whether it is products, services or a mixture of the two, should be captured somewhere at the business level.

Of course, this list is provided only for illustration purposes and is not intended to be an exhaustive list of business requirements. An example of a business context is provided in the case study chapter.

Business requirements should drive every project in the organisation and, as such, every project must contribute either directly or indirectly to meeting the business requirements. If a project exists that does not meet the business requirements, then serious questions must be asked as to whether it should go ahead or not, or whether the business requirements for the organisation should be revisited. Having a good set of business requirements is a good way to provide a basis for ensuring that the business stays on track and maintains its focus.

Business requirements may refer to something that needs to be done or to some sort of constraint on the business. Business requirements may, therefore, reflect 'functional' or 'non-functional' requirements at a business level. See sections 4.2.1.2 and 4.2.1.3 for a full description of these two terms.

4.2.1.2 The 'Functional Requirement'

Functional requirements, in their essence, yield some sort of observable result to someone, or something, who is using the system. By their very definition, functional requirements 'do' something and result in some sort of function being performed. Functional requirements are usually what are referred to when people misuse the term 'user requirements'. See the section on contexts for a full discussion on this matter. Examples of functional requirements include the following:

- 'Do *X*'. A functional requirement often looks like a direct instruction to do something.
- 'Provide service *X*'. A functional requirement may state that a particular service must be provided to a set of stakeholders.
- 'Deliver *X*'. A functional requirement may state that a particular product or artefact must be delivered to a set of stakeholders.

Functional requirements tend to have verbs that reflect direct action, such as 'do', 'provide', 'build' and so on. It is often possible to distinguish between functional and non-functional requirements based on the verbs that are used to describe them.

This list is not intended to be exhaustive but to provide some simple examples of the types of needs that may be represented by functional requirements.

Functional requirements should be represented in such a way that they may be realised by any number of different system or technologies. Ideally, all requirements should be solution-independent and free to be realised in any way that can be imagined by the requirement owners. In real life, however, it is rarely the case that all requirements can be specified in such a broad way, and there may be limitations on the way that requirements may be realised. When such limitations exist, we say that a requirement is being constrained by one or more 'Non-functional Requirement'.

4.2.1.3 The 'Non-functional Requirement'

A 'Non-functional Requirement' constrains or limits in some way, the way that a functional requirement may be realised. It should be noted that the term 'constraint' is often used rather than 'Non-functional Requirement'. This is perfectly valid but is deliberately not used in this book as the formal term on the ontology. This is because the term 'constraint' is one of the key constructs in the SysML notation and therefore to avoid unnecessary confusion, the term 'Non-functional Requirement' is used.

It is essential to identify and understand the non-functional requirements that exist in a system. Non-functional requirements are sometimes treated as secondary requirements that are not as important as functional requirements. Although this may be the case in some instances, overall it is the satisfaction of these non-functional requirements that decide whether the project is successful or not. Non-functional requirements are directly linked to functional requirements, which means that they *must* be considered as important as the functional requirements.

Requirements taxonomies in general focus on classifying the non-functional requirements. This is a symptom of the complexity added through the development of non-functional requirements. Examples of non-functional requirements that may be used to populate a requirement taxonomy include the following:

- 'Meet standard *X*'. In most real systems there are non-functional requirements that insist that a particular standard must be complied with. The term 'standard' is used here in its broadest sense and may include international standards, industry standards, in-house standards, work instructions, procedures and so on. Such non-functional requirements are often referred to as 'quality requirements'.
- 'Comply with legislation *X*'. In many real systems there may be a non-functional requirement that insists that a particular piece of legislation or law must be complied with. Obviously, these requirements are usually mandatory as laws tend not to be optional! Such non-functional requirements are often referred to as 'legal requirements'.
- 'Use technology *X*'. In many systems there may be a non-functional requirement that states that a specific technology must be used to realise the system. For example, it may be that a specific modelling must be used (such as SysML) for modelling the system, that a specific set of electronics must be used, that a specific manufacturer must be used, that a particular programming language must be used and so on. These non-functional requirements limit the way that a requirement may be implemented and, therefore, these are often referred to as 'implementation requirements'.
- 'Ensure performance *X*'. In many systems, such as safety-critical and real-time systems, there are non-functional requirements that dictate minimum or maximum limits on the performance of the system. Examples of this include a statement that the system must respond between two time limits, that a system must be able to cope with certain data capacity, that the business must operate within a specified profit margin and so on. Such non-functional requirements are often referred to as 'performance requirements'.
- 'Ensure size *X*'. Most physical systems have non-functional requirements that specify the physical size of some aspect of the system. Example of this include ensuring that a cabin is big enough to fit an operator into, that a vehicle will fit into a tunnel, that a computer is small enough to fit into a pocket and so on. These sorts of size-limiting non-functional requirements are often referred to as 'physical requirements'.
- 'Ensure that the system will work with system *X*'. In almost every real-life system there is some sort of interface with another system. This may be a physical, an electronic, a visual, an aural and any number of other types of interface. Non-functional requirements are used to describe how the system under development must work with other systems. Examples of this include ensuring that a plug on a lead fit into a pre-existing socket, that a communication system uses the same protocol as an existing system, that software runs on a particular operating system and so on. Such non-functional requirements are often referred to as 'interface requirements'.
- 'Ensure system can be operated by *X*'. Any system that involves human needs to consider certain human factors, such as usability, comfort, temperature and

so on. Examples of this are ensuring that the system has buttons that can be seen in the dark, that the system operates within defined temperature ranges, that the number of GUIs (graphical user interfaces) is below a certain number and so on. These non-functional requirements are often referred to as 'human factor' requirements. It should be noted that human factor requirements are notorious difficult to quantify and that there is an entire discipline of study related to human factors.

This list is not intended to be exhaustive, but it should provide an indication of just how complex these non-functional requirements can be. In fact, one of these requirements may result in another, e.g. a human factors requirement that describes the comfort of an environment may contain a quality requirement that specifies a standard that must be met.

When considering non-functional requirements, it is worth remembering that by its own definitions each will constrain another requirement. This is important to understand as there is a tendency to simply list non-functional requirements and not to think about what they are actually constraining. This is discussed again in section 4.2.6.

4.2.2 The 'Requirement Description' concept

The 'Requirement Description' is an essential element of any requirements engineering approach, but it is one that is very often misunderstood. The Requirement Description is exactly what it says – it is a description of a requirement. The 'Requirement Description' is *not* the requirement itself but is a way to describe the abstract concept of the requirement. This is needed because, as described in the previous section, there are a number of features that make up a 'good' requirement and it is the 'Requirement Description' that captures these. This relationship is shown in Figure 4.3.

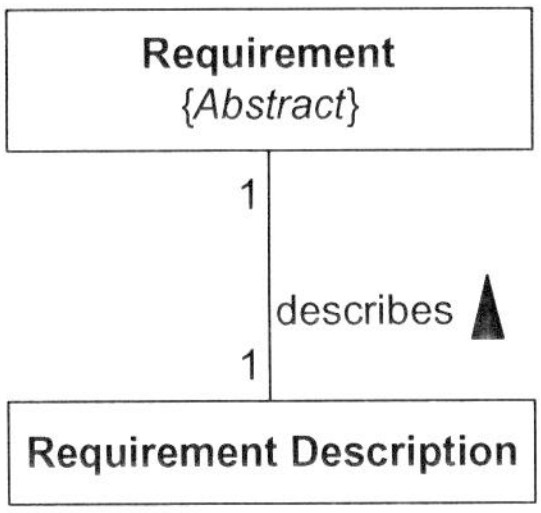

Figure 4.3 The relationship between the 'Requirement Description' and the 'Requirement'

Figure 4.3 presents a subset of the ontology showing that each 'Requirement Description' describes a 'Requirement'.

Again, it must be stressed that the 'Requirement Description' is not the requirement but a representation of it. Of course, each 'Requirement' must have a 'Requirement Description', but the fact that each requirement has an associated description does not mean that the requirement itself is understood – in fact far from it. The 'Requirement Description' provides a high-level description of each 'Requirement', but this does not mean that the 'Requirement' has been given meaning. A 'Requirement' has not been given a meaning until it has been put into context. See the section on use cases for more on this.

The Requirement Description has the following uses:

- *Describe the requirement.* This often provides many people's first view of the requirement.
- *Provide a basis for traceability.* Traceability is essential for systems engineering, and requirements engineering is no different. It must be possible to trace from any element in the development right back to the Requirement Descriptions and, hence, the requirement Source Elements. Once this traceability is established, it is possible to trace back to requirements and forward to the rest of the development. One of the advantages of adopting a model-based approach is that traceability is 'free' in a model, as it is an inherent feature of a good model.
- *Provide a basis for measurement.* Very often it is desirable to measure requirements, whether it is a crude measurement, such as the number of requirements, or a more sophisticated measure, such as a complexity measure. Such measurements may be applied to individual Requirement Descriptions or to the requirements set as a whole.
- *Provide a basis for contractual agreements.* The requirements will often form part of the contractual agreements on a project, and it is usually the Requirement Descriptions that are used for this. Clearly, traceability and validation are important issues here.

All Requirement Descriptions describe requirements and must be abstracted from source information, which is discussed in section 4.2.3.

4.2.2.1 Typical properties of a requirement

It is important that each individual requirement is described in a consistent and accurate way. It is usual to identify a number of features or properties for each requirement. Although there is no definitive set of these properties, there are some common, best-practice properties that occur in most approaches. For example, the SysML suggests two properties that must be defined as an absolute minimum: the identifier and the description. In reality, there should be more than these two, but they are intended as a start point only and the SysML specification recommends that more are defined. Some examples of attributes that may be considered are shown in Figure 4.4 [4].

Figure 4.4 shows the concept of a requirement represented graphically as a SysML requirement block and the properties of that requirement are shown as the list of properties:

- 'Absolute ref' – provides a mechanism for traceability that the requirement can always be identified, regardless of how much the description may evolve or change

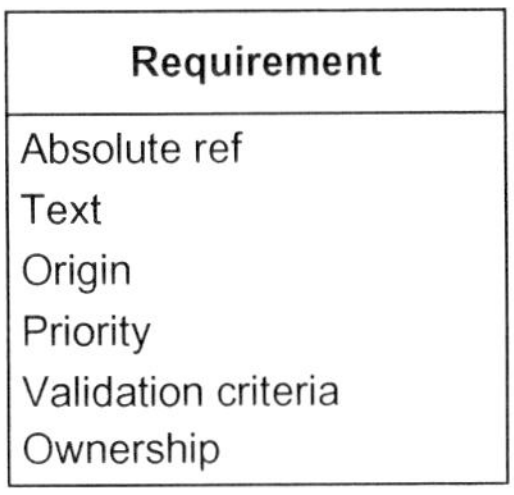

Figure 4.4 Properties of a requirement

- 'Text' – forms the main description
- 'Origin' – a reference to the stakeholder that originated the requirement
- 'Priority' – usually one of a defined set of values, such as 'essential', 'desirable', 'bells and whistles' or 'unknown'
- 'Validation criteria' – the criteria that reference a set of Validation Views
- 'Ownership' – a reference to a stakeholder who is responsible to ensure that the requirement is satisfied

Of course, these properties match the essential elements of describing a requirement that were discussed previously.

4.2.3 The 'Source Element' concept

All requirements must originate from somewhere and this is where the 'Source Element' comes into play. Figure 4.5 shows how the 'Source Element' relates to the other concepts in the ontology.

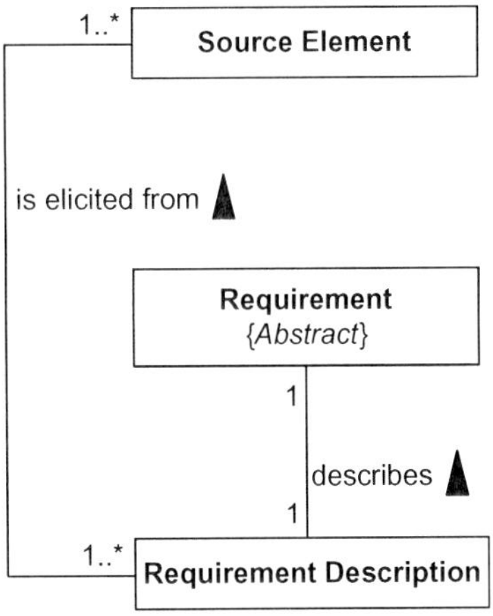

Figure 4.5 Relationship between 'Source Element' and other elements

Figure 4.5 shows that one or more 'Requirement Description' is elicited from one or more 'Source Element'. Note that the 'Source Element' is still traceable to the conceptual 'Requirement' but that this is achieved via the 'Requirement Description'.

For a 'Source Element' to be usable, it must be a configurable item. This means that it must be identifiable in terms of its version number and findable in terms of its location.

In terms of what a 'Source Element' actually is, the list of examples is almost endless as a 'Source Element' can be just about anything. Examples of Source Elements include the following:

- *Requirement lists*. Sometimes people generate lists of (what they think are) requirements. These are usually just statements concerning the requirements that have not been analysed in any way and are, therefore, classified as requirement sources rather than requirements.
- *Conversations*. Possibly the most informal of requirement sources, but one of the most common and often most useful. The main problem with conversations is how to make them configurable items. This can be achieved through making notes or even making recordings of comments.
- *Emails*. Again, quite informal but emails can be invaluable sources for requirements. Emails can be simply saved or their contents pasted into a document.
- *Workshop outputs*. An incredibly powerful way of obtaining requirement sources is through holding workshops. The format can be almost anything, ranging from text notes to sketches, models and even video and audio recordings.
- *Business requirements*. It has already been stated that business requirements drive functional requirements and this makes them ideal requirement sources.
- *Standards*. Love them or hate them, standards have a massive impact in most systems, and they can also be excellent sources for requirements. They can be sources for any type of requirement although they are typically viewed as non-functional requirements.
- *Existing systems*. No system exists in isolation, and one system always interacts in some way with another system. When these systems are known and understood, they often provide non-functional requirements that constrain the new system, such as interface requirements.
- *Specifications or designs*. Requirements engineering activities are typically carried out towards the beginning of a project, but it should be remembered that the requirements engineering process can be applied at any point in the development life cycle. Indeed, the requirements can and should be revisited throughout the life cycle. As such, it is often the case that specification and designs, whether from the system being developed or other systems, can provide valuable insights for new requirements.
- *Information sources, such as books and articles*. Again, no system exists in isolation and just about anything can be the inspiration behind a requirement. These can be books and articles, videos, paintings, music – in fact anything that makes people think.

- *Higher level requirements*. It must be stressed that requirements engineering is a highly iterative activity and can be carried out at any level of abstraction of a system. Of course there will be high-level system requirements, but requirements will exist at every level, such as subsystem, assembly and component levels. As such, higher level requirements often provide a source for lower level requirements.

This list is by no means exhaustive and can include just about anything. The key point here is that requirements can be abstracted from almost anything but, regardless of what they are, it is essential that there is traceability established between these sources and the Requirement Descriptions.

4.2.4 The 'Rule' concept

When describing any requirement, there is a lot of room for ambiguity and misinterpretation. To minimise these problems, it is quite common to see a number of rules defined that are applied to describing requirements.

This idea of a 'Rule' is shown in Figure 4.6.

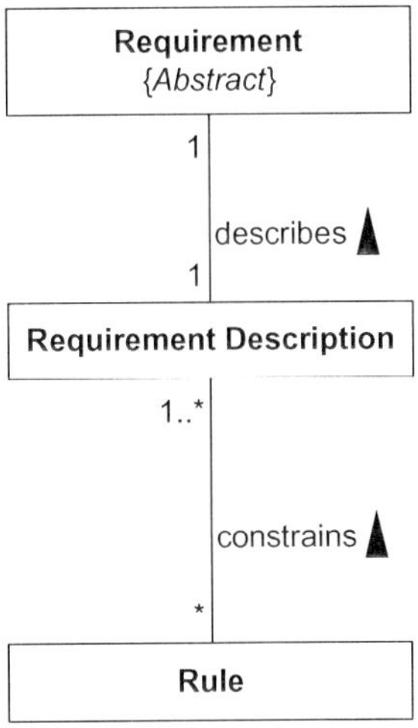

Figure 4.6 Defining a 'Rule'

Figure 4.6 shows that one or more 'Rule' constrains one or more 'Requirement Description'. These rules may apply to the requirement itself or, more usually, to the properties of a requirement.

Sometimes these rules apply to the way that the wording in a Requirement Description must be applied. One of the best examples of wording rules concerns the use of words such as 'shall', 'may', 'should' and 'can'.

The IEEE has a set of guidelines that describes all sorts of requirements for writing standards, which contains some useful information concerning the use of these words. The following extract is taken from the 'IEEE Standards Style

Manual' [5] (note that the bold text here has been added by the authors to highlight pertinent parts of the text):

> The word **shall** is used to indicate mandatory requirements strictly to be followed in order to conform to the standard and from which no deviation is permitted (**shall equals is required to**). The use of the word must is deprecated and shall not be used when stating mandatory requirements; must is used only to describe unavoidable situations. The use of the word will is deprecated and shall not be used when stating mandatory requirements; will is only used in statements of fact.
>
> The word **should** is used to indicate that among several possibilities one is recommended as particularly suitable, without mentioning or excluding others; or that a certain course of action is preferred but not necessarily required; or that (in the negative form) a certain course of action is deprecated but not prohibited (**should equals is recommended that**).
>
> The word **may** is used to indicate a course of action permissible within the limits of the standard (**may equals is permitted to**).
>
> The word **can** is used for statements of possibility and capability, whether material, physical or causal (**can equals is able to**).

The extract here provides good guidance on the use of language that can be embedded into rules that constrain the way in which the requirement can be described. In this case, and using the terminology introduced previously in this chapter, these rules would apply to the 'Description' property of the 'Requirement Description'.

Another example of a rule concerns the complexity of the text description that is being used to describe the requirement. There are many best practice complexity measures that can be used to assess the complexity of a sentence or paragraph of text. One of the most widely used is the Flesch Reading Ease Score [6] that indicates how easy text is to read. This is measured according to the following formula:

$$206.835 - 1.015(\text{total words}/\text{total sentences}) \\ - 84.6(\text{total syllables}/\text{total words})$$

This formula results in a score between 0 and 100, where a high score indicates that the text is easy to read, whereas a low score indicates that the text is more difficult to read. Again, this rule would be applied to the 'Description' property of a 'Requirement Description'.

Another example of a different type of rule may be where the value of a property of a 'Requirement Description' is constrained by the values that it can take on. For example, this may take the form of an enumerated list where a property of 'Priority' may only take on one of four values: 'Essential', 'Desirable', 'Bells and whistles' or 'Unknown'. The value that a property may take on may also be constrained by defining a special 'type' for the value, e.g. the value of an 'Owner' property may be constrained to be defined as a reference to one of a number of stakeholders.

The previous three examples show how various properties of Requirement Descriptions may be constrained by rules, but these rules may also apply to the set of Requirement Descriptions. For example, the total number of requirements may be measured.

The four examples of rules that are shown here were chosen to make a specific point: the rules themselves can take on many different forms and can be applied to either the properties of the Requirement Descriptions or the Requirement Descriptions themselves. These rules can be realised in many ways, ranging from simple text-based constraints, to SysML parametric constraints, to formal methods.

4.2.5 The 'Context' concept

The idea of the 'Context' is fundamental to the approach taken in this book and, hence, it is very important that the concept is well understood. In its simplest form, a context may be thought of as a 'point of view'. It is essential, however, that it is well understood from what point of view each context is taken. It is possible to view the requirements of a system from any number of different points of view, so it is essential that the origins for these points of view are well defined. Figure 4.7 shows two of the more common contexts that are possible.

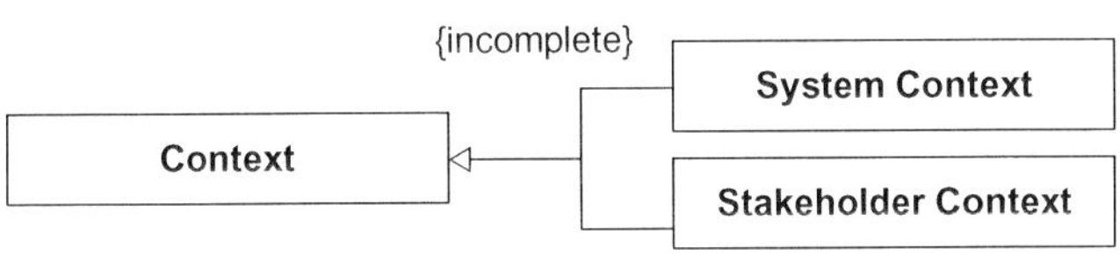

Figure 4.7 Definition of two types of 'Context'

Figure 4.7 shows that there are two types of 'Context', which are the 'Stakeholder Context' and the 'System Context'. The diagram also shows that the full set of contexts shown here is incomplete (indicated by SysML {incomplete} constraint), so there are many other types of context that are not shown here.

4.2.5.1 Stakeholder context

The first type of context that is considered is the 'Stakeholder Context'. The stakeholder context is a set of point of views that is defined by looking at a set of requirements from the point of view of different stakeholders. Key to getting this right is being able to identify what the various stakeholders are. A stakeholder is defined as a role of any person or thing that has an interest in a system. Each stakeholder is realised by a name or set of names, which is associated with the role. The following points should always be remembered when considering stakeholders:

- A stakeholder represents the *role* played by 'something'.
- This 'something' is associated with the stakeholder role and represents how the stakeholder manifests itself for the system.

- This 'something' may be a person or thing. A stakeholder is not always manifested by a person and very often it will be another system, an organisation, a standard and so on.
- A single stakeholder may have multiple 'somethings' associated with it.
- A single 'something' may have multiple stakeholders associated with it.

To identify and reason about stakeholders, it is usual to classify them into a number of different groups. Figure 4.8 shows a generic set of stakeholders in the form of a classification hierarchy, or taxonomy.

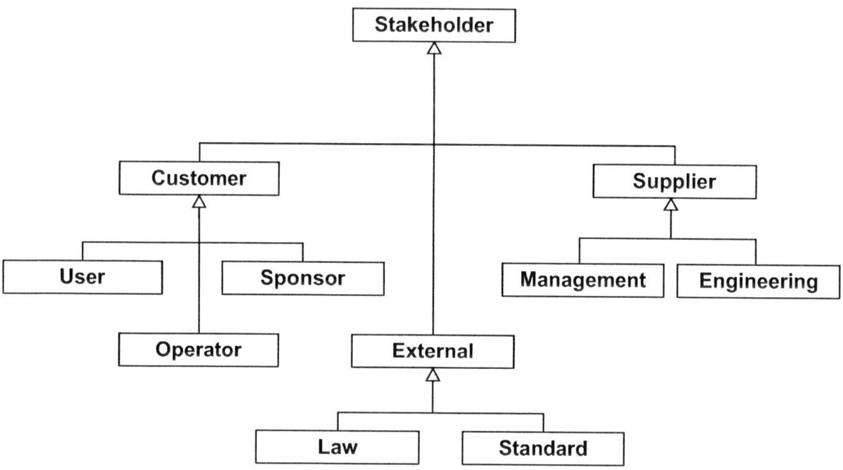

Figure 4.8 Generic set of stakeholders

Figure 4.8 shows a generic set of stakeholders. There are following three broad types of 'Stakeholder':

- 'Customer', which describes all roles that may make use of the system or service or that the supplier needs to keep satisfied. In many cases, the supplier may have some influence over the customer and is in a position to discuss issues and agree on compromises with them.
- 'External', which describes the roles that must be satisfied by the supplier but that cannot be negotiated with in any way. In other words, the supplier has no influence on external stakeholder roles.
- 'Supplier', which describes all the roles that are involved with the development and delivery of the product or service.

This broad categorisation is quite common and can be seen in a number of different standards – see (ISO 15505) and (ISO 15288) for examples – and is a good start point for helping to identify a set of stakeholders.

A number of other sub-types of stakeholder have also been identified on the diagram. To explain these special types in a coherent fashion, it is a good idea to consider the stakeholders involved with an example system. Let us consider, therefore, a passenger airline system as a basis for explaining the various stakeholder roles.

For the 'Customer' stakeholder roles, the following will happen:

- The 'User' role will be realised by any number of passengers who are the end users of the airline. In almost all cases, this will be a list of names that can be specified to realise a single role. In other words, a single stakeholder role is not limited to a single name and may have any number of names associated with it in order to realise that role.
- The 'Operator' role will encompass all the employees or associates of the airline, ranging from ticket sales, to the pilot and stewards, to the air-traffic controllers. Again, this is an example of many names that will realise a single role. Note here that the role does not imply in any way how it is realised. For example, the role of ticket sales may be realised by a person (or people) or by an automated system. The stakeholder role does not care how it is realised and is only concerned with the definition of the role.
- The 'Sponsor' role will represent all the people or businesses that have a financial stake in the airline enterprise. Again, this will be realised by many names in reality. Note here, however, that some of the names may represent organisations or businesses, rather than individual names.

This makes the case quite nicely for a single stakeholder role having any number of names associated with it. It is also possible, however, to have a single person that takes on any number of roles. Imagine a person who is a pilot for their main job. When working, this person takes on the role of a pilot, which will be a type of 'Operator'. When the person is not working, they may go on holiday and they may travel on an aeroplane. In this situation, the same person that usually takes on the role of pilot (when working) will take on the role of passenger (which will be a special type of 'User'). Also, that person may have shares in the company, and this means that the same person will sometimes take the role of shareholder, which will be a type of 'Sponsor'. In this example, it is possible for a single person to take on multiple roles, which is why it is essential to avoid individual or company names when it comes to specifying stakeholder roles.

For external roles, the following will happen:

- The 'Law' role will include all legal roles that may exist. Examples of this role include individual piece of legislation, treaties or agreements, national or international laws or any industry-specific mandatory requirements. This will be realised by all the laws and legislation that apply to the world of air travel.
- The 'Standard' role will encompass all the standards that may apply to the airline. This will range from international standards, to country-specific standards, to industry standards, right down to in-house processes, guidelines and work instructions.

Note in the example here that none of the roles discussed in the above list is realised by people, but they are 'things'. In some cases, these things may be pieces of paper (documents) or may even be virtual documents (such as work instructions on the Internet).

For supplier stakeholder roles, the following will happen:

- The 'Manager' roles will be realised by all the managers involved in the project. This will include, e.g., project managers, operational managers, and configuration managers.
- The 'Engineer' role will be realised by all the engineers involved in the project, such as software engineers, hardware engineers, electrical engineers, and mechanical engineers.

Once the stakeholders have been identified, it is possible to create a set of contexts. When considering contexts, the following basic facts should be considered:

- The stakeholders identified here form the basis for defining a number of contexts.
- Each of these stakeholders will have their own point of view or context.
- Each context will be used to express the requirements of system from the point of view of a single stakeholder, in the form of 'use cases' – explained in the next section.
- Each context will potentially conflict with other contexts, as they represent different point of view.
- All contexts must be consistent with one another to form a model. By looking at the various contexts together and applying the modelling techniques discussed in Chapters 2 and 3, it is possible to identify conflicts, overlaps, similarities and so on.

The identification of stakeholders is one way to help to define a number of contexts that form the heart of the requirements model. These contexts lead to the definition of use cases.

4.2.5.2 System contexts

The second type of context that is considered is a set of contexts based on a system or, more specifically, the level of hierarchy of a system. This type of context is particularly relevant where a system is being developed that can be broken down into many subsystems, assemblies and components. This is very often the case where manufacturing systems are concerned, such as in the automotive and rail industries.

Figure 4.9 shows a simple, generic hierarchy of a system that is used to discuss the concept of the system context.

Figure 4.9 shows a generic hierarchy where a 'System' is made up of one or more 'Subsystems', each of which is made up of one or more 'Assemblies', each of which is made up of one or more 'Components'. The terms used here are intended to be generic and used to represent the different levels that exist in a system – in reality, different terms are used to reflect the industry or organisation. The hierarchy itself can have any number of levels on it – in this case there are four levels in the hierarchy, but this will vary depending on the system.

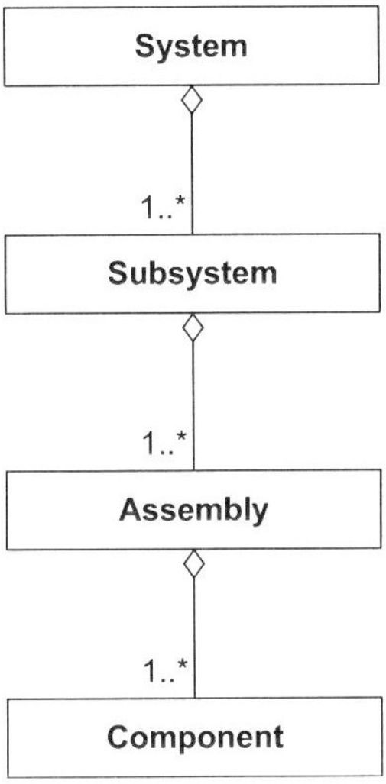

Figure 4.9 A simple, generic system hierarchy

When considering such a system, it is quite common to have a number of different types of requirements defined that exist at the various levels in the hierarchy. This is shown in Figure 4.10.

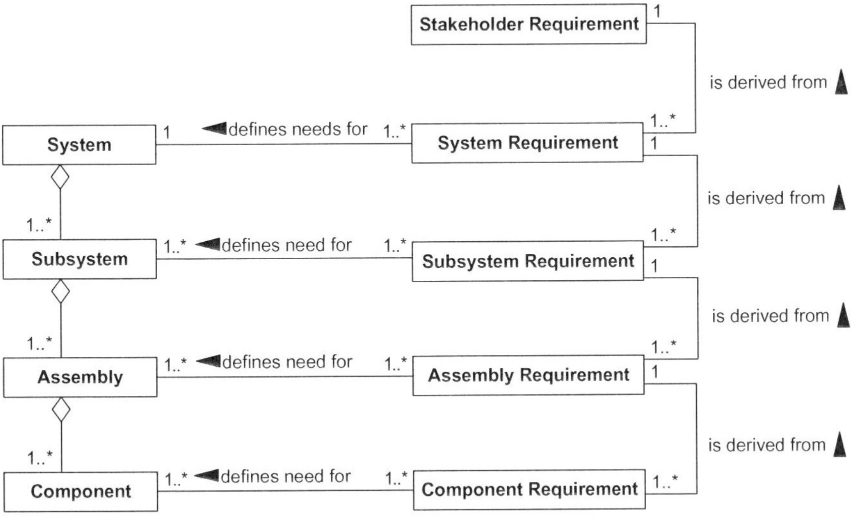

Figure 4.10 Requirements at different levels of system hierarchy

Figure 4.10 shows different types of requirement that exist at different levels of hierarchy. The highest level requirement shown here is the 'Stakeholder Requirement' that has been discussed in the previous section. It is then quite common to derive lower level requirements from these which are shown as one or more 'System Requirements'. Further lower level requirements are then derived from these that reflect the levels of the hierarchy, which are one or more 'Subsystem Requirement' is derived from each 'System Requirement', one or more 'Assembly Requirement' is derived from each 'Subsystem Requirement' and one or more 'Component Requirement' is derived from each 'Assembly Requirement'.

By considering the different levels of hierarchy, it is possible to identify a context at each level. In other words, each level of the hierarchy has its own point of view. To illustrate this, consider the example of a car as described below:

- At the 'System' level, the contexts will be from the point of view of various systems. In this example, these systems will be the car itself and the other systems may be other cars, the environment, drivers, passengers and so on. These system requirements are derived from the initial stakeholder-level requirements and will form the overall requirements for the car.
- At the 'Subsystem' level, the contexts will be from the point of view of various major elements that make up the car. In this example, these major elements may be the chassis, the engine, the transmission system and so on. So, the engine will have its own context that must be consistent with the contexts of the chassis and the transmission system that it must interact with. These subsystem-level requirements will be derived from the higher level system-level requirements.
- At the 'Assembly' level, the contexts will be from the point of view of the major assembly elements. So, if the engine is taken as the subsystem, this may be broken down into various assemblies, such as the fuel assembly, the cylinder assembly and so on. The fuel assembly, therefore, will have its own context, as will the cylinder assembly and so on. These assembly-level requirements will be derived from the higher level subsystem-level requirements.
- At the 'Component' level, the contexts will be from the point of view of the major components that make up the assemblies. So, in the case of the fuel assembly, components may be the injection needle, the injection timer, the component harness and so on. The injection needle will, therefore, have its own context, as will the other components. These component-level requirements will be derived from the higher level assembly-level requirements.

Once the system hierarchy levels have been identified, it is possible to create a set of contexts. When considering contexts, the following basic facts should be considered:

- The system hierarchy levels identified here form the basis for defining a number of contexts.
- Each one of these system hierarchy levels will have their own point of view or context.
- Each context will be used to express the requirements of system from the point of view of a single system hierarchy level, in the form of 'use cases' – as explained in the next section.
- Each context will potentially conflict with other contexts because they represent different point of view.

- All contexts must be consistent with one another to form a model. By looking at the various contexts together and applying the modelling techniques discussed in Chapter 3, it is possible to identify conflicts, overlaps, similarities and so on.

The identification of system hierarchy levels is one way to help to define a number of contexts that form the heart of the requirements model.

4.2.5.3 Other contexts

In the two previous examples, we have discussed two common ways to identify the contexts of the system, but there are many more contexts that may exist, depending on the nature of the system or project that is being considered. Other examples of contexts include the following:

- Business contexts, whose use cases describe business requirements, mission statements, business goals and so on.
- Project contexts, whose use cases describe requirements that relate to delivering the project on time, within budget and so forth.
- Programme contexts, whose use cases describe programmatic requirements that represent strategy, program goals and so forth.
- System of systems contexts, whose use cases represent requirements that do not exist in any single system, but exist for the system of systems.
- Context by life cycle stage, whose use cases represent requirements for a specific stage in a project. For example, a missile will have a different set of requirements depending on whether it has been fired or not – this is particularly relevant where safety is concerned.
- Context by system state, whose use cases represent requirements that are dependent on a specific state of the system. An example of this may be a set of use cases that are only relevant when the system is in 'normal operating mode' or 'emergency mode'.

This list is not intended to be exhaustive, but it should provide an indication of just how complex the whole area of system engineering is, and drive home the point of how crucial it is to consider the system from different points of view, rather than just having a single flat view.

4.2.6 The 'Use Case' concept

The concept of a use case is one that is very often misunderstood. Many people assume that a use case is the same as a requirement, but this is not the case. The definition of a use case used in this book is that a *use case is a requirement that has been given meaning by putting it into context.* Any single requirement may be interpreted in different ways depending on the point of view, or context, that it is viewed from. The previous section has description of how a number of contexts may be identified for any system, and it is the use case that looks at each requirement in these different contexts. This is illustrated in Figure 4.11.

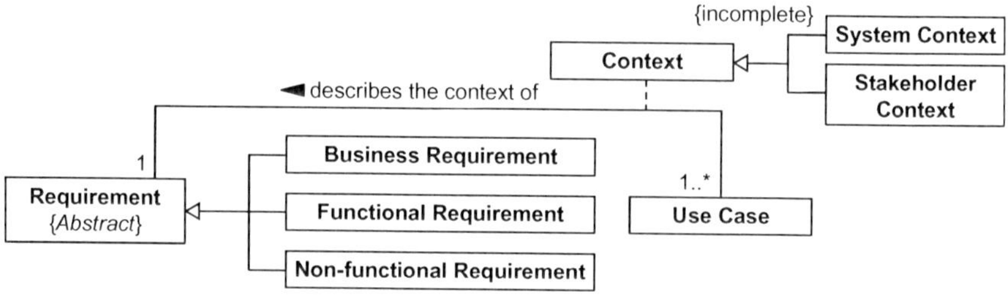

Figure 4.11 Definition of a use case

Figure 4.11 shows that one or more 'Use Case' describes the context of each 'Requirement' via the 'Context'.

To illustrate the concept of the use case, let us consider the two examples that have been used in section 4.2.5.

In the example of the passenger airline, imagine that there is a requirement, and hence a Requirement Description, that is defined as 'Save money'. This seems quite straightforward and easy to understand as almost everybody can understand the idea of saving money. However, depending on the role that we are playing (which stakeholder we are), the requirement will take on different meanings (different use cases) as discussed below:

- From the passenger's point of view, this may be interpreted as save money on the fare paid to make a journey. Therefore, the requirement is 'Save money', the context is from the point of view of the 'User' stakeholder and the use case may be 'save money on cost of fare'.
- From the airline owner's point of view, this may be interpreted as save money on the cost of providing a flight, in terms of fuel costs, staff costs, food costs and so on. Therefore, the requirement is 'Save money', the context is from the point of view of the 'Sponsor' stakeholder and the use case may be 'save money on the cost of providing a flight'.

This simple example just goes on to show how the context considered can completely alter the meaning of the original requirement. It should also be noted in the example here that the two use case will potentially conflict, as the passenger doesn't want to pay much money, but the owner doesn't want to spend any money either. By considering the various use case, it is possible to identify any areas of potential conflict, overlapping, gaps in understanding and so on.

In the example of the car system, imagine that there is a requirement, and hence a Requirement Description, to 'conserve fuel'. Again, this may seem quite straightforward, but depending on the context, this time, based on the level of hierarchy, it will take on different meanings as discussed below:

- From the driver's point of view, this may be interpreted as conserving fuel to save on running costs of the car. Therefore, the requirement is 'Conserve fuel', the context is from the point of view of the 'System' level and the use case may be 'conserve fuel to minimise the running costs of the car'.

- From the fuel injection unit's point of view, this may be interpreted as conserving fuel to optimise the engine performance, which was derived from the engine's requirements. Therefore, the requirement is 'Conserve fuel', the context is from the point of view of the 'Assembly' and the use case may be to 'conserve fuel to optimise engine performance'.

In the example here, the use cases do not conflict, but they do have quite different, yet complementary, meanings.

When it comes to demonstrating that the original requirements can be met, known as 'validation', it is the use cases that must be validated, which, in turn, will validate the original requirements. This is discussed more in the next section concerning 'scenarios'.

4.2.7 The 'Scenario' concept

The previous section discussed how the original requirements must be given meaning by putting each into context using use cases. It is essential that we can demonstrate that we have met the original requirement, but this must be achieved by demonstrating that we can satisfy the relevant use cases. The way that use cases are validated is by considering various 'scenarios' for each use case.

A scenario is defined in this book as *an exploration of a 'what if' for a use case*. Each use case gives rise to a number of different situations that may arise when it is being satisfied. This is shown in Figure 4.12.

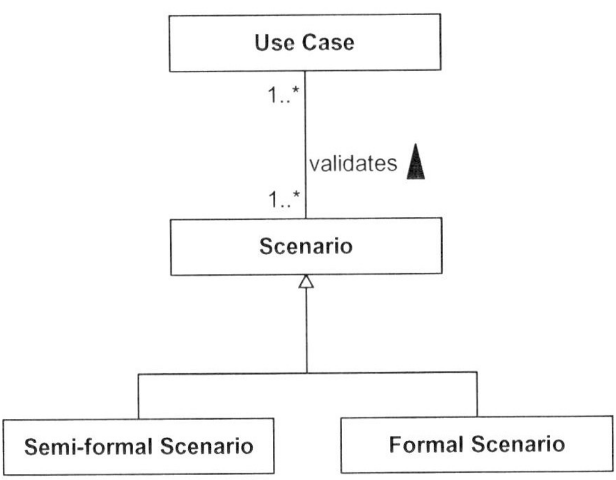

Figure 4.12 Definition of a scenario

Figure 4.12 shows that one or more 'Scenario' validates one or more 'Use Case' and that there are two types of 'Scenario': the 'Semi-formal Scenario' and the 'Formal Scenario'.

The semi-formal scenarios are realised by SysML sequence diagrams that show interactions between elements in the system. These interactions will usually exist at two levels:

- Stakeholder-level scenarios that treat the system as a black box and analyse the interactions between the stakeholders and the system.
- System-level scenarios that look at the interactions between system elements within the system.

These scenarios allow a use case to be analysed by considering different 'what ifs' and representing them with the sequence diagram.

The formal scenarios will be realised by SysML parametric constraints and their usages. This allows a more mathematical-based approach to be taken for understanding the use cases. The parametric usages are connected together into different networks that allow 'what if' analysis and are particularly powerful when considering trade-offs.

All these scenarios may be used for two purposes:

- Understanding, to allow each use case to be analysed.
- Validation, to allow each use case to be satisfied in a demonstrable way.

Scenarios provide a very powerful mechanism that is essential for any requirements exercise.

4.3 The ontology revisited

Figure 4.13 shows the full ontology with all of the concepts that have been discussed so far. Together, these concepts form the heart of the approach to context-based requirements engineering (ACRE), which is the subject of this book.

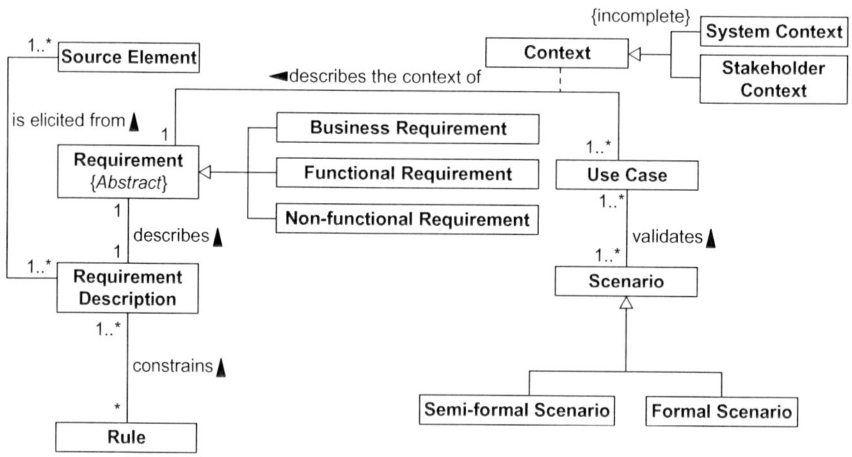

Figure 4.13 The ACRE ontology

Figure 4.13 shows the ACRE ontology and identifies and defines the concepts and terminology that is used throughout this book. There is, however, another aspect of the ontology that provides another powerful mechanism – the relationships. The relationships on the ontology provide the basis for traceability throughout the whole system. Traceability is very important for any system development, but when modelling is applied properly, the traceability is inherent in the model.

4.4 Summary

This chapter has introduced the model-based requirements engineering ontology that identifies and defines the concepts, terms and their inter-relationships. The ontology forms the heart of the ACRE, which is described in this book.

The ontology provides the basis for all the views that make up the ACRE framework (described in the next chapter) and enforces the modelling rigour necessary for a successful project.

References

1. Holt J. *A Pragmatic Guide to Business Process Modelling*. 2nd edn. London: BCS Publishing; 2009
2. Dickerson C.E., Mavris D.N. *Architecture and Principles of Systems Engineering*. Boca Raton, FL: CRC Press; 2009
3. Holt J., Perry S. *Modelling Enterprise Architectures*. London, UK: IET; 2010
4. Stevens R., Brook P., Jackson K., Arnold S. *Systems Engineering – Coping with Complexity*. Upper Saddle River, NJ: Prentice Hall; 2008
5. IEEE. Available from http://www.science.uva.nl/research/csa/Presentations/2005Style.pdf [Accessed 9 August 2011]
6. Flesch. Available from http://en.wikipedia.org/wiki/Flesch%E2%80%93Kincaid_readability_test [Accessed 9 August 2011]

Chapter 5

A framework for model-based requirements engineering

I got all choked up and I threw down my gun
And I called him my pa, and he called me his son,
And I came away with a different point of view

Johnny Cash, 'A boy named Sue'

5.1 Introduction

The principles behind model-based systems engineering (MBSE) have been discussed briefly in Chapter 1. Key to the success of MBSE is to have three main enablers, which are shown in Figure 5.1.

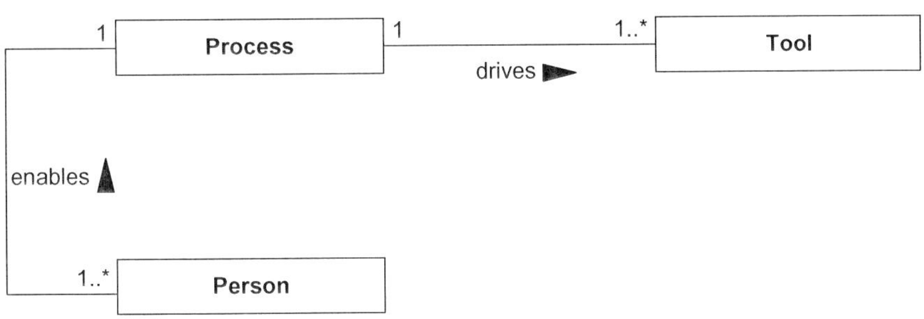

Figure 5.1 The three aspects of MBSE

The main concepts required for the approach to context-based requirements engineering (ACRE) have been introduced in the previous chapter in the form of the ACRE ontology. This chapter looks at how the ontology may be visualised by considering a number of different views that make up the ACRE framework.

5.2 Introducing the framework

The concepts introduced in the ontology must be realised in some way. Different subsets of the ontology are identified and then the elements within each subset are visualised using views. The collection of these views makes up the ACRE framework. This framework identifies and defines a set of views that are used to help us identify, define, understand, document and communicate requirements. There are some very pragmatic reasons why the definition of such a framework is important:

- *Coverage*. It is important that the whole of the ontology is realised. Each view considers a small set of the ontology and the totality of the views covers the whole ontology.
- *Rigour*. By generating all the views, applying the appropriate rules for each view and ensuring consistency, we produce a true requirements model. This true model provides the rigour for ACRE. Realising all the views provides the highest level of rigour, whereas realising only some of the views provides less rigour. This means that the approach is flexible for projects of different levels of rigour.
- *Approach*. The approach defines how we do things, i.e. what is the process that we follow to realise the ACRE views.
- *Flexibility of scale*. The framework defines a number of views but, depending on the type of project being undertaken, not all views need to be realised. This ability to realise some or all views makes the ACRE very flexible in terms of the size of the project.
- *Flexibility of realisation*. The views required by ACRE may be realised in any number of different ways. The approach promoted in this book is primarily through using the SysML notation, but any suitable notation may be used to realise the views. In the same way, any suitable tool may also be used.
- *Integration with other processes*. The framework allows integration with any other systems engineering processes provided that the information views for the processes are known. This allows ACRE to be used with many other methodologies and systems engineering approaches [3].
- *Automation*. The framework provides the basis for automating ACRE using sophisticated systems engineering tools. One of the main benefits of an MBSE approach is that it saves a lot of time and effort as many of the process artefacts may be automatically generated.

All of these points will be covered in greater detail in Chapter 6, where implementing ACRE on real projects is discussed.

5.3 Introduction to the framework views

The ACRE framework comprises a number of views. These views are identified in Figure 5.2.

Figure 5.2 shows that there are six main ACRE views that are needed according to the framework:

- *Source Element View*. This view contains all the source information that is required in order to get the requirements right.

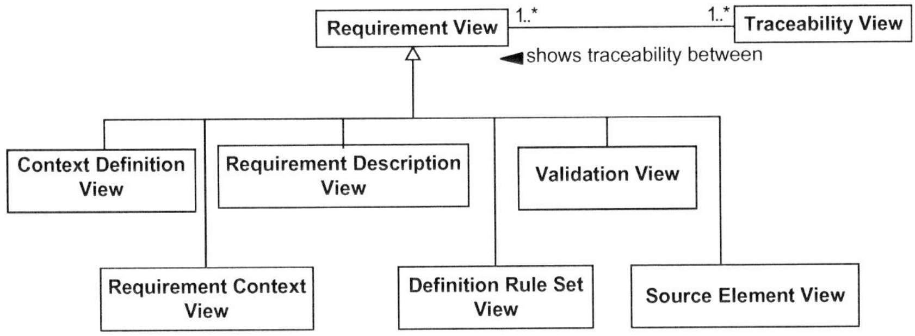

Figure 5.2 The ACRE framework views

- *Requirement Description View*. This view contains structured descriptions of each requirement. These requirements are considered individually and will usually have a number of attributes, or features, associated with each one.
- *Definition Rule Set View*. This view contains rules that may have to be applied to each requirement definition. For example, these may be complexity rules in the form of equations of more general text-based rules.
- *Requirement Context View*. This view takes the requirements and gives them meaning by looking at them from a specific point of view. This is known as putting the requirements into context and forms the basis of the approach presented in this book.
- *Context Definition View*. This view identifies the points of view that are explored in the Requirement Context View. These points of view, or contexts, may take many forms including stakeholders and levels of hierarchy in a system.
- *Validation View*. These views provide the basis for demonstrating that the requirements can be met or complied with in some way. These views can be informal (such as scenarios at various levels of abstraction) or may be formal (such as mathematical-based representation).

Alongside these core ACRE views there is an additional set of support views known as Traceability View. These views allow traceability between different elements of the model to be explicitly shown. These traceability links may exist between views or between elements within views.

To use these views effectively it is important to keep in mind the following:

- Each view is understood in terms of why the view is necessary.
- Which areas of the ontology are being focused on in the view is understood.
- Which other views are associated with the view is understood.
- How each view may be visualised using graphical notations, text, mathematics, etc. is understood.

Each of the views introduced here is described in these terms, and the framework is built based on these definitions and discussions.

It should be stressed that not all views are always necessary on a given project as this will depend upon the level of rigour of the project, the scale of the project and the tools available. This is discussed in more detail in Chapter 7.

5.4 The views

This section looks at each of the ACRE views that was introduced in the previous section and describes them according to the criteria that were also described in the previous section.

These views are not presented in any particular order, as it is the process that is being followed which will define this. Examples of such processes can be found in Chapter 7.

5.4.1 The Source Element View

5.4.1.1 Rational

The Source Element View contains all relevant source information that is required to get the requirements right. It is essential that the origin of all requirements is known and this is what this view allows us to define. This view is used primarily as a mechanism to establish traceability in the system and provide links between the requirements and any other aspect of the system.

As to what these Source Elements can be, the list is almost limitless! A Source Element can range from a single utterance to a full system specification or design. Examples of some typical Source Elements are discussed below:

- *Conversations*. This may be an informal way to elicit requirement but, nonetheless, it is an approach that is used very often. It is important that these conversations can be captured in some way, even if it is just a rough set of notes. With the increased power of mobile phones and PDAs, it is now often a simple matter to make short voice recordings and attach them to a document. Indeed, there are many applications that are aimed specifically at doing exactly this.
- *Emails*. People are forever making suggestions and changing existing ideas using email. In many projects where there are no formal requirements, emails can be a rich source of information.
- *Informal documents*. Informal documents such as logbooks, whiteboards, and post-it notes can be a good source for requirements. The use of applications on mobile phones and PDAs can be invaluable here, with some applications being specifically aimed at capturing notes, images, handwriting from notes and so on. In fact, when combined with voice and video capture, these applications can generate very rich documents that contain many types of media.
- *Formal requirements documents*. One of the most obvious sources for requirements is formal requirements documents. These are often held in a requirement management tool or in an office document. Caution should be exercised, however, as it is all too easy to mistake well-documented and

well-presented requirements for good requirements. Even when a requirements set is presented in a beautiful-looking document, it does not mean that the requirements are understood or in any way rigorous or correct.

- *Systems specifications, designs, etc.* In addition to formal requirements documents, it is also quite common to use the specification, or even the design of a system, to elicit requirements. This is particularly common when the system is broken down into a hierarchy of subsystems, assemblies and components, where the specification of one element in a system may very well form requirements for another element in the system.
- *Processes.* The processes associated with a system, whether they are the development, management or the operational processes may constrain the requirements in some way. For example, development processes may constrain some of the requirements taken from the project's point of view, in terms of the way that the project may be run. Another example may be the operational processes that exist in a system that is being updated or automated. It may be that the operational processes, from a human operator point of view, cannot be changed; therefore, this may constrain the requirements.
- *Existing systems.* Existing systems, updates to systems and legacy systems that interact in some way with the new system will need to be interfaced to, which brings a whole set of interface requirements and constraints along with it.
- *Brainstorming sessions.* This is related to two of the previous points concerning conversations and informal documents. Again, the use of smartphone and PDA technology can be a real bonus here. The outputs of brainstorming sessions tend towards being quite unstructured, such as whiteboards and flipcharts with dumps of ideas and thoughts on them.
- *Structured workshops.* Structured workshops may be similar to brainstorming sessions, but tend to have more well-defined goals and desired outcomes for the session. Also, a structured workshop often has a more formal agenda with specific discussion points on them. Formal documents are often produced as a result of a structured workshop and, yet again, smart phones and PDAs may be a boon for these sessions.
- *Standards, laws and best-practice models.* In most real-life projects there are a number of standards or laws that must be complied with. As to the level of compliance, this depends on the nature of the standard or law and the priority in the context of the system. For example, laws tend to be mandatory, whereas some standards are and others are not. The same holds true for best-practice models, such as found in books, publicly available specifications and so on.

The list presented here simply shows some typical Source Elements and is not intended to be exhaustive. Do not feel constrained in any way by this list, but use it as a start point to consider where each source may come from.

5.4.1.2 The ontology

The Source Element View focuses on the subset of the original ACRE ontology that is shown in Figure 5.3.

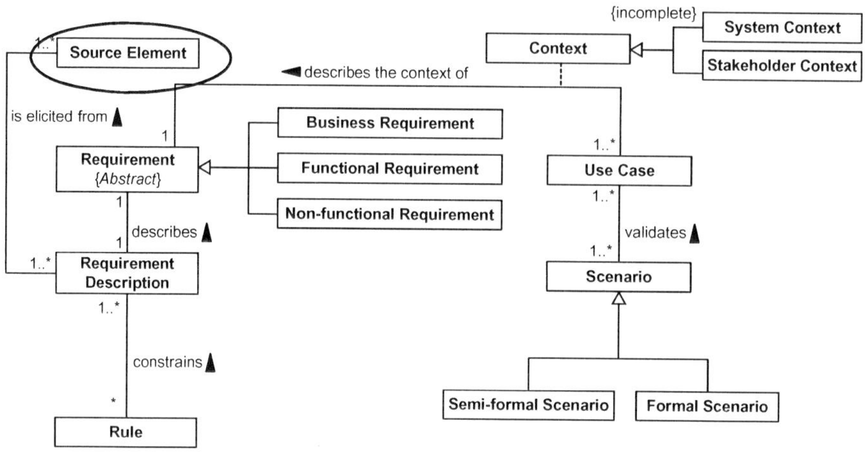

Figure 5.3 Subset of the original ACRE ontology for the Source Element View

The Source Element View focuses on the 'Source Element' from the original ontology, as shown here. This view is perhaps the simplest of all the views in the framework, as reflected in its simple structure, shown in Figure 5.4.

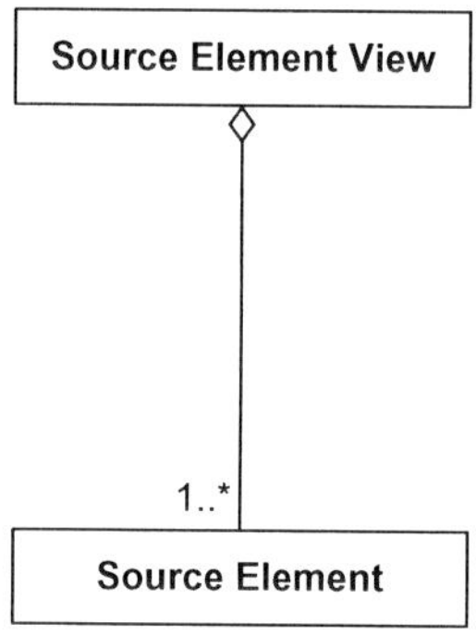

Figure 5.4 Structure of the Source Element View

Figure 5.4 shows that the 'Source Element View' is made up of one or more 'Source Element'. Due to the varied nature of the structure and format of the Source Elements, and given the fact that this view is really just a collection of elements that can be linked back to, the structure of this view is very simple.

It should be borne in mind that the structure of the Source Element itself may be highly complex, such as the case when the Source Element is a system specification. However, this view is used primarily as a basis for traceability and as such the information in the view may be thought of as a list-like collection of elements.

5.4.1.3 Relationships with other views

The Source Element View is related to another view in the framework, hence consistency checks that must be applied to ensure that these relationships are valid (Figure 5.5).

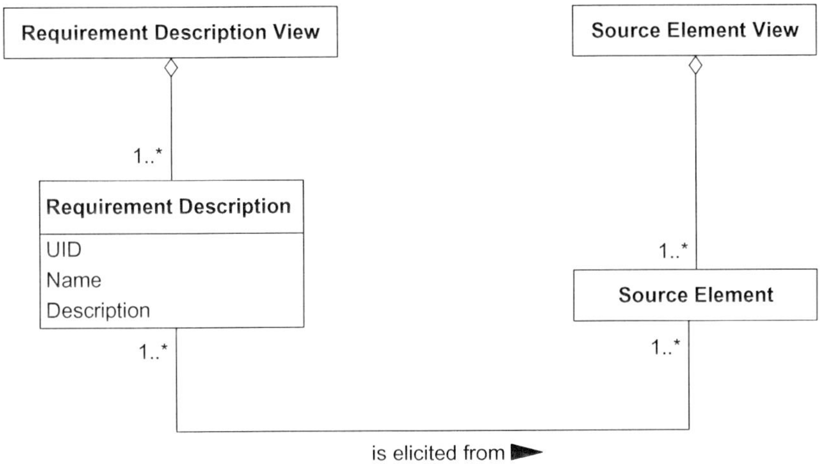

Figure 5.5 Relationships between Source Element View and the rest of the framework

These relationships are defined as the *Requirement Description View*. The Requirement Descriptions in the Requirement Description Views are related back to the Source Elements. Each Requirement Description must be related to at least one Source Element and *vice versa*. The following rules must be enforced:

- Each Source Element in the Source Element View must be traceable to one or more Requirement Description in the Requirement Description View.
- Each Requirement Description in the Requirement Description View must be traceable to one or more Source Element in the Source Element View.

To maximise the benefits of a true MBSE approach, these rules should be automated rather than manually applied to the model.

5.4.1.4 Visualising the view

This view may be realised in a number of ways. The two main mechanisms considered here are as follows:

- A list
- A set of blocks in the model

Figure 5.6 shows that an example of this may be realised using blocks in a SysML block definition diagram.

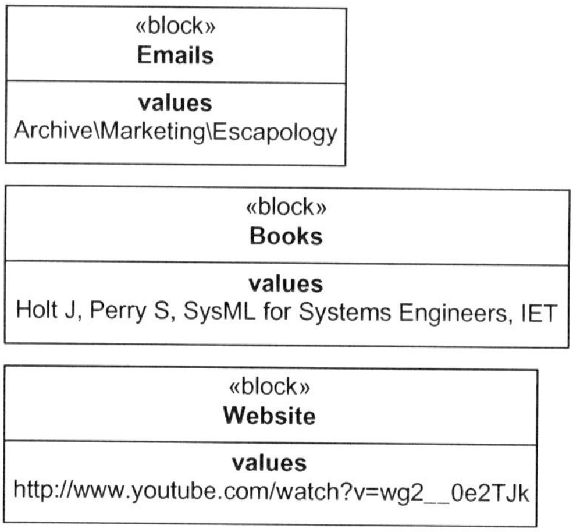

Figure 5.6 Example visualisation of Source Element View using SysML block definition diagram

Figure 5.6 shows a collection of blocks that are used in a very simple fashion. Each block is used only as a reference point to specific external Source Element. The example here shows a number of different types of Source Element such as:

- references to a number of Source Element emails;
- references to a number of Source Element books; and
- references to a number of Source Element websites.

It can easily be argued that the use of a SysML diagram here is 'overkill' and that the Source Elements could simply be listed, but there is a good reason the way this has been done. One of the benefits of a model-based approach is that traceability is inherent in the model. The traceability can only exist for elements that are either part of the model or explicitly linked to the model, and Figure 5.6 serves this single but important purpose to provide an explicit link between external Source Elements and the model itself.

There are other ways that this view can be visualised. Any way is potentially okay, providing that it meets the criteria laid out in this section.

5.4.1.5 Discussion

The Source Element View is both the simplest view in terms of its structure and the loosest in terms of what it may look like. This view is present to provide solid traceability back to the rest of the system model.

There is a tendency in system engineering for the requirements to be developed using very good techniques, but then for these requirements to bear no relation to the rest of the system model. The best requirements in the world are worth nothing if they cannot be related to the rest of the system model. Unfortunately, this is nowhere near as uncommon as it should be.

5.4.2 Requirement Description View

5.4.2.1 Rational

This view contains structured descriptions of each requirement. These requirements are considered individually and will usually have a number of attributes, or features, associated with each one.

The main purpose of this view is to describe each individual requirement according to a pre-defined set of attributes. These attributes vary depending on the process that is being followed, the industry that the work is being carried out in, any standards or best-practice models that may be used and any other number of factors.

This view is primarily used for managing the requirements of a system and is often the basis of implementation for many of the commercial requirements management tools that are in the market today.

Each Requirement Description provides a non-contextual description of the requirement. When a requirement is put into context, it is known as a 'use case' and hence there is a very strong relationship between the Requirement Descriptions and the use cases from the Requirement Context Views.

5.4.2.2 The ontology

The Requirement Description View focuses on the subset of the original ACRE ontology that is shown in Figure 5.7.

The Requirement Description View focuses on the 'Requirement Description' from the original ontology, as shown here. This view is one of the simplest (bearing in mind that 'simple' does not imply 'easy') of all the views in the framework, along with the Source Element View, as reflected in its simple structure shown in Figure 5.8.

Figure 5.8 shows that the 'Requirement Description View' is made up of one or more 'Requirement Description'. It is important to remember that the 'Requirement

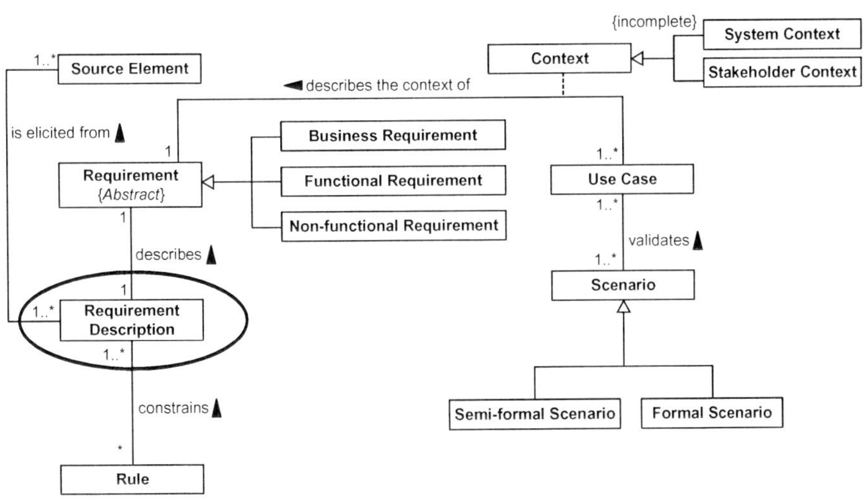

Figure 5.7 Subset of the original ACRE ontology for the Requirement Description View

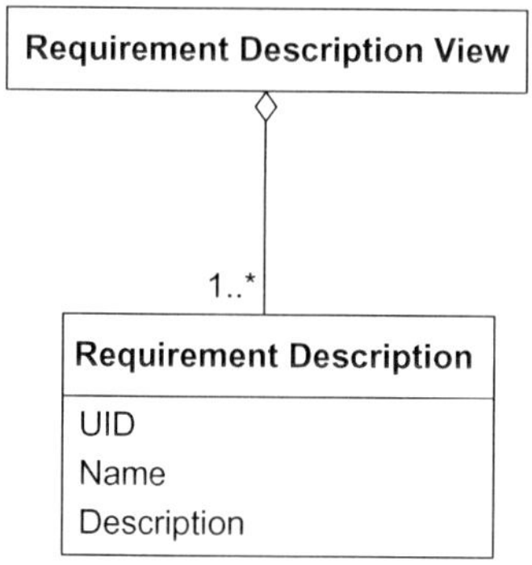

Figure 5.8 Structure of the Requirement Description View

Description' shown here is *not* the requirement itself but an abstraction of that requirement that describes it using a number of features or attributes. This figure also shows the following generic set of attributes:

- *UID*. UID is the unique identifier for the requirement. Requirements on any real project evolve during the life cycle of the project and hence the names and descriptions of requirements may change beyond all recognition. To ensure that each requirement can always be located, regardless of how it has evolved, each requirement should have a unique identifier. This can be used to locate the Requirement Description and also is used as a basic mechanism for traceability in the system. The UID is used primarily by automated services, such as software management tools, as it will be non-memorable to most people who read it.
- *Name*. The name is a simple label that should have some intuitive meaning that can be used to identify the requirement by human eyes. In some ways, this serves the same role as the UID, but from a human point of view. It should be remembered, however, that the name may evolve as the project progresses, but the UID may not.
- *Description*. The description is a piece of text that describes the requirement. This should be kept as simple and unambiguous as possible, but in reality this is often ignored and long, verbose descriptions are often defined.

This is a very brief list that is based on the SysML modelling standard and should be seen as an absolute minimum set of attributes. In any real-life system, several other attributes, such as the following, should also be considered:

- *Origin*. Any requirement must have an origin – it must have originated from somewhere. This information is recorded as part of the Requirement Description. This source will be taken directly from the context identification. For example, this may be a stakeholder or a level of hierarchy on the systems, or any other source for the context.
- *Priority*. Each requirement will often have a level of priority associated with it. For example, it may be decided that a requirement may be 'essential' in that it

is always needed, 'desirable' in that it would be strongly recommended that it is met, or 'bells and whistles' in that it is a nice idea, time permitting. Of course, a number of levels may be defined with any definition for each level.
- *Verification criteria*. It is essential that each requirement can be proven to work – verification. The verification aspect is often omitted, whereas the validation criterion is usually deemed essential.
- *Validation criteria*. It is essential that for each requirement it can be proven that it does what it is supposed to do – validation. The verification aspect is often omitted, whereas the validation criterion is usually deemed essential. The validation criterion is related directly to the Validation Views.
- *Ownership*. Each requirement must be owned and hence have a stakeholder who is responsible for delivering the requirement. This attribute should reference one of the stakeholders identified in the system directly. Note that this attribute differs from the 'Origin' attribute despite the fact that they both are often referring to stakeholders. The 'Origin' is related to the origin of the requirement, whereas the 'ownership' relates to who is responsible for delivering the requirement.

There are many other attributes that may be considered for a requirement, as the list shown here is only intended to show a generic best-practice list that may need to be tailored for specific applications.

5.4.2.3 Relationships with other views

The Requirement Description View is related to other views in the framework, hence consistency checks that must be applied to ensure that these relationships are valid.

Figure 5.9 shows that the Requirement Description View is related to the following other views from the framework:

- *Source Element View*. Each Requirement Description must be traceable back to a Source Element. This is essential for traceability in the system and for enforcing validation of each requirement.
- *Definition Rule Set View*. The attributes that make up the Requirement Description may be checked by applying a number of rules. For example, complexity rules may be enforced on the 'Description' attribute to ensure that the text is legible.
- *Requirement Context View*. Each Requirement Description will be related to a number of use cases. Each use case takes one or more requirement and gives them meaning by putting it into context.

These relationships may be enforced by applying the following rules:

- Rules, when they exist, must apply to a Requirement Description.
- Each Requirement Description must relate back to a Source Element.
- Each Requirement Description must be related to at least one Use Case.
- The Requirement Description Views must relate to a Requirement Context View.
- Each Requirement Description must have a full set of attributes defined.

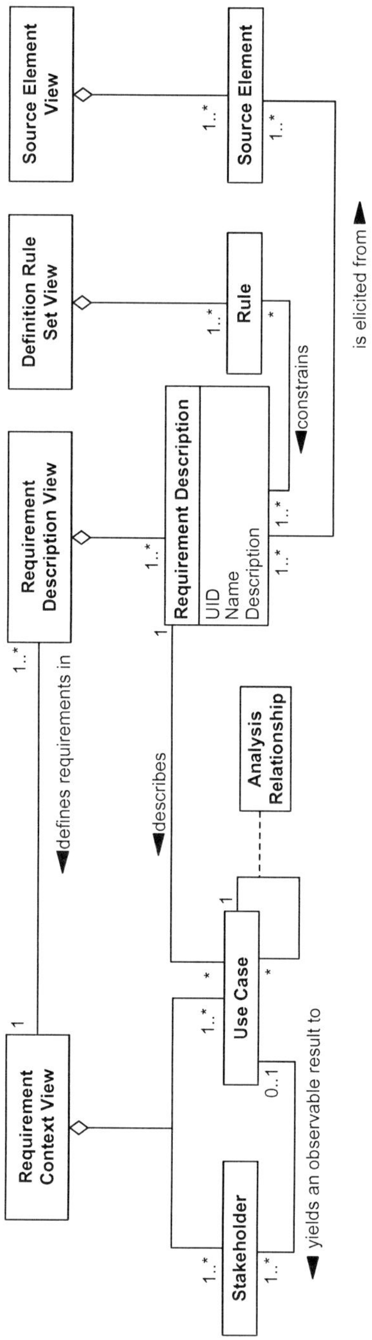

Figure 5.9 Relationships between the Requirement Description View and the rest of the framework

Rules, such as the following, are also needed for each of the additional attributes:

- Rule – The 'Origin' attribute of any requirement must refer to a context element.
- Rule – the 'Validation criteria' attribute must reference Validation Views.

To maximise the benefits of a true MBSE approach, these rules should be automated rather than manually applied to the model.

5.4.2.4 Visualising the views

This view may be realised in a number of ways, including the following:

- Using a text-based description or table
- Using a requirements management tool
- Using SysML as part of the model

An individual Requirement Description is shown in Figure 5.10 using the SysML «requirement» block.

Figure 5.10 shows a SysML representation of a single Requirement Description. Note that the property values have been defined here to show the 'UID', the 'Name' and the 'Description'. In this example, only a basic set of properties has been shown.

This is quite a simple representation and the equivalent visualisation using a text table is shown in Figure 5.11.

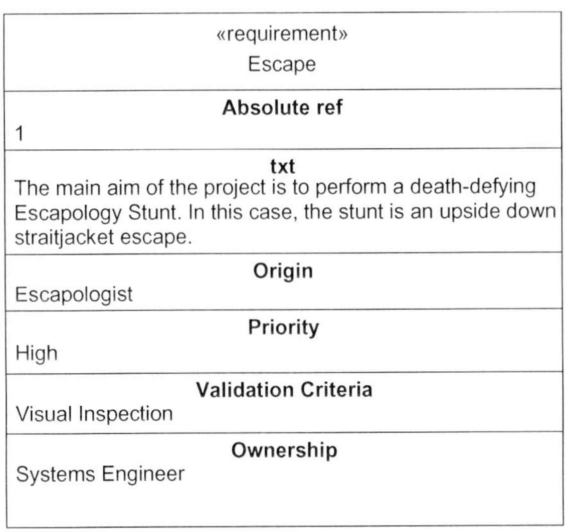

Figure 5.10 A single Requirement Description showing defined properties

Requirement Description	Escape
Absolute ref	1
Txt	The main aim of the project is to perform a death-defying Escapology Stunt. In this case, the stunt is an upside down straitjacket escape.
Origin	Escapologist
Priority	High
Validation criteria	Visual inspection
Ownership	Systems Engineer

Figure 5.11 Representing a Requirement Description using a text table

Upon comparing the tabular and the SysML visualisations, it is evident that the two are almost identical, which is expected with such a simple construct. Any other visualisation, such as the one using a requirements management tool, will also look very similar. This may lead the question of why bother to use the SysML modelling notation when a table will achieve the same ends. Again, the answer to this is one of following the MBSE philosophy which brings along with it a number of benefits. One of the main benefits here is that as the information is captured in the model, given a set of sharp tools, the view can actually be visualised using almost any representation by generating a new view artefact from the model. An example of this is provided towards the end of this chapter.

5.4.2.5 Discussion

It should be noted here that the Requirement Descriptions in this view have a flat structure, whether they are visualised using lists or SysML requirements. This is quite deliberate. There is a temptation amongst requirements engineers to group requirements based on their functionality or type and to create a requirements taxonomy (classification hierarchy). This goes fundamentally against the context-based approach. One of the key points of the context-based approach is that requirements should be grouped together based on context and *not* functionality or type.

Many historical approaches to requirements engineering and many requirements documents have been obsessed with putting requirements into strict classifications. At a high level, many documents will often split up 'functional' and 'non-functional' requirements. This is fine if the requirements are being described, but holds no meaning in the real world. In the real world, and by the very nature of a non-functional requirement, they constrain functional requirements. It is essential, therefore, that these constraining relationships are identified. These constraining relationships can only be defined when looked at in context. The same holds true for inclusions, extensions and generalisations – it is the context that gives them meaning.

The same idea holds for when people want to group requirements because they appear to have similar functionality. This is fundamentally wrong because the functionality of a given requirement may change depending on the context.

Requirements must be grouped according to their context, which forms the basis of the whole 'context-based requirements engineering' approach advocated in this book. If requirements are classified based only on type or functionality, it would be as well to group them alphabetically.

5.4.3 Definition Rule Set View

5.4.3.1 Rational

This view contains rules that may have to be applied to each requirement definition. For example, these may be complexity rules in the form of equations of more general text-based rules.

When defining individual Requirement Descriptions, it is often desirable to put constraints on either the values or the measurable aspects of the values. These constraints are enforced by applying one or more rule to the Requirement Description values. These rules must be defined somewhere and the relevant attributes of the Requirement Descriptions must be identified. The purpose of this view is to capture this information.

5.4.3.2 The ontology

The Definition Rule Set View focuses on the subset of the original ACRE ontology that is shown in Figure 5.12.

This view is concerned with the 'Rule', one or more of which constrains one or more 'Requirement Description'. Note that the rules are related to the Requirement Descriptions, rather than the requirements themselves, as the Requirement

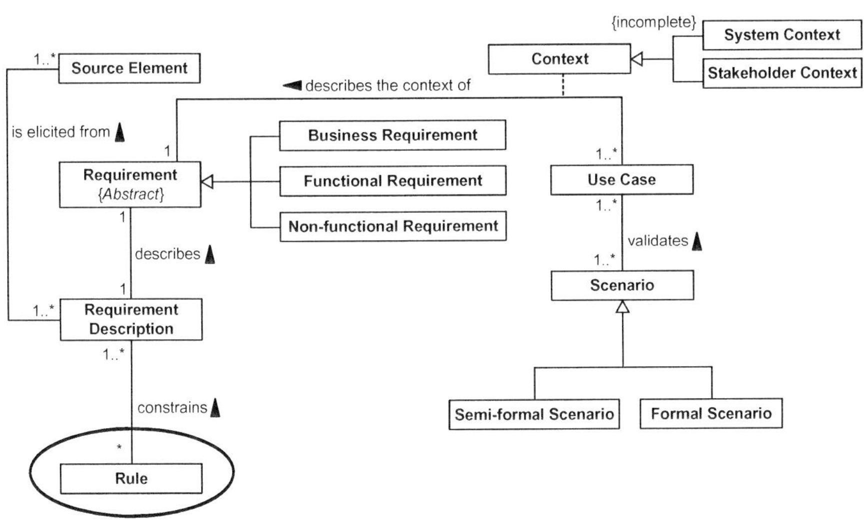

Figure 5.12 Subset of the original ACRE ontology for the Definition Rule Set View

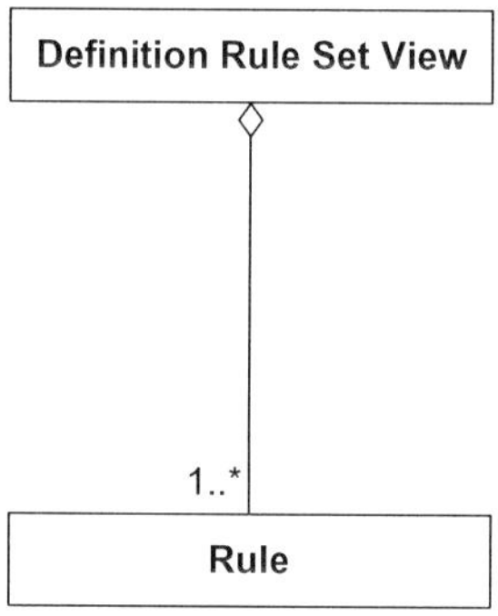

Figure 5.13 Structure of the 'Definition Rule Set View'

Description is one of the ways that the abstract 'Requirement' is manifested in the model (Figure 5.13).

It is shown in the figure that the 'Definition Rule Set View' is made up of one or more 'Rule'. Each 'Rule' is, in turn, made up of a 'Rule Definition' and a 'Parameter Set':

- The 'Rule Definition' defines the rule itself in some form. These rules may take a number of different forms, e.g. equations, heuristics, enumerated lists, tables, graphs and so on. A number of examples of these are discussed later.
- The 'Parameter Set' defines the elements that are used by the rule itself, in other words the parameters of each rule. This allows the rules to be applied in a uniform and consistent manner.

To illustrate how these rules may be applied, consider the set of attributes that have been defined in the previous section for each Requirement Description:

- *UID.* A rule here may take the form of a heuristic that states that each 'UID' must be unique. This may be deemed as a self-evident rule, but it may also be that the rule is applied across a set of project, has a specific format for a pre-fix for the UID, has a specific format for the UID itself (such as a time or date element) or any other aspect of the UID.
- *Name.* A rule here may specify that each 'Name' must have a maximum number of words in it, contain only nouns and so on.
- *Description.* The rule here may limit the number of words in the description, the number of sentences, the ratio between words and sentences and so on. These rules may often take the form of complexity measurements, an example of which is discussed.
- *Origin.* A rule here may be quite straightforward in that it insists that the 'Origin' is related directly back to a stakeholder that has been defined in one of the other views.
- *Priority.* The rule associated with the 'Priority' may be quite simple in that the value is constrained by an enumerated list – a pre-defined set of values that is available to the editor of the Requirement Description. In the previous section,

it was discussed how this may take on values such as 'Essential', 'Desirable' or 'Bells and whistles'. It may also be the case that additional values are defined, such as 'unknown' or 'undefined', to provide a default value for the attribute.

- *Verification*. A rule here may state that each 'Verification' must be linked to a set of scenarios, or some other part of the model.
- *Validation*. A rule here may state that each 'Validation' must be linked to a set of scenarios, or some other part of the model.
- *Ownership*. A rule here may be quite straightforward in that it insists that the 'Origin is related directly back to a stakeholder that has been defined in one of the other views.

It is also possible to define rules that apply across all attributes, such as one that states that all attribute values must be completed before approval for the Requirement Description can be obtained. This need not apply to all attributes, but may be limited to a subset of them. Again, the point here is that the rules are completely flexible and can be applied as the requirements engineers or process authors see fit when defining the requirements process.

5.4.3.3 Relationships with other views

The Definition Rule Set View is related to other views in the framework, hence consistency checks that must be applied to ensure that these relationships are valid.

From Figure 5.14, it is evident that one or more 'Rule' from the 'Definition Rule Set View' constrains one or more 'Requirement Description' from the 'Requirement Description View'.

These relationships result in the following process rules:

- Rule – Each rule must apply to at least one Requirement Description attribute or the Requirement Description itself.
- Rule – Each Requirement Description may be constrained by zero or more rules.

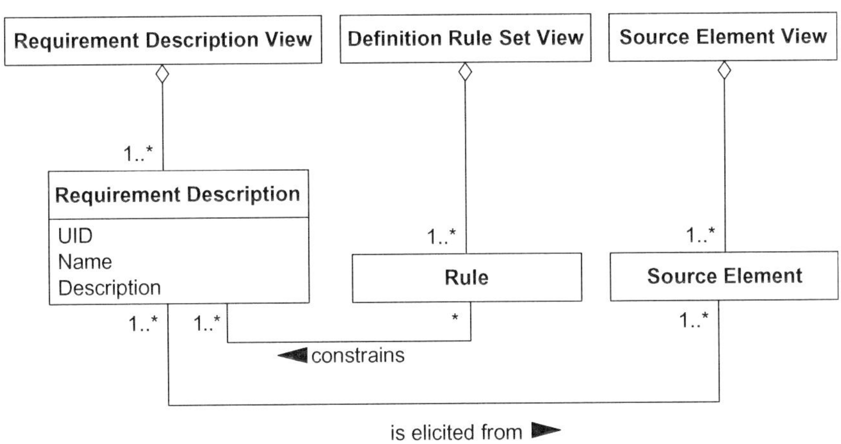

Figure 5.14 Relationships between the 'Definition Rule Set View' and the rest of the framework

To maximise the benefits of a true MBSE approach, these rules should be automated rather than manually applied to the model.

5.4.3.4 Visualising the view

There are a number of ways that the rules defined in this view may be specified, including mathematical specifications, SysML parametric constraints, textual descriptions and so on.

The example below shows how a set of rules that apply to the 'Description' attribute of the 'Requirement Description' may be specified.

Figure 5.15 shows how a rule may be defined using a SysML parametric constraint. The rule that is being defined forms part of a standard set of rules for complexity measurement of sentences.

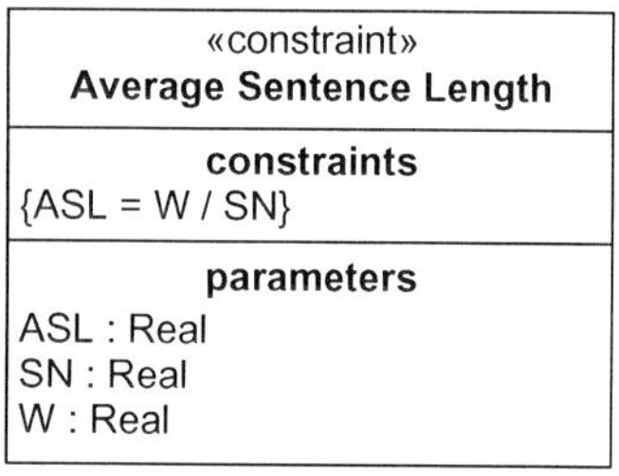

Figure 5.15 Example of a rule defined as a SysML parametric constraint

The rule here is made up of three main elements:

- *The rule name*. This is shown in the top box in Figure 5.15 and has the word "constraint" above it to indicate that this is a SysML parametric constraint construct. The name for the rules in this case is defined as 'Average Sentence Length', which is a standard measurement for text.
- *The rule definition*. This is shown in the middle box and has the word 'constraints' above it to indicate that this is part of the standard SysML parametric constraint construct. The rule itself is shown in curly brackets and, in this instance, is defined in terms of a mathematical equation. The various parameters that are used as part of the equation are defined in the next box.
- *The parameter definitions*. This is shown in the lowest of the three boxes and has the word 'parameters' above it to indicate that this is part of the standard SysML parametric constraint definition. In this instance, there are three parameters: 'ASL', which is the 'Average Sentence Length' and is a 'Real' number, 'SN', which is the 'Syllable Number' and is an 'Integer' and 'W', which is the 'number of words' and is also an integer.

Clearly, these rules must be verified, which can be done in any way seen fit, such as using a mathematical proof, a software package and so on.

It is often the case that the rules themselves are made up of a number of rules. This is the case in the example shown in Figure 5.16, in which a complete set of rules for performing the text description complexity measure is also shown.

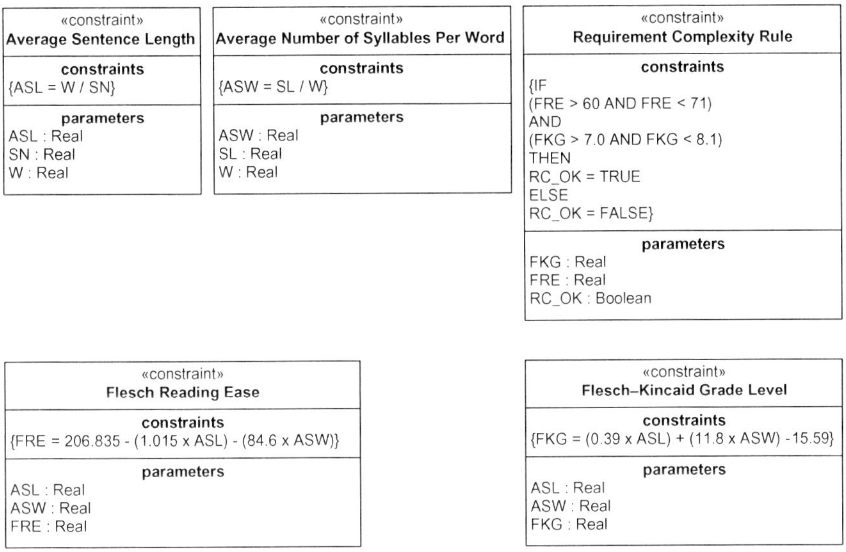

Figure 5.16 Complete set of rules for measuring the complexity of text

Figure 5.16 shows the complete set of rules for calculating the complexity of text descriptions. These rules are based on the so-called Flesch–Kincaid grade level test and the Flesch reading ease test [1,2].

Where a number of rules are defined that are dependent on one another, e.g. in the order that they must be calculated, it is also important to show a network of these rules and their inter-dependencies. This can be achieved in SysML by using the parametric usage diagram; an example of which is shown in Figure 5.17.

Figure 5.17 shows how the rules that have been defined can be strung together in a network to provide the ultimate end result that is required for the complexity measure. There are four main rules that are used to calculate the complexity using mathematical equations ('Average Sentence Length', 'Flesch Reading Ease Test', 'Average Number of Syllables Per Word' and 'Flesch–Kincaid Grade Level Test') and then a single rule in the form of a heuristic ('Requirement Complexity Rule') that is used to make the decision of whether or not the attribute satisfies the ultimate rule or not.

5.4.3.5 Discussion

The Definition Rule Set View is quite an odd view in the way that it is frequently used. In many cases, this view is simply never considered, and the individual Requirement Descriptions can be written in an *ad hoc* fashion provided that they are in the correct place in the requirement document.

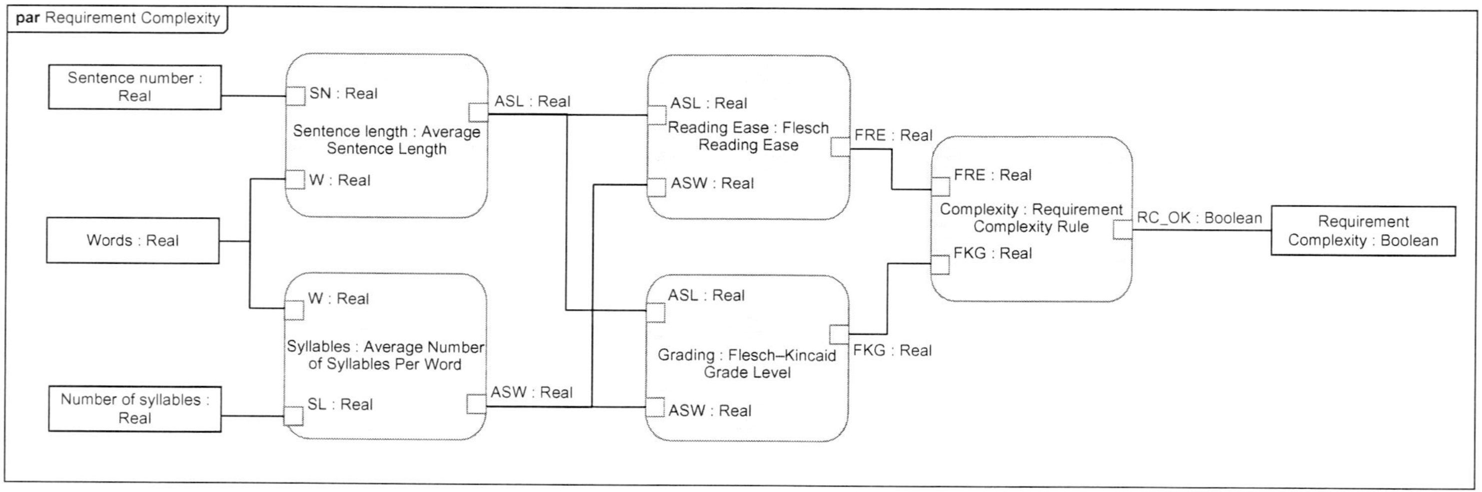

Figure 5.17 Example of use of complexity rules

Another way that this view is often badly used is when its importance is overemphasised. It must be made clear that this view is very important, particularly when the project is concerned with a mission-critical or safety-critical system. However, in many cases all that is presented for a complete set of requirements is the Requirement Description View (or equivalent), and provided that the rules in the Definition Rule Set View have been applied, this is deemed as acceptable. Again, it should be stressed that the main emphasis should be on the context of the requirements and hence the set of all the views.

5.4.4 Requirement Context View

5.4.4.1 Rational

This view takes the requirements and gives them meaning by looking at them from a specific point of view. This is known as putting the requirements into context and forms the basis of the approach presented in this book.

The requirements have only been described so far by defining a number of Requirement Descriptions in the Requirement Description View. This is all well and good and an essential part of any requirements exercise, but this is by no means complete. The problem arises that these Requirement Descriptions may be interpreted in different ways depending on the viewpoint of the reader of the Requirement Description. It is essential then that each requirement is looked at from different points of view or in different contexts. It will also be found that different contexts are concerned with different sets of requirements, all of which will be related together in some way. When a requirement is put into context it is known as a 'use case' and by considering these uses case and the relationships between them and other use cases or stakeholders, it is possible to generate a complete point of view or context.

These contexts may be based on a number of elements, such as stakeholders or levels of hierarchy in a system.

5.4.4.2 The ontology

The Requirement Context View focuses on the subset of the original ACRE ontology that is shown in Figure 5.18.

Figure 5.18 shows that the Requirement Context View is primarily concerned with one or more 'Context' of the system that is concerned with using one or more 'Use Case' to describe the context of a 'Requirement'.

Figure 5.19 shows that the 'Requirement Context View' is made up of one of more 'Stakeholder' and one or more 'Use Case'. The definition of a use case in SysML is that a 'Use Case' yields an observable result to one or more 'Stakeholder' either directly or indirectly. Use cases are related together by one or more 'Analysis Relationship' and it is these relationships that provide the true meaning of the set of use cases or context.

A number of these Requirement Context Views exist, the number of which is determined by the information contained in the 'Context Definition View' that is described later in this chapter.

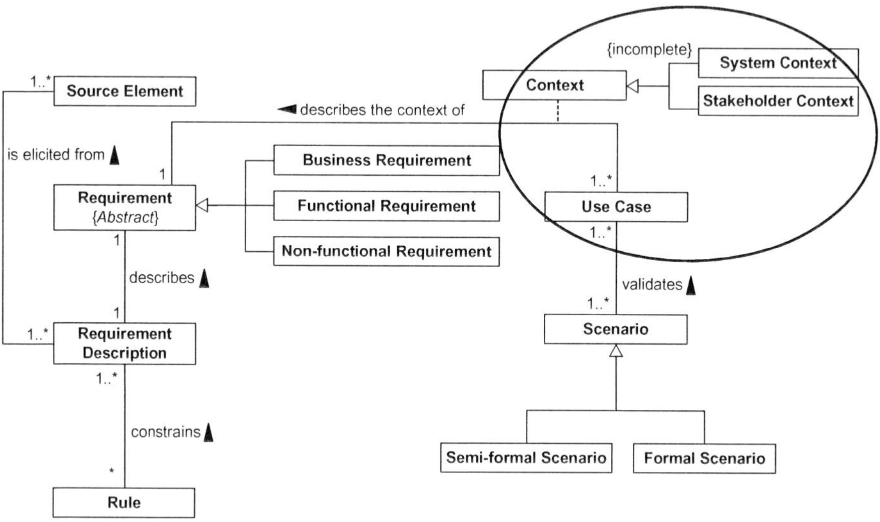

Figure 5.18 Subset of the original ACRE ontology for the Requirement Context View

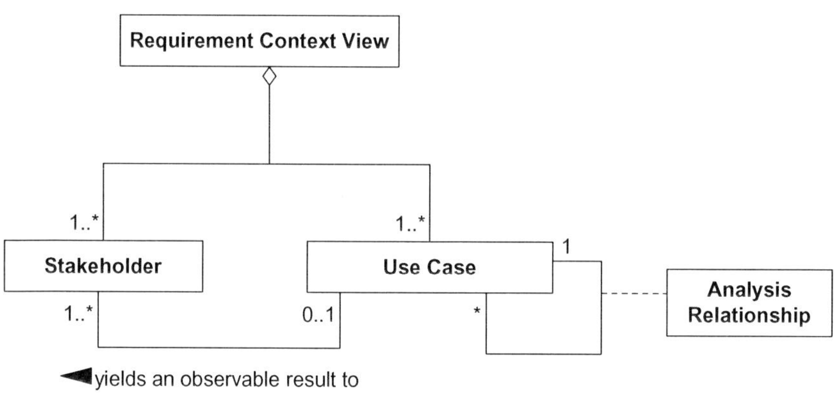

Figure 5.19 Structure of the 'Requirement Context View'

5.4.4.3 Relationships with other views

The Requirement Context View is related to other views in the framework, hence consistency checks must be applied to ensure that these relationships are valid.

It can be seen from Figure 5.20 that one or more 'Requirement Description View' describes the requirements in each 'Requirement Context View'. This reflects the very strong relationship between each 'Requirement Description' and one or more associated 'Use Case'. One or more 'Validation View' validates each 'Use Case'. This is a very important relationship as every requirement in the system

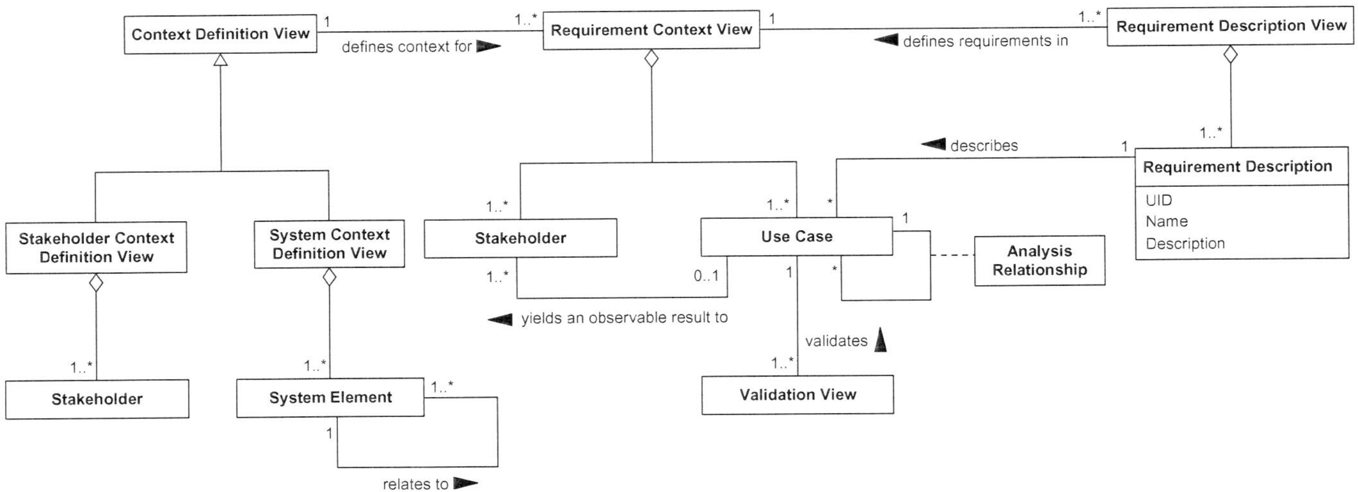

Figure 5.20 Relationships between the Requirement Context View and the rest of the framework

must be demonstrated to be achievable and that it has been satisfied, which is the purpose of the Validation Views. Note, however, that it is the use cases that are the subject of the validation, rather than the Requirement Descriptions. This is because a single Requirement Description may be interpreted in a number of different ways depending on the context of the requirement.

A 'Context Definition View' defines the context for one or more 'Requirement Context View' which may result in there being quite a number of Requirement Context Views.

It is clear from Figure 5.20 that the Requirement Context Views form the heart of the whole framework possessing, as they do, direct relationships to almost all other views in the framework. This is only to be expected as the whole approach advocated and described in this book is context-based requirements engineering, so it should be no great surprise how important this view is.

The following rules should be applied:

- Each Requirement Context View must have a related element on a Context Definition View that defines the context.
- Each Use Case must be related to at least one Requirement Description.
- Each and every Requirement Description must have at least one Use Case.
- Each Stakeholder on the Requirement Context View must have an associated Stakeholder on a Stakeholder Context Definition View.
- Each Context Definition View must be related to at least one Requirement Context View.
- Each Use Case must be related to either another Use Case or a Stakeholder.
- Each Use Case must have at least one Validation View associated with it.

To maximise the benefits of a true MBSE approach, these rules should be automated rather than manually applied to the model.

5.4.4.4 Visualising the view

There are many number of ways that the Requirement Context View may be visualised, such as text, free-form diagrams or a more structured approach. The view is visualised in SysML using the use case diagram, which has been a cornerstone of requirements engineering for many years.

Figure 5.21 shows a SysML visualisation of the Requirement Context View using a use case diagram. In this figure, the use cases from the ontology are shown as SysML use cases. This is slightly confusing as the term use case is being used twice here, once to refer to the concept of a use case (a requirement in context) and once to refer to the SysML element (the ellipse on the diagram).

5.4.4.5 Discussion

This view focuses on the contexts of the requirements that are described in the Requirement Description View and, as such, it forms the heart of the whole approach advocated in this book.

The Requirement Context View really does enforce the need for applying and effective modelling notation to requirements engineering as it demonstrates the

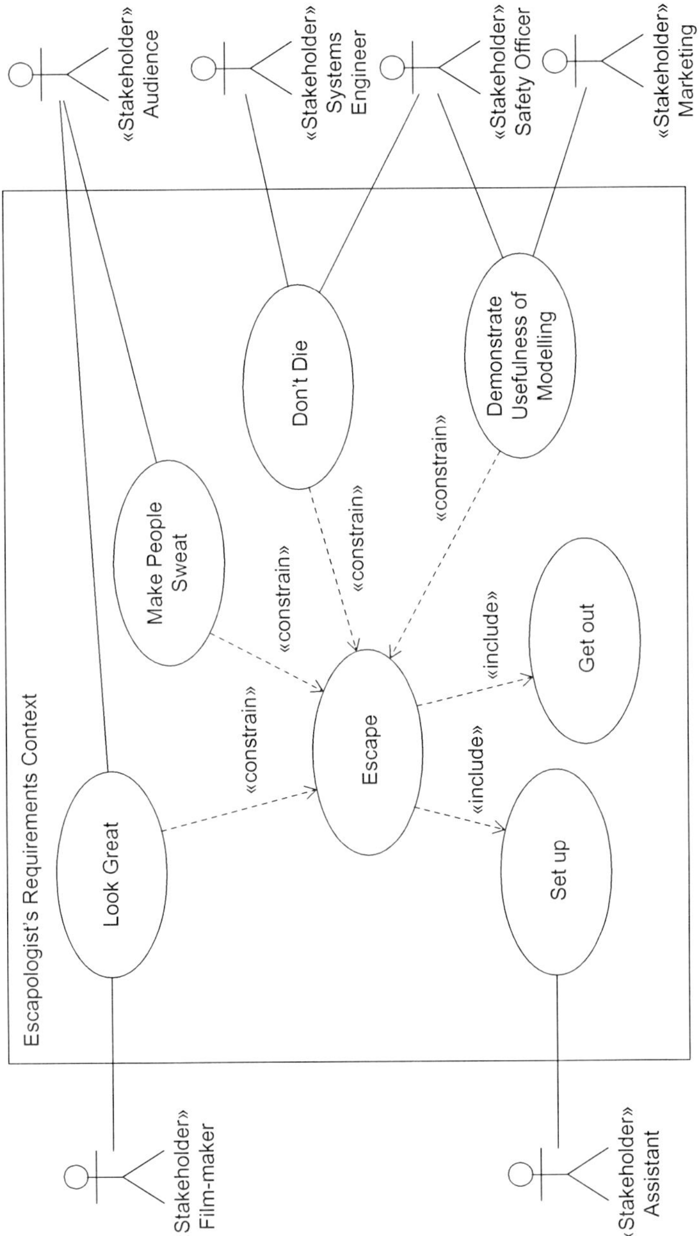

Figure 5.21 Example visualisation of the 'Requirement Context View' using a SysML use case diagram

sheer number and also the complexity of the relationships between the various views in the framework and the elements within each view.

5.4.5 Context Definition View

5.4.5.1 Rational

This view identifies the points of view that are explored in the Requirement Context View. These points of view, or contexts, may take many forms including stakeholders and levels of hierarchy in a system.

5.4.5.2 The ontology

The Source Element View focuses on the subset of the original ACRE ontology that is shown in Figure 5.22.

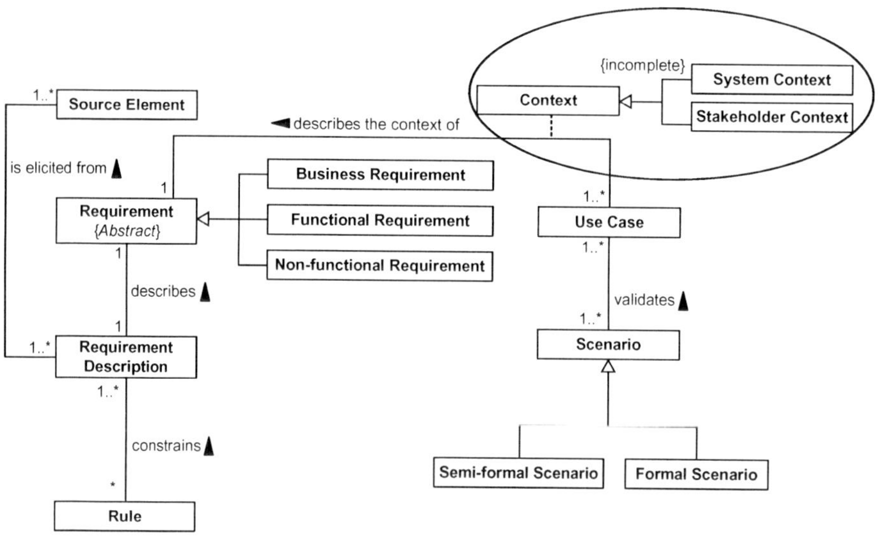

Figure 5.22 Subset of the original ACRE ontology for the Context Definition View

The area of the ontology that is focused on for this view is the various types of 'Context'. Only two types of context are shown here: 'System Context' and 'Stakeholder Context', although many more types of context are possible, as discussed in Chapter 2.

Figure 5.23 shows the following two types of 'Context Definition View':

- The 'Stakeholder Context Definition View' that is made up of one or more 'Stakeholder'. This view identifies a number of stakeholders in a classification hierarchy that are used as a basis for defining contexts.
- The 'System Context Definition View' that is made up of one or more 'System Element', each of which relates to one or more other 'System Element'. This view identifies a number of system elements, usually in a structural hierarchy, that are used as a basis for defining contexts.

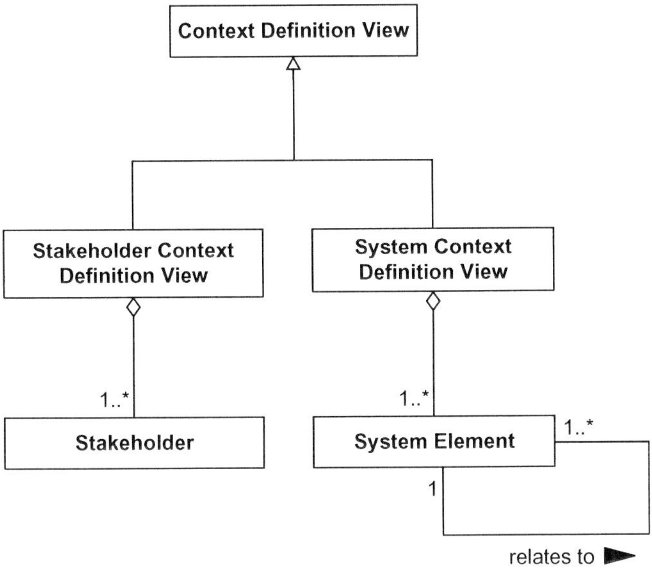

Figure 5.23 Structure of the 'Context Definition View'

It should be kept in mind that there may very well be more different types of view identified here, which will depend on the application of the modelling and the project. The two shown above are two of the most common types.

5.4.5.3 Relationships with other views

The Context Definition View is related to other views in the framework, hence consistency checks must be applied to ensure that these relationships are valid (Figure 5.24).

The main relationship shown here is between Context Definition View and one or more 'Requirement Context View' because the main purpose of Context Definition View is to define the contexts. As a consequence of this, there will be several Requirement Context Views for each Context Definition View. There will also be other relationships that are not shown on the diagram, in Figure 5.24, from the 'Stakeholder Context Definition View'. Stakeholders are used and referenced in several other views, including the following:

- Stakeholders will appear as SysML actors on the Requirement Context View, as the stakeholder forms an important part of this view.
- Stakeholders will be referenced from the Requirement Description View where attributes of 'Origin' and 'Ownership' are defined. The values of these two attributes are actually a set of one or more stakeholders, which must be taken from this view.

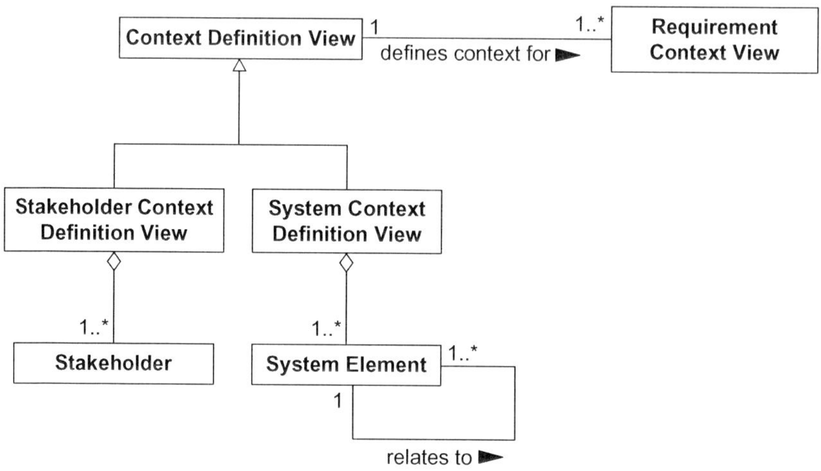

Figure 5.24 Relationships between the Context Definition View and the rest of the framework

The following rules may be applied to these views:

- Each element in each Context Definition View ***may*** define an individual Requirements Context View.
- Each Stakeholder on a Stakeholder Context Definition View ***may*** appear as a Stakeholder on a Requirement Context View.

To maximise the benefits of a true MBSE approach, these rules should be automated rather than manually applied to the model.

5.4.5.4 Visualising the view

The Context Definition Views can be realised in a number of ways, including the following:

- Text descriptions
- Informal diagrams to show hierarchy
- SysML block definition diagrams to show the system hierarchy

Figures 5.25 and 5.26 show the SysML visualisation of both the Stakeholder Context Definition View and the System Context Definition View.

Figure 5.25 shows how the Stakeholder Context Definition View may be realised using a SysML block definition diagram. The stakeholders on this view are usually shown as taxonomy, or classification hierarchy, using the SysML generalisation (has types) relationship. This allows a number of categories of stakeholders to be defined.

Each of these stakeholders potentially have its own context, hence they require the creation of a Context Definition View.

Figure 5.26 shows how the System Context Definition View may be realised using a SysML block definition diagram. The various system elements are shown

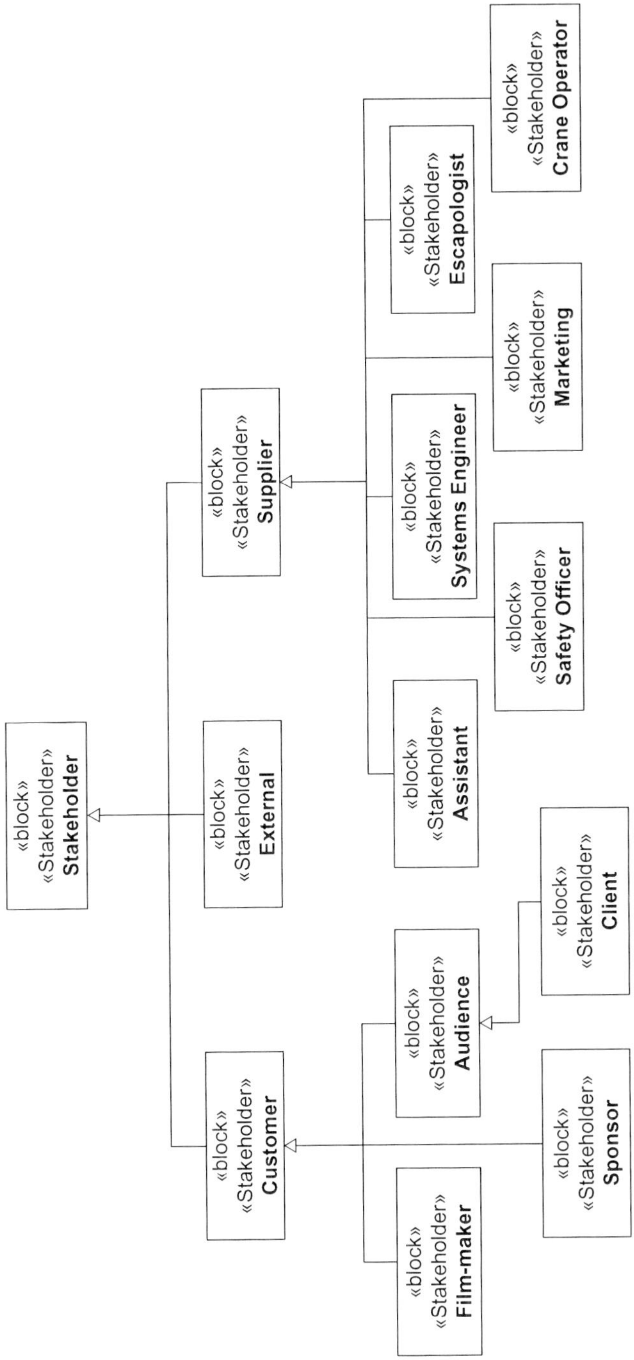

Figure 5.25 SysML representation of the Stakeholder Context Definition View

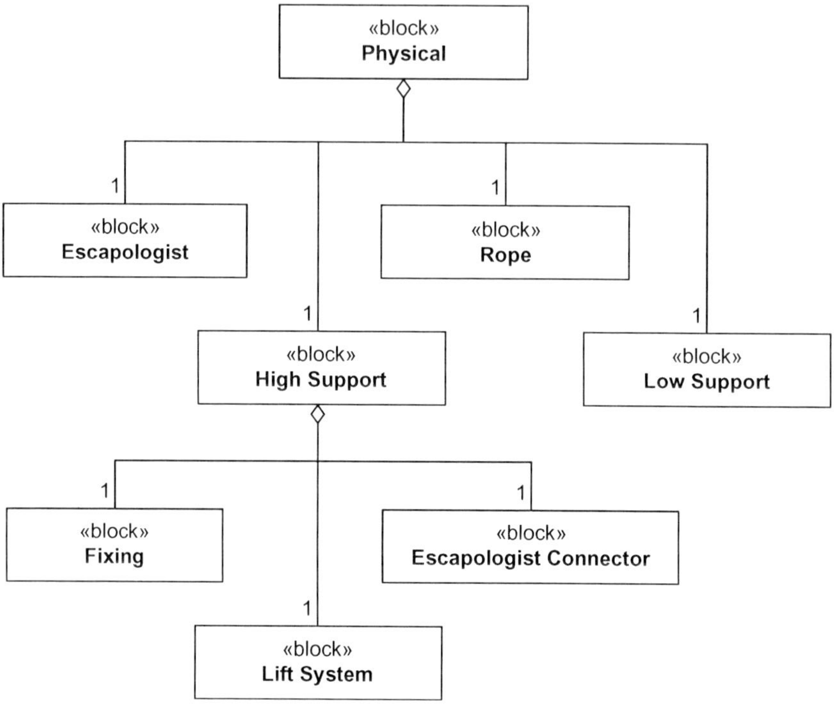

Figure 5.26 SysML representation of the System Context Definition View

as SysML blocks and are expressed in the form of a structural hierarchy using the SysML aggregation (is made up of) relationship.

Each of these system elements potentially have its own context, hence they require the creation of an associated Context Definition View.

5.4.5.5 Discussion

The Context Definition Views are crucial to the whole context-based approach because they are the views that allow us to identify the various contexts that are relevant for the system or project at hand. The views can look deceptively simple because they can be very difficult to get right. Once the stakeholders have been identified and the Context Definition Views have been created, the original stakeholders will often be refined. Any change to stakeholders may also require another iteration of the Context Definition Views.

5.4.6 Validation View

5.4.6.1 Rational

These views provide the basis for demonstrating that the requirements can be met or complied with in some way. These views can be informal (such as scenarios at various levels of abstraction) or may be formal (such as mathematical-based representation).

5.4.6.2 The ontology

The Validation View focuses on the subset of the original ACRE ontology that is shown in Figure 5.27.

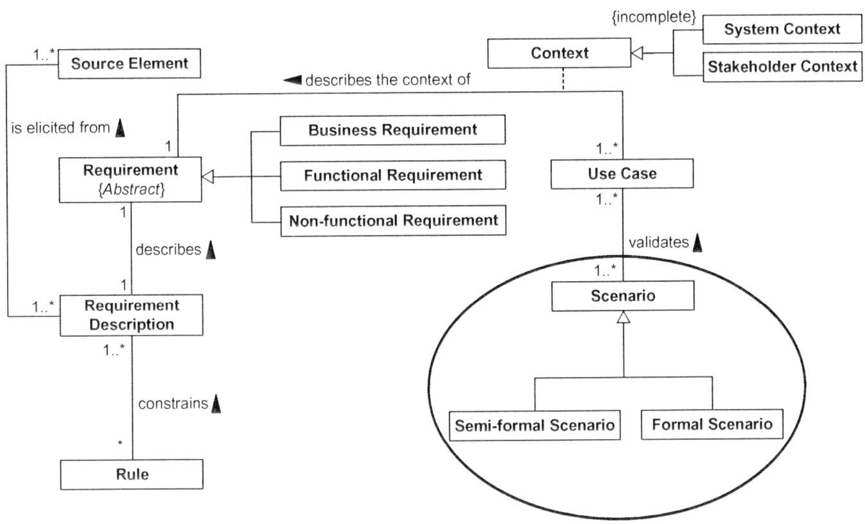

Figure 5.27 Subset of the original ACRE ontology for the Validation View

This view focuses on the scenarios that are used to validate the use cases and, hence, the original requirements. There are two main types of scenario that are considered here:

- *Semi-formal scenarios*. These scenarios explore various 'what-if?' situations by considering the relationships between entities in the system, e.g. by looking at how the various stakeholders interact with the system in order to satisfy a particular use case. These scenarios are visualised using, primarily, SysML sequence diagrams.
- *Formal scenarios*. These scenarios explore various 'what-if?' situations by considering how the values of various properties vary and, hence, impact the system. These scenarios are visualised using SysML parametric constraints and their associated diagrams.

These scenarios result in a number of different views being necessary, as shown in Figure 5.28.

Figure 5.28 shows that there are three types of 'Validation View', which are the 'Stakeholder Scenario View', the 'System Scenario View' and two types of 'Constraint Validation View'. Each of these is explored in more detail in the following sections.

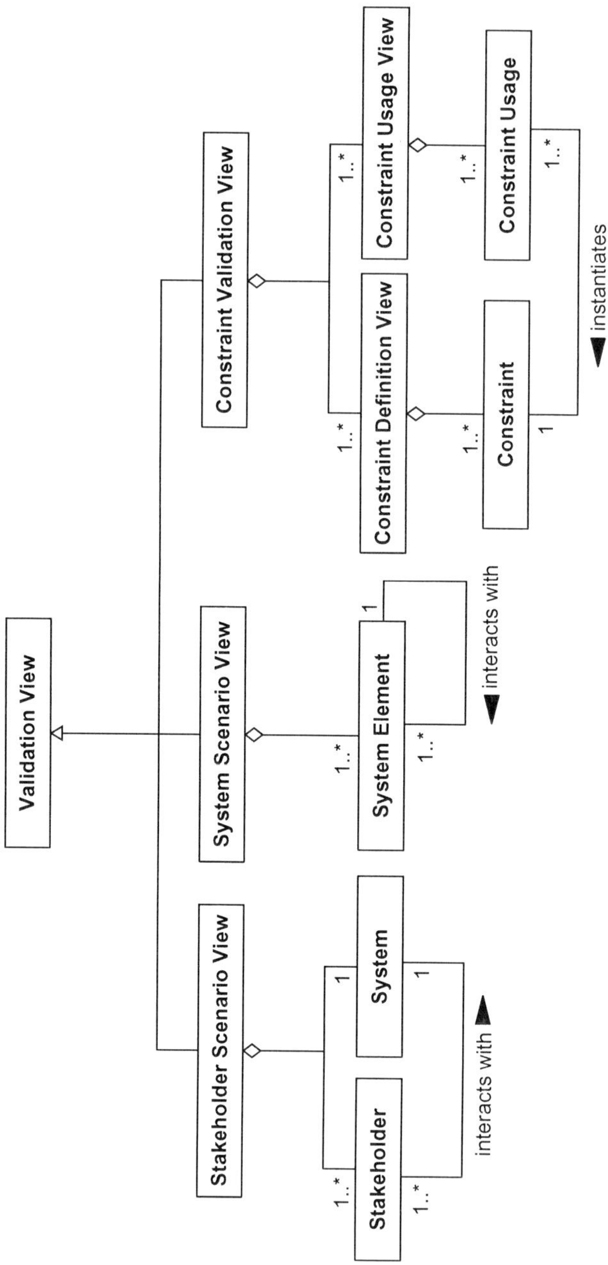

Figure 5.28 Structure of the various Validation Views

5.4.6.3 Relationships with other views

The Validation View is related to other views in the framework, hence consistency checks must be applied to ensure that these relationships are valid.

Figure 5.29 shows that one or more 'Validation View' validates each 'Use Case'. The validation of all the original requirements is achieved through validating each use case.

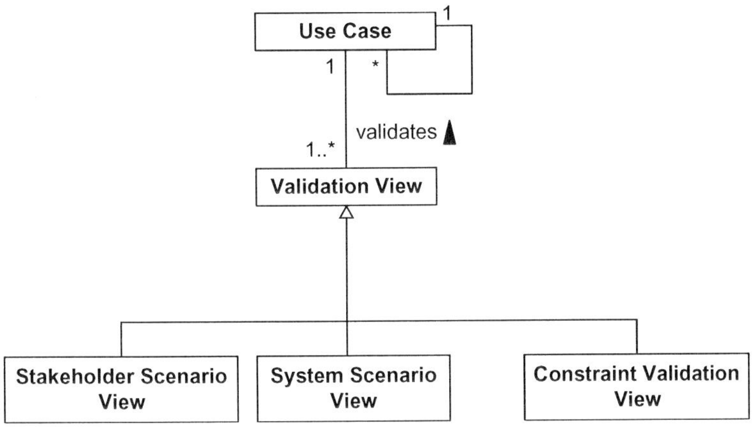

Figure 5.29 Relationship between 'Validation View' and the rest of the framework

The following rules may be applied:

- Each Use Case ***must*** have one or more Validation View associated with it.
- Each Constraint Validation View must use properties that exist on the System Context Definition View.

To maximise the benefits of a true MBSE approach, these rules should be automated rather than manually applied to the model.

5.4.6.4 Visualising the 'Stakeholder Scenario View'

The first of the Validation Views that will be considered is the 'Stakeholder Scenario View'. This view is concerned with looking at scenarios using the SysML sequence diagram. The sequence diagram looks at the interactions between sets of life lines, as discussed in Chapter 3, and the life lines in this view will be the system itself and a number of actors.

To generate a scenario, the first step is to select a specific use case from a context. Next, the context itself, or the system, is visualised using a single life line. The stakeholders that relate to the selected use case are then identified, by seeing which actors relate to the use case, either directly or indirectly. An example of this is shown in Figure 5.30.

Figure 5.30 shows a scenario that has been described for the use case 'Escape' from the 'Escapologist' context.

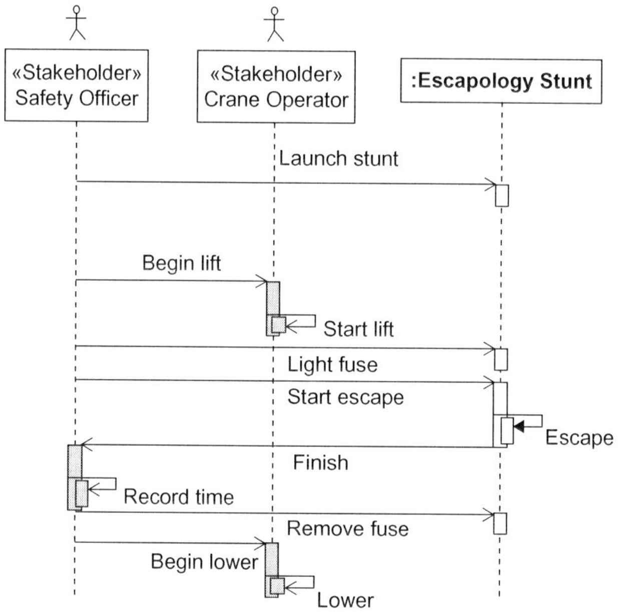

Figure 5.30 Example of a 'Stakeholder Scenario View'

5.4.6.5 Visualising the 'System Scenario View'

This view also looks at scenarios and also visualises them using the SysML sequence diagram. In this case, however, the life lines are realised by system elements and the interactions between them are analysed for a specific use case.

These system scenarios must be consistent with the higher level Stakeholder Scenario Views, should they exist. One way to think about these two types of scenario and the differences between them is to think of the stakeholder scenarios as 'black box' scenarios, where the system is treated as a single entity with no details on what goes on inside. The system scenarios, however, may be thought of as 'white box' scenarios, where the inner workings of the system are considered by looking at the system elements and the interactions between them.

An example of a System Scenario View is shown in Figure 5.31.

Figure 5.31 shows the same scenario that was considered for the 'Stakeholder Scenario View' in Figure 5.30, but this time using the 'System Scenario View'. In essence, the figure shows exactly the same information, but from two different points of view. Although the Stakeholder Scenario View focuses on the interactions between the stakeholders and the system, the System Scenario View focuses on the interactions between system elements within the system.

5.4.6.6 Visualising the 'Constraint Validation View'

The 'Constraint Validation View' actually has two views associated with it, the 'Constraint Definition View' and the 'Constraint Usage View'. The constraint

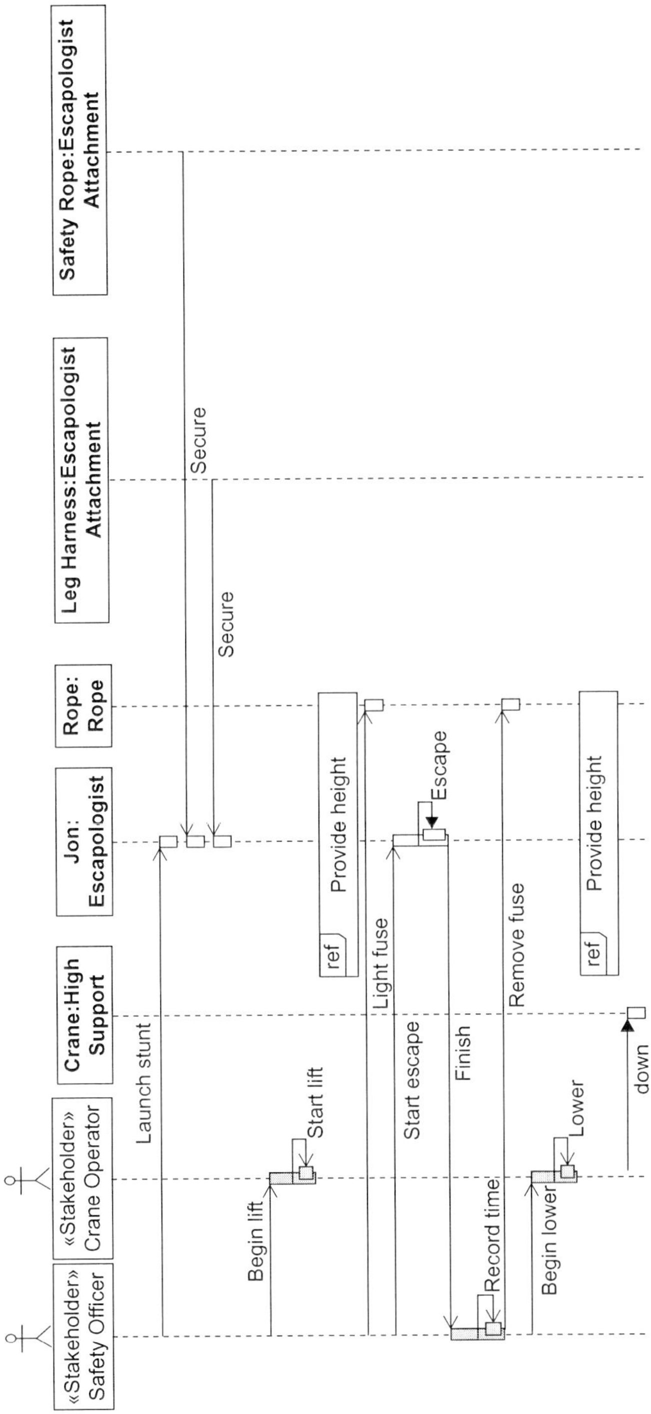

Figure 5.31 Example of a 'System Scenario View'

Validation Views allow formal scenarios to be considered. Although the previous semi-formal scenarios looked at interactions between various elements, the formal scenarios allow different properties of the system to be measured and reasoned about depending on the use case.

The approach taken is the same as previously, in that a single use case from a context is chosen and then a scenario is created. This time, the scenarios will be realised by a series of parametrics that can be reasoned about. This reasoning will take the form of applying equations, logic, heuristics, look-up table and any number of other mathematical-based techniques. These techniques are defined using the SysML parametric constraints, as shown in Figure 5.32.

Figure 5.32 shows an example of a set of parametric constraints that have been defined. Note how some of these constraints are basic mathematical operators ('Plus (2)', 'Minus (4)', 'PercentageIncrease'), one is a standard trigonometry equation ('Hypotenuse') and two are logical operators that form a rule ('AndDecision', 'GreaterThanDecision').

These parametric constraints form a library of calculations that can be applied to the system, but, as yet, how they are applied to the system has not been defined. This usage of the parametric constraints is shown in the 'Constraint Usage View' below.

5.4.6.7 Visualising the 'Constraint Usage View'

The 'Constraint Usage View' shows how the parametric constraints defined in the 'Constraint Definition View' are applied to the model itself. This view is visualised using the SysML parametric usage diagram.

Figure 5.33 shows how the parametric constraints that have been defined previously may be applied to the system elements. The parametrics themselves are connected together in the form of a network, and then the parameters that are required for each parametric usage are taken either from other usages or directly from properties on the model.

5.4.6.8 Discussion

The various Validation Views are an essential part of the framework. Requirements are not requirements if they cannot be satisfied. To satisfy a requirement, it is necessary to understand its context by considering its use cases. These use cases may then be validated by the scenarios. The original requirements, therefore, are validated via their use cases.

The Validation Views have many uses. The obvious use is that they allow us to demonstrate how we can satisfy each requirement. As a result of this, it is possible to use these Validation Views as a basis for acceptance tests. Acceptance tests are the only means by which the end customer can assess whether or not the project has been successful and are based solely on the original requirements.

These Validation Views are also used to ensure that the use cases are actually correct. Not only do they allow us to satisfy the requirements but they also force us to understand the use cases and, hence, the source requirements better. This understanding allows conflicts to be identified, gaps to be spotted, overlaps and

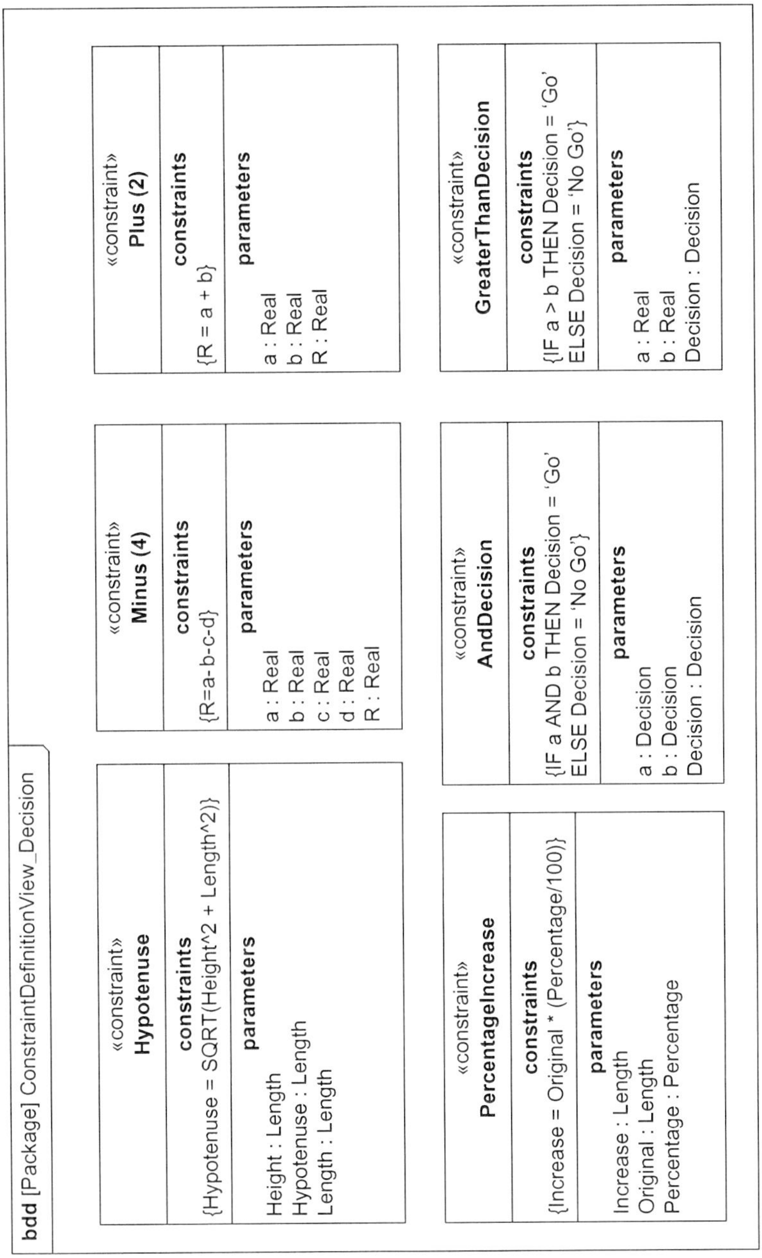

Figure 5.32 Example of a 'Constraint Definition View'

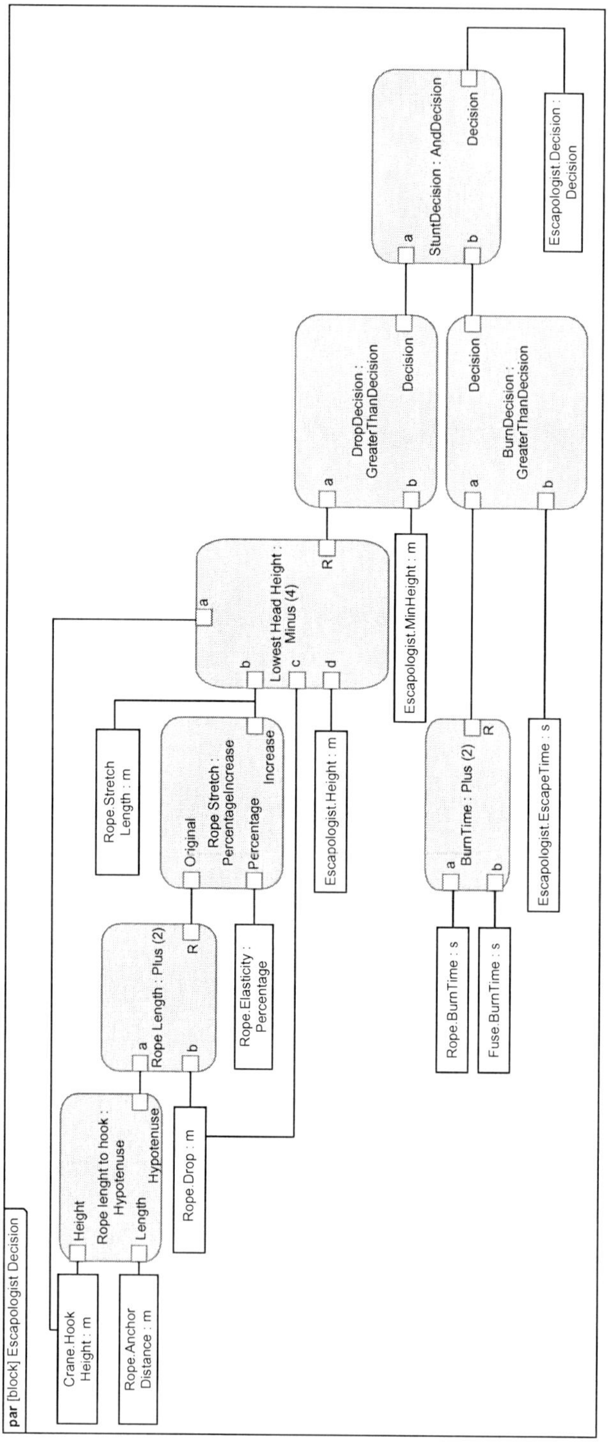

Figure 5.33 Example of a 'Constraint Usage View'

identical requirements to be highlighted and all sorts of other powerful analysis techniques to be implemented. See Chapter 3 for a full discussion on such analysis techniques.

Another use for these validation scenarios is to allow us to reason about the requirements at a business level. Business analysts often use terms like 'measure of effectiveness' (MOE) and 'measure of performance' (MOP). The clue to the relationship between these measures and the scenarios is the word 'measure'. When using parametric constraints, this is exactly what we are doing. As a result of this, the formal Validation Views can be used to provide both measures of effectiveness and measures of performance. In the same way that the semi-formal scenarios can be applied at either the stakeholder level or the system level, the same is true for the formal validation. By applying formal scenarios at the stakeholder level, we can provide measures of effectiveness, and by applying them at the system level, we can provide measures of performance.

Applying the same thought processes, it is also possible to perform trade studies on the requirements set. This takes a very similar form, where different formal scenarios are considered and the results then analysed.

5.5 The Traceability Views

The views that have been looked at so far form the heart of the context-based approach; however, a key part of any requirements engineering endeavour is to provide traceability both to and from the original requirements. This is essential for both quality and validation reasons and provides a level of rigour and, hence, confidence to any requirements set.

Establishing traceability can be a long, tedious and error-prone activity, especially when requirements must be traced by hand. There are many types of relationships that may exist between requirements, as discussed in Chapter 3, and defining what these are and where they exist is no simple task. One of the big benefits of adopting a model-based approach is that all of this traceability is inherent in the model.

There are two levels of traceability relationships that exist in the model, the 'implicit' and 'explicit' relationships. The implicit relationships are the ones that are inherent in the modelling language itself. One of the advantages of using a standard model notation, such as the SysML, is that there is an underlying meta-model beneath the notation that specifies exactly how each of the modelling elements relates to one another. The explicit relationships are those that are not inherent in the modelling notation but that are dependent on the application of the modelling. These relationships can be identified directly from the ontology and the framework.

In the area of requirements engineering, these traceability relationships can take on many different forms. For example, just consider the 'Source Element' and the number of different things that can be a Source Element. These range from conversations, to higher level requirements, to specifications, to entire systems. It is often necessary, therefore, to define exactly where the traceability relationships exist. It is possible to trace between almost any system element and any element in the framework. It is also often desirable to trace between the views themselves. This is shown in Figure 5.34.

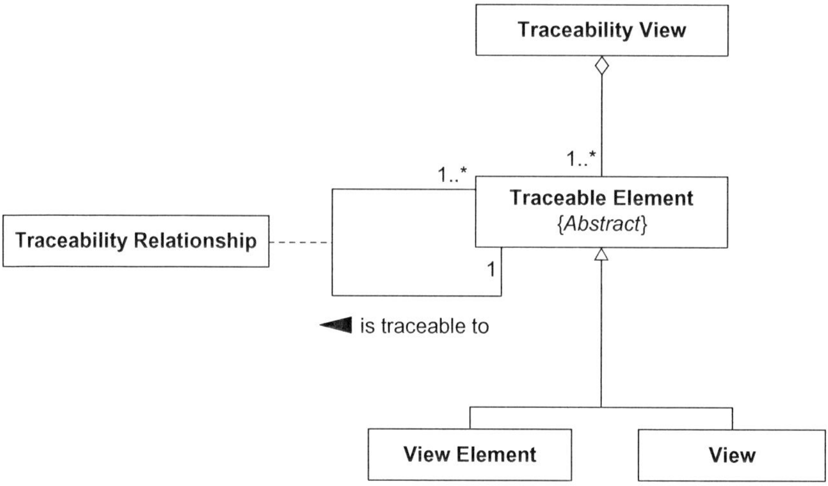

Figure 5.34 The 'Traceability View'

Figure 5.34 shows that the 'Traceability View' is made up of one or more 'Traceable Element', of which two types are listed below:

- The 'View Element' that represents any element that exists within the ontology.
- The 'View' that represents any view from the framework.

Each 'Traceable Element' is traceable to one or more other 'Traceable Element' via a 'Traceability Relationship'.

5.5.1 Relationships to other views in the framework

The very nature of this view is to relate other views and view elements together; therefore, it can potentially relate to all of the core views in the framework.

5.5.2 Visualising the view

This view can be visualised using any number of different techniques, such as the following:

- Tables – Simple tables to show relationships between elements.
- Spread sheets – Similar to the previous point, but using a spread sheet.
- In a database, using a requirements management tool – Many requirements management tools allow traceability to be established using the underlying database and then visualised in different ways.
- SysML diagrams – Any relationships may be defined in the model.

As has been mentioned several times already, when a true model is produced then the traceability is inherent and, therefore, many of these views may be automated. This is discussed in more detail in the next chapter.

5.6 Conclusions

This chapter has introduced the ACRE framework that comprises a number of views. The views are based on realisations of the ontology that was discussed in the previous chapter.

The next step is to look at how the framework may be used to implement the ACRE philosophy on real projects and to discuss the 'people, process and tools' aspects of MBSE.

References

1. Flesch R. 'A New Readability Yardstick'. *Journal of Applied Psychology*. 1948;**32**:221–33
2. Kincaid J.P., Fishburne R.P. Jr, Rogers R.L., Chissom B.S. *Derivation of New Readability Formulas (Automated Readability Index, Fog Count and Flesch Reading Ease Formula) for Navy Enlisted Personnel*. Research Branch Report 8-75, Millington, TN: Naval Technical Training, U. S. Naval Air Station; 1975
3. ISO 15288. *Systems Engineering – Systems Life Cycle Processes*. London, UK ISO Publishing; 2004

Part III

Realisation

Chapter 6
Case study

The case has, in some respects, been not entirely devoid of interest.

Sir Arthur Conan Doyle (1859–1930), Sherlock Holmes

6.1 Introduction

This chapter discusses the application of model-based requirements engineering by looking at a case study. This brings together and uses many of the concepts and terminology defined in the ontology and framework in Chapters 4 and 5 as well as using the notation and modelling techniques described in Chapters 2 and 3.

This case study exercises each of the concepts and views of the ACRE framework, as shown in Figure 6.1, which are discussed in detail in Chapters 4 and 5.

Iteration is a necessary part of modelling and as such all of the models that follow have been developed in an iterative manner. Iteration is one of the fundamental principles of modelling; a model is never 'right first time'. Confidence in a model is gained by ensuring consistency between the diagrams forming the views; to achieve this, partly completed diagrams may need to be updated based on information gained from populating other views within the framework.

An additional complexity is the reuse of the framework at different system levels. Figure 6.2 shows three levels that may be defined: stakeholder, system and subsystem:

- Stakeholder refers to the role of a person, place or thing that will be affected by the system.
- System refers to the main deliverable, which may be split down into component parts.
- Subsystem refers to each of the component parts of the system.

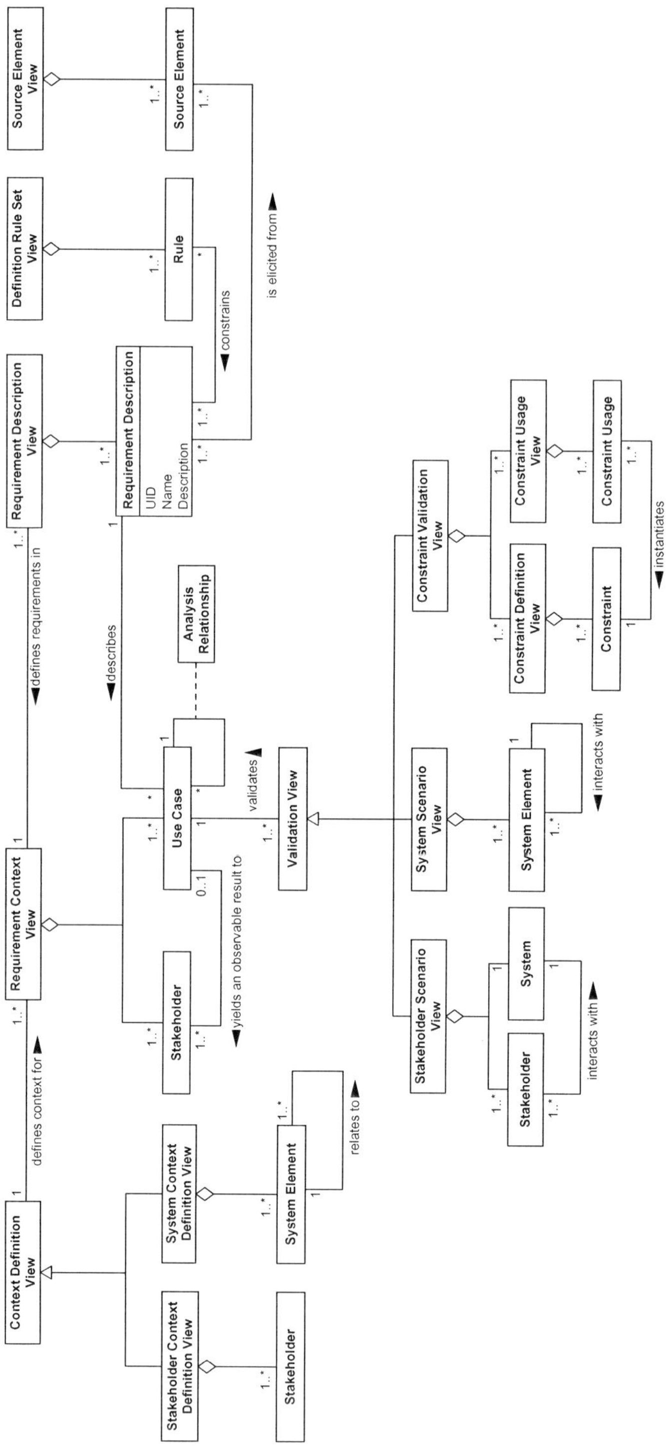

Figure 6.1 Framework

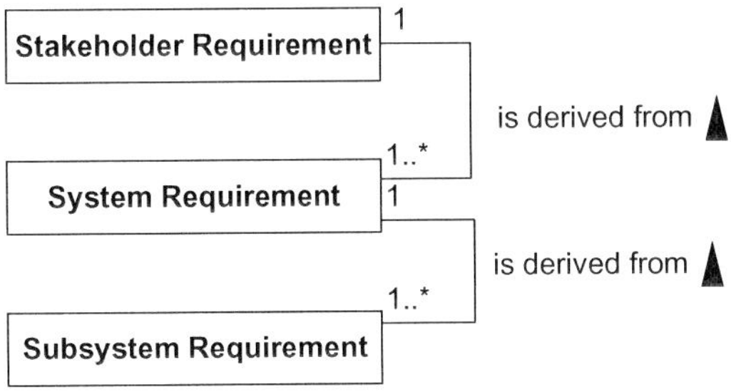

Figure 6.2 System hierarchy

The framework may be applied at each level, and traceability relationships are used to capture the dependencies between levels, represented as Traceability Views.

Because of the high level of iteration, it is impossible to say which diagram was 'finished' first and even in which order that were completed. In reality, the views may be produced in a specific order, but this will depend on the exact process that has been followed. In practice, it is rare to complete a single view completely and then move onto the next one. Typically, partial views are created and then updated and enhanced when other views are considered.

Figure 6.3 shows an example of how this iteration manifests itself.

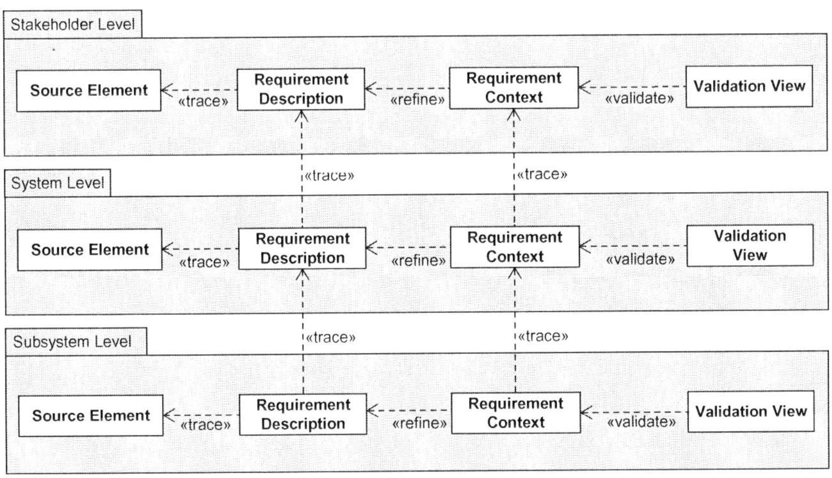

Figure 6.3 Framework iteration

Figure 6.3 shows that each of the views and elements in the ACRE framework may be used at each level: stakeholder, system and subsystem. The traceability provides the relationships between each level. The remainder of this chapter discusses the case study that is used to illustrate each and every view in the ACRE framework. Each section highlights the area of the framework on which it is based, discuss the development of each view highlighting the advantages of different representations and expose the traceability that has been captured during the development of the view.

6.2 Background

In terms of safety-critical systems, there are many, many examples that can be used. For the purposes of this book, the decision was made to include an example that would focus on a safety-critical application that could be fully modelled for the case study. This had to be a real system, not just a theoretical one, and should be non-industry specific, so as to appeal to all readers of the book.

Many people claim to have great faith in their approach to systems engineering, but the question is would these same people actually put their life on the line to demonstrate this? This question is often asked at workshops and conferences relating to safety-critical systems, which will also be addressed here.

In a book published earlier, the authors have used an example from the world of escapology to model a famous death-defying escape in order to illustrate modelling concepts. The same example – the coffin escape – is used in Chapter 3 to illustrate the basic modelling notation. This example has been used widely and has proved to be very popular, so the question arose, could we do something similar in real life to provide the example for this book? Therefore, an example from the world of escapology was sought where we could actually implement the system based on a system model built from a set of requirements defined using the ACRE approach.

The world's most famous escapologist is the now-legendary Harry Houdini. Houdini was an outstanding showman who had many skills, but he is probably most remembered as being as an escapologist. Houdini is one of the few people *ever* to have their name in the Oxford English Dictionary as an official word.

The stunt that is almost synonymous with Houdini is the straight jacket escape which, it was decided, would be an ideal candidate for the example for this book. The idea would be that one of the authors would take the role of the escapologist, whilst the other two would take the role of the systems engineer. The escapologist would model the escape and train himself up to become a competent escapologist for straight jacket escapes (competency scopes available on request). The systems engineers would perform all the systems modelling and would implement the system based on this model.

The systems engineers had decided that the escapologist would be raised into the air on the end of a rope to perform the escape. To spice events up a little, they also decided to set fire to rope – this was purely to make the model more interesting, so the escapologist was told.

So it was on a bright afternoon of April 2010 that the escapologist turned up to a location in the United Kingdom, complete with straight jacket, to perform the escape. The implementation of the system consisted of an industrial crane, a length of rope, a 2-minute fuse, a lighter, a timer and a camera crew.

The following case study is the actual model that was used for this stunt.

6.3 The Source Element View

The Source Element View highlighted in Figure 6.4 is used to capture the origin of requirements. This view is used primarily to establish traceability.

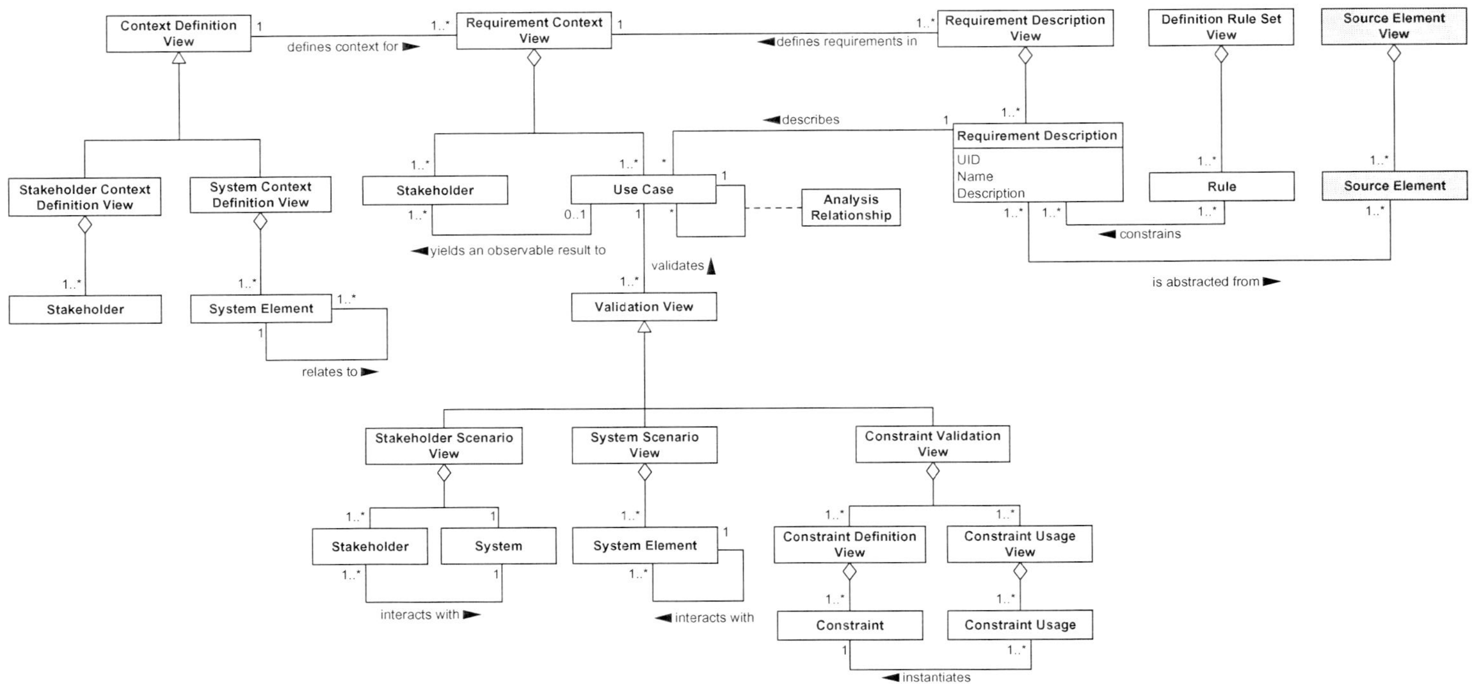

Figure 6.4 Framework – Source Element View

Figure 6.4 highlights the Source Element View and the Source Elements from the ACRE framework. To carry out a straight jacket stunt, a number of sources of information, as with any project, need to be considered. These sources included the following:

- Technical reference material such as Houdini's books on how to escape.
- Discussions regarding the overall intention of the project such as emails.
- Legal reference material such as health and safety standards and laws.

In this case, the sources identified initially were the following:

- Emails – three emails discussing the outline of the stunt.
- Books – two books including the technical aspects of SysML and escapology.
- Standards – two safety critical equipment standards providing concepts for risk management.

There are many ways that these sources could be recorded, as discussed in Chapter 5. In this example we consider four approaches, including text-based, tables and block definition diagrams.

- Escapologist email
- System engineer email
- Marketing email
- Houdini book [1]
- Houdini video [2]
- IEC 61508 [3]
- ISO 31000 [4]

A simple text-based list, as shown above, may well provide the first iteration of the Source Elements and on small project may provide enough information to be a complete Source Element View. However, on larger projects this needs to be recaptured in an engineering tool, which will allow traceability relationships to these Source Elements to be captured.

The blocks in Figure 6.5, each of which represents a Source Element as indicated by the «Source Element» stereotype, provide reference information which allows the original information to be found in an archive location, as with the emails, in a book as with the SysML for systems engineering reference [5] or on a DVD as with the Houdini video [2]. This ability to provide referencing is useful when the information cannot be captured in a central model. In this case, the Source Elements can be traced as they are captured in a SysML block definition diagram which allows traceability relationships to be added as discussed in the Chapter 3.

Figure 6.6 shows a Source Element Email captured as a SysML block within the model. When considering each Source Element, the level of information held can be important to the task being completed. In some security-based situations, it may be necessary to ensure information is not communicated but that the Source Element can still be traced. The approach taken in Figure 6.5 would support this need by not showing the text within the source. In this case study, the detail associated with the source provides important information and must be communicated. It provides the first set of information that informs the stakeholder requirements.

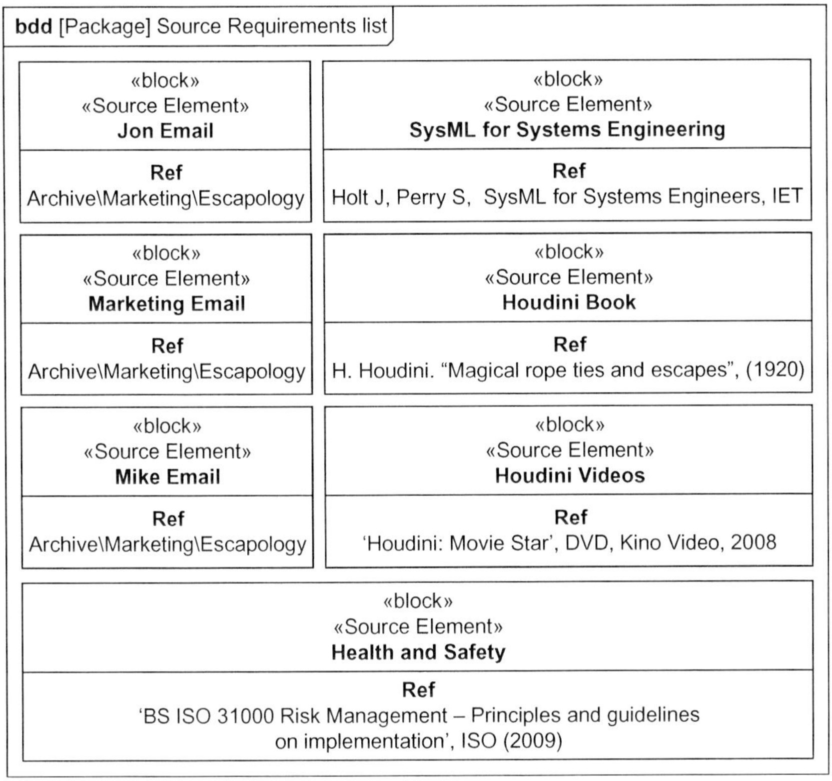

Figure 6.5 Referencing sources

In Figure 6.6, the descriptive text within the block is shown. It represents an email that forms part of the discussion leading up to the definition of the stunt. However, it may represent any of the types of source information discussed in Chapter 4. If required, more information could be added to the block, capturing identifiers, longevity, relation to other sources, history etc.

Although providing a model of the sources may be very useful for capturing source detail and traceability, it may not suit every stakeholder. Using a model-based approach enables the delivery of information captured in a model to be delivered in many different ways.

Table 6.1 shows a tabular representation of Source Elements; it is an alternative way to view Source Elements, which is an extension of the initial text-based list. The difference here is that this table has been automatically generated from the information held inside each of the three Source Element blocks representing emails, as shown in Figure 6.5 and expanded in Figure 6.6. The automatic generation of tables is a feature of many tools and can be used very effectively to deliver model-based information in a more traditional format.

These views are all similar in nature due to the type of information being held. When collating source information, there are two important points to consider. The first is whether the Source Element can be traced? If the Source Element

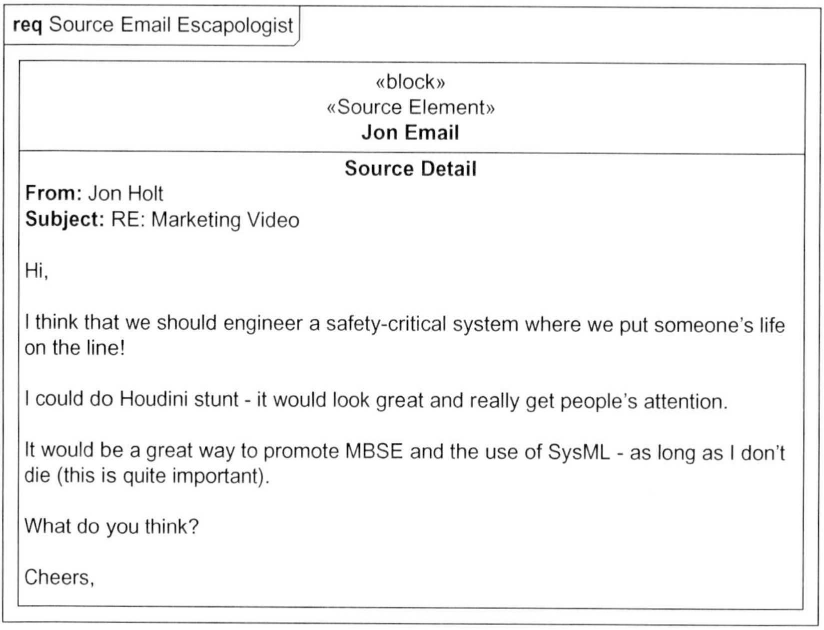

Figure 6.6 Source email

Table 6.1 Source emails

Name	Txt
Jon Email	From: Jon Holt Sent: 2 March 2010 08:32 Subject: RE: Marketing Video Hi, I think that we should engineer a safety-critical system where we put someone's life on the line! I could do Houdini stunt – it would look great and really get people's attention. It would be a great way to promote MBSE and the use of SysML – as long as I don't die (this is quite important). What do you think? Cheers, Jon
Marketing Email	From: Marketing Sent: 1 March 2010 18:22 To: Jon Holt Cc: Mike Brownsword; Simon Perry Subject: Marketing Video

Table 6.1 Continued

Name	Txt
	Hi Jon, Can you start to think about doing a video as part of a new marketing initiative? We want to promote our tools and services for SysML in particular so that we can generate some more leads. Above all, we want it to be interesting. Have a think and let me know. Cheers, Hedley
Mike Email	From: Mike Brownsword Sent: 2 March 2010 08:34 To: Marketing; Jon Holt Cc: Simon Perry Subject: RE: Marketing Video Hi guys, Hang on a second – if we are going ahead with this, then we can't have Jon doing the safety modelling. Let me, Simon and Matthew do all the system modelling and Jon can model the stunt itself. That way we can make sure that all the equipment is safe and that we have a chance of keeping him alive. All the best Mike

cannot be traced, it will not provide any justification for the existence of requirements or system design. It will also turn the analysis of any impact from that source into a manual task which will need to be repeated every time any changes are considered. Second, the level of granularity needs to be considered. In the examples shown here, some provide the full text of the Source Element, whereas others do not. In cases where the original text, or portions of it, needs to be referenced, the complete text may need to be captured rather than a simple reference.

6.4 The Requirement Description View

The Requirement Description View is used to capture the structured descriptions of each requirement. These requirements are considered individually and usually have a number of properties, or features, associated with each one.

Figure 6.7 shows the ACRE framework with the Requirement Description View highlighted.

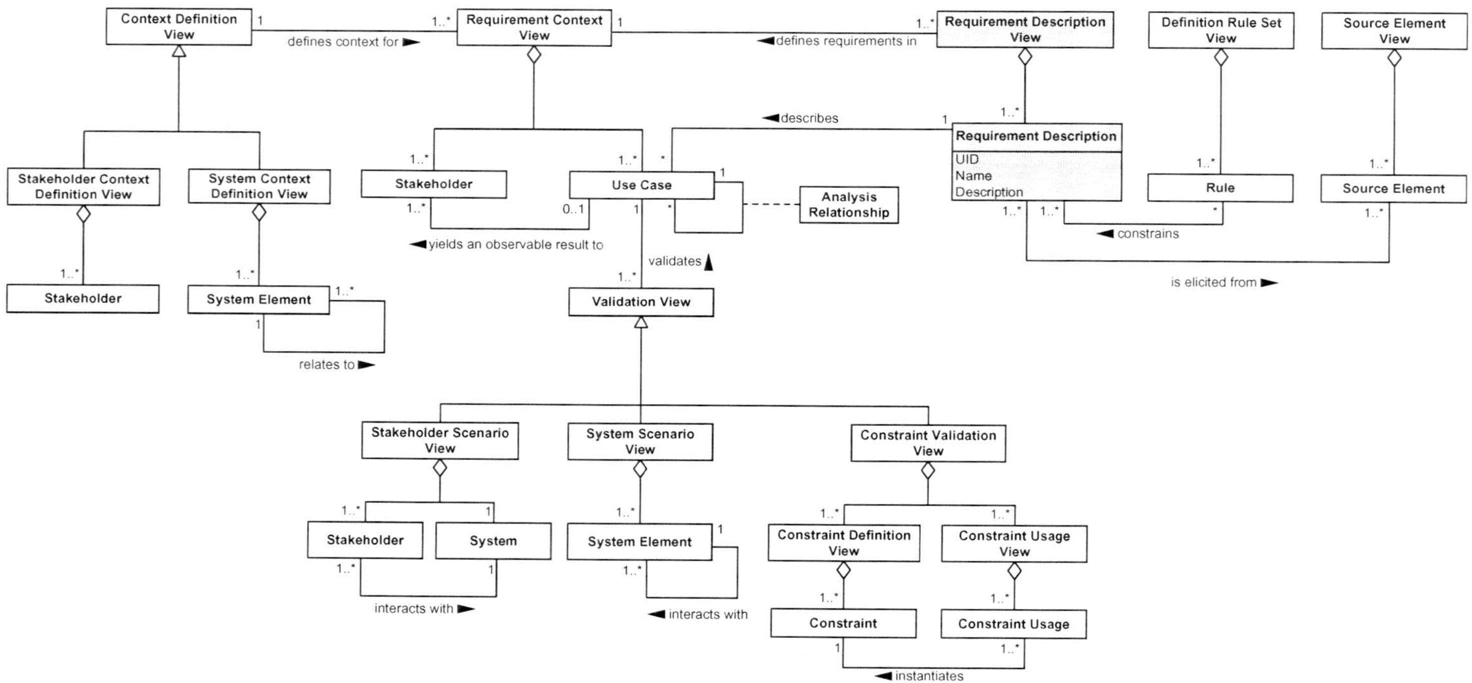

Figure 6.7 Framework – Requirement Description View

To understand the requirements rather than the sources of information, the description or explanation of each requirement must be considered in the form of Requirement Descriptions. To ensure that there is a formality to the requirement and that it can be found in the future, all Requirement Descriptions should also have a number of properties, such as 'Name' and 'Unique Identifier (UID)'. Other properties can be added to the Requirement Description as discussed in Chapters 3 and 5.

This section considers two ways to capture Requirement Descriptions – the first using a SysML requirements diagram and the second a traditional table-based view.

The requirement diagram in Figure 6.8 shows a number of Requirement Descriptions using the SysML construct of a requirement block. Requirement diagrams can be used to show the nesting relationship of requirements, an example of which is shown between 'Get out' and 'Remove straight jacket'. The diagram also allows different levels of information to be shown, e.g. in the 'Don't die' Requirement Description, a text description associated with the requirement is also displayed using the 'txt' compartment on the block.

Table 6.2 shows a textual representation of the requirements, typical of what may be found in commercial requirements management tools. Many industrial modelling tools also create tables in this form, based on the information in the model. This has the advantage of being able to ensure that system traceability is directly connected to the requirement whilst enabling table-based viewing and editing, if required.

The requirements considered so far are stakeholder requirements that represent what a stakeholder needs or would like the system to do, rather than system

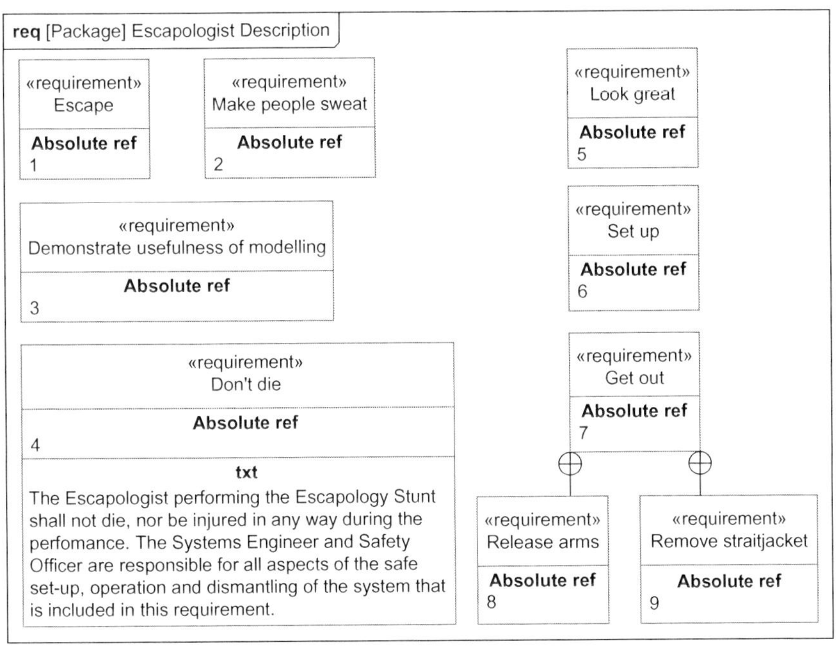

Figure 6.8 Requirements description diagram

Table 6.2 Requirements description table

Absolute ref	Name	Description	Priority
3	Demonstrate usefulness of modelling	The escapology stunt shall demonstrate the usefulness of model-based systems engineering that can be used by Marketing	Medium
4	Don't die	The escapologist performing the escapology stunt shall not die, nor be injured in any way during the performance. The systems engineer and Safety Officer are responsible for all aspects of the safe set-up, operation and dismantling of the system that is included in this requirement	High
1	Escape	The main aim of the project is to perform a death-defying escapology stunt. In this case, the stunt is an upside down straight jacket escape	High
7	Get out	The escapologist shall escape from the straight jacket in a timely and safe manner	High
5	Look great	The escapologist performing the escapology stunt shall look good from the point of view of the audience and the film-maker	Low
2	Make people sweat	It is essential that the excitement experienced by the audience is maximised while still ensuring that the requirement to minimise risk to escapologist is met	Medium
8	Release arms	The escapologist will release both arms whilst still wearing the straight jacket	Medium
9	Remove straight jacket	The escapologist will remove the jacket fully. This has to be done after the arms have been released	High
6	Set up	The escapology stunt shall be set up by the Assistant	High

requirements, and, as such, there is a question regarding the relevance of the text. Many stakeholders describe their requirements in terms of system requirements as they often think about what they want the system to achieve. However, there may be many stakeholders with varying levels of influence over the final system, all generating requirements. These requirements have the potential to conflict, and the prioritisation of stakeholders may preclude a specific requirement from being incorporated in the final system. It would be better for each stakeholder to describe the problem they perceive rather than what they believe the system should do. In this case, the requirement 'The escapology stunt shall be set up by the Assistant' may be better described as 'The escapologist needs the escapology stunt to be set up by a qualified and competent individual who understands the configuration of the escapology stunt as well as the processes to be employed during the escape.' The explanation of the problem seen by the escapologist may then open up a number of alternative solutions.

It is often important to be able to return to the stakeholder who originated the requirements and so it would be useful to capture this information within the requirement. As mentioned earlier, many other pieces of information are often captured as part of the Requirement Description.

Figure 6.9 shows an expansion of two Requirement Descriptions which includes other properties such as 'Priority', 'Origin' and 'Validation Criteria'.

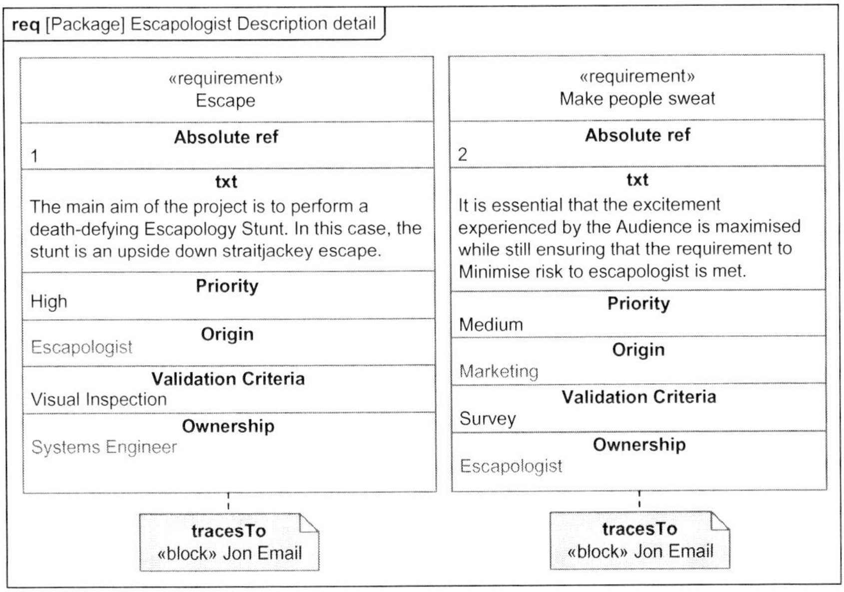

Figure 6.9 Expanded requirements description

Within a strong modelling environment, many of these attributes are linked to elements already defined within the model, further supporting consistency within the model.

6.4.1 The Traceability View (Requirement Description View to Source Element View)

The highlighted Traceability View may be used to allow traceability between different elements of the model to be explicitly shown. These traceability links may exist between views or between elements within views.

Figure 6.10 shows the Traceability View and also highlights the Requirement Description View and the Source Element View which will be traced to each other. Traceability is often limited to relating system and subsystem requirements or system and stakeholder requirements together. Limited use of traceability in this way does not realise the true value of traceability because, in this case, it can only be used to analyse changes from stakeholder requirements into system requirements. Traceability has much wider implication and can be used effectively to support analysis of coverage, review, change and impact. Within the ACRE framework, traceability enables the realisation of the relationships between many framework elements. As with other areas, traceability is used repeatedly to build up a full picture of the relationships, not only between stakeholder and system requirements but also all the way through the framework from source to validation.

Table 6.3 shows a matrix of the traceability between the stakeholder requirements (shown across the top of the table) and the Source Elements (shown on the left), which have been discussed in the previous two sections. Having sight of this traceability enables the impact of changes in source material on the stakeholder

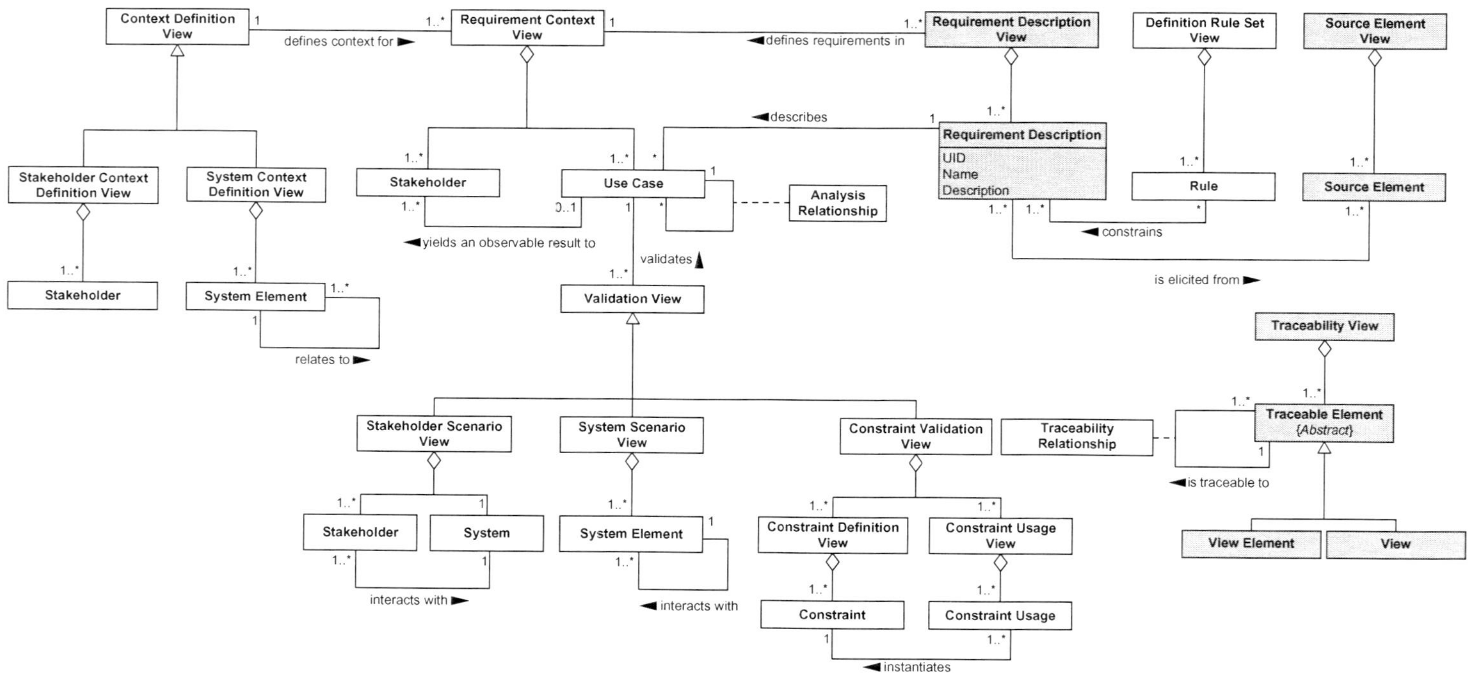

Figure 6.10 Framework – Traceability View

Table 6.3 Requirements trace to source

Traces to \ Traces from	«requirement» Remove strait jacket	«requirement» Release arms	«requirement» Get out	«requirement» Make people sweat	«requirement» Look great	«requirement» Escape	«requirement» Demonstrate usefulness of modelling	«requirement» Set up
«block» Houdini Book	trace	trace	trace					
«block» Jon Email				trace	trace	trace		
«block» Marketing Email							trace	
«block» Mike Email							trace	trace

requirements to be considered. Such changes may derive from any of the Source Elements, such as changes to standards and stakeholder's comments.

The example shown here is a relatively simple one, with few rows and columns in the table. In a straightforward example like this, it is feasible to generate the table manually. However, for larger projects with many requirements and sources, this manual development of matrices is simply not realistic. In these cases, there are many tools that can be used to capture traceability, including requirements management tools, traceability tools and modelling tools. When capturing traceability in tools, there are a number of issues to be addressed, such as the number of relationships, the role of the person adding the traceability, the views of the traceability which are produced and the possible use of multiple tools. The use of tools is discussed in detail in Chapter 7.

As in Table 6.3, many tools are able to automatically generate matrices to aid in the visualisation of the traceability. This autogeneration enables the traceability information to be stored in a modelling tool whilst still allowing the information to be viewed in standard office tools.

6.5 The definition rule set view (testing the Requirement Descriptions)

The Definition Rule Set View contains the rules that may have to be applied to each Requirement Description.

Figure 6.11 shows the ACRE framework with the definition rule set view highlighted. Applying rules to the Requirement Descriptions provides confidence in the requirement set as a whole. These rules may focus on the actual words used in the Requirement Description properties, such as text descriptions, the status of the requirement, coverage of source information and many other areas. This may be as straightforward as defining a list of forbidden words. Some words are

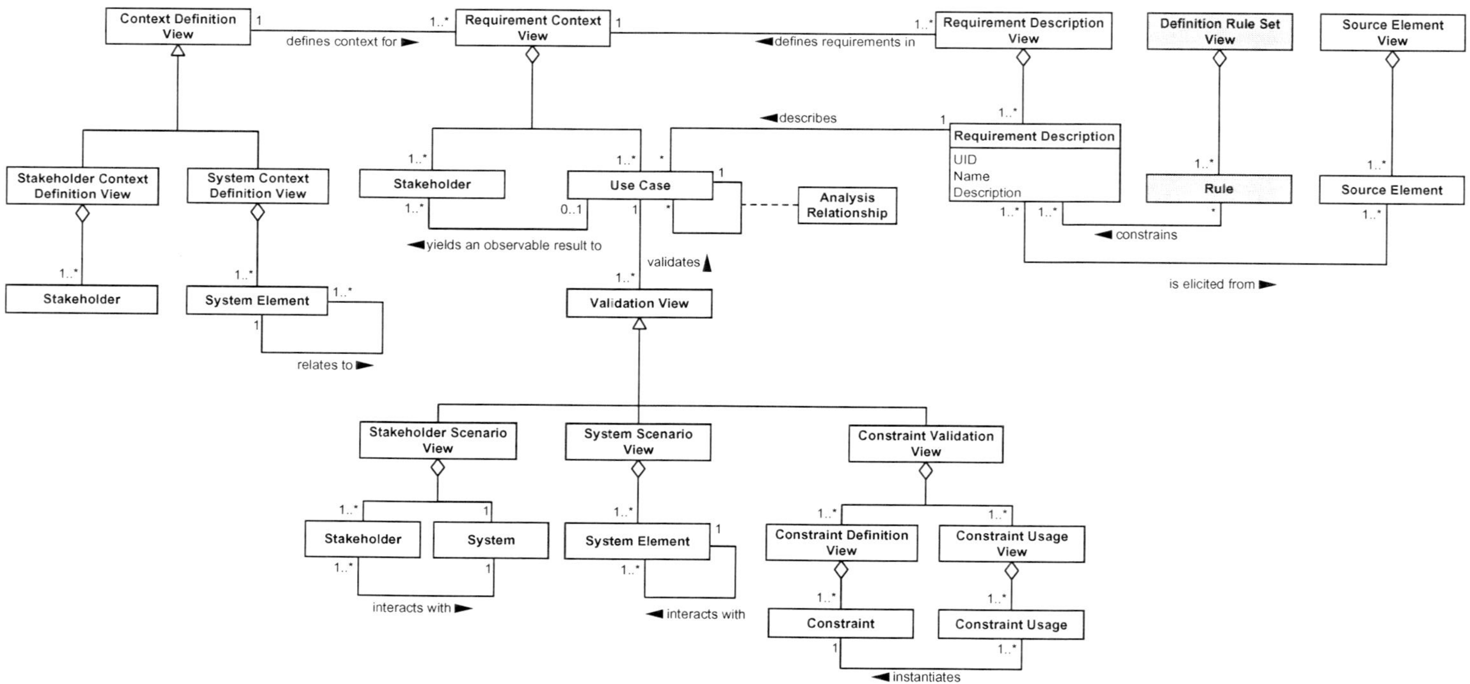

Figure 6.11 Framework – definition rule set view

notoriously ambiguous and, as such, are frowned upon in the world of requirements engineering. Examples of some of these forbidden words include the following:

- *Quickly*. This is considered to be ambiguous as it doesn't provide any time-frame or idea of what quick may be.
- *Reasonable*. The subjective nature of what is reasonable makes this a dangerous word to use.
- *Minimum*. It is also subjective and needs quantitative clarification.
- *Maximum*. Similar to minimum, this needs clarification to be considered unambiguous.

It should be kept in mind that the values of the properties that form the Requirement Description are used as part of the validation process. Therefore, it is essential that each Requirement Description is as unambiguous as possible.

Another example of a typical type of rule is one that calculates complexity. Complexity rules may come in many forms, such as equations or more general text-based rules. Metrics can be defined specifically for a project or organisation. However, many standard metrics exist such as the Flesch–Kincaid complexity metrics [6] that are already established and may form a good off-the-shelf library of readymade metrics. These metrics are based on the desire to understand the complexity of text [7] and have since been specialised to apply to requirements.

Figure 6.12 shows how metrics of this sort can be captured in a model-based environment using the SysML construct of a parametric constraint as discussed in Chapter 3. These rules can take on many forms, including mathematical and logical operators, both of which are shown in the figure. The benefit of having metrics defined in this way is that they can be reused in a number of parametric networks within this model as well as models in other projects.

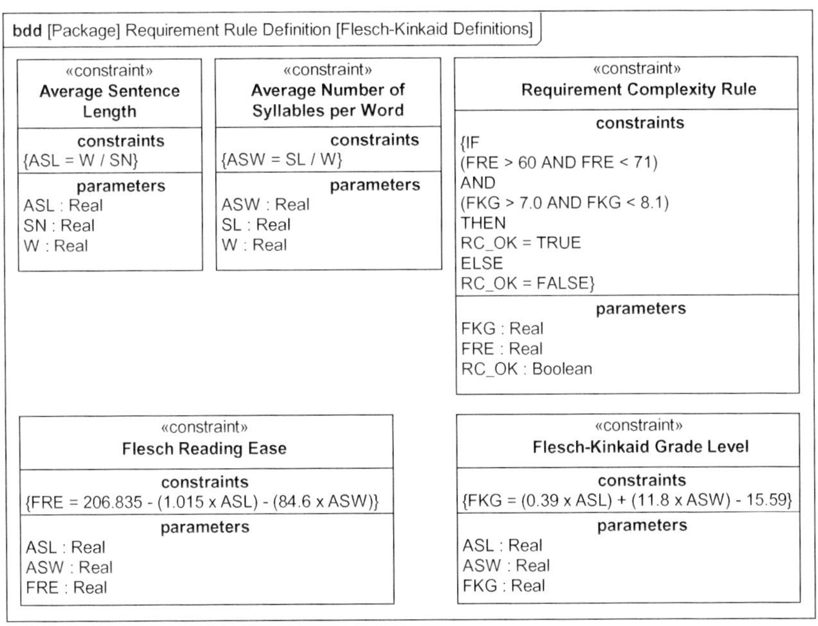

Figure 6.12 Rule definition

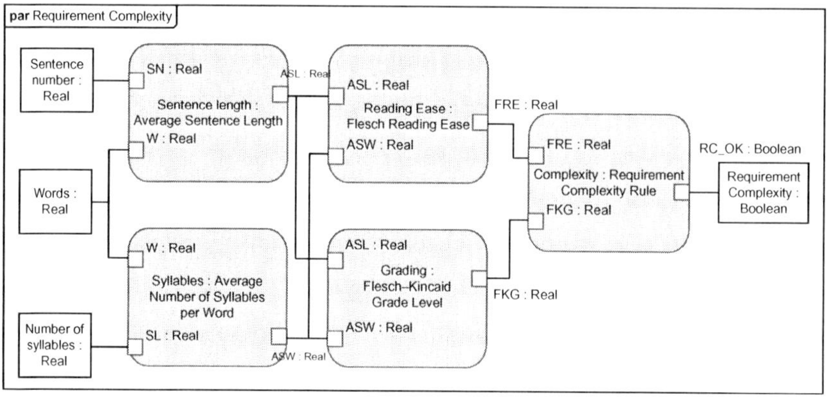

Figure 6.13 Rule usage

Table 6.4 Complexity metrics

Absolute ref	Average sentence length	Average syllable per word	Reading ease	Grade level	Complexity
1	12.5	1.44	72.3	6.2	False
2	25	1.84	25.7	15.8	False
3	17	2.00	20.3	14.6	False
4	22.5	1.85	27.9	14.9	False
5	20	1.65	46.9	11.6	False
6	10	1.60	61.3	7.1	True
7	14	1.57	59.6	8.4	False
8	12	1.58	60.7	7.7	True
9	7	2.00	30.5	10.7	False

Figure 6.13 shows how the constraints defined in Figure 6.12 can be connected together in a parametric network. This particular network can be used to calculate the complexity of each Requirement Description based on the Flesch–Kincaid Requirement Complexity Rule. The metrics are applied to the 'txt' attribute of each Requirement Description to ensure that the text descriptions lie within acceptable limits of readability.

Table 6.4 shows the results of applying the Flesch–Kincaid set of metrics to the Requirement Descriptions in Table 6.2. Each row of the table corresponds to a requirement, with columns for the following:

- 'Absolute ref' – the unique number associated with the Requirement Description.
- 'Average sentence length' – the average number of words in each sentence.
- 'Average syllable per word' – the average number of syllables in each word.
- 'Reading ease' – the result of the Flesch Reading Ease test (higher numbers mean more readable).
- 'Grade level' – the result of the Flesch–Kincaid Grade level (lower numbers mean readable at a younger age).

- 'Complexity' – whether the requirement passes the complexity test defined in the Requirement Complexity Rule constraint on Figure 6.12.

The Requirement Complexity Rule is calculated from the Flesch Reading Ease and the Flesch–Kincaid Grade Level, which in turn are calculated from the average sentence length and average syllables per word.

When reading any metric, it is important to know what the scale represents. In the case of the Flesch Reading Ease a 100-point scale is used which is interpreted as the higher the score, the easier the text is to understand. The Flesch–Kincaid Grade Level refers to the U.S. school grades, e.g. a score of 8 means that an eighth grader (13–14 years old) should be able to understand the text. The Requirement Complexity Rule applies boundaries to both of these, looking for scores between 60 and 71 for the Flesch Reading Ease test and between 7 and 8.1 for the Flesch–Kincaid Grade Level test. The result of applying this metric is that a number of requirements fail because they are too complex. Requirement 'Absolute ref 3' for example fails due to the low reading ease and high-grade level scores. Conversely, the Requirement Description identified by 'Absolute ref 1' fails because it is not complex enough.

This metric shows that only two of the requirements descriptions are written at the correct complexity level. Therefore, the requirements should have been changed in order to pass all the metrics. However, doing this raises the question as to whether or not this would still capture the meaning intended by the stakeholder.

Table 6.5 shows two Requirement Descriptions identified by 'Absolute ref 2' and 'Absolute ref 3', which failed the complexity metric. It can be seen that the text has a very low reading ease and a very high-grade level, causing the requirements to fail the complexity check. Therefore, the aim is to reduce the length of each word and increase the number of sentences to improve the score.

Table 6.5 Requirements showing failed metric

Absolute ref	**Description**	**Average sentence length**	**Average syllable per word**	**Reading ease**	**Grade level**	**Complexity**
3	The escapology stunt shall demonstrate the usefulness of model-based systems engineering that can be used by marketing	17	2.00	20.3	14.6	False
2	It is essential that the excitement experienced by the audience is maximised while still ensuring that the requirement to minimise risk to escapologist is met	25	1.84	25.7	15.8	False

Table 6.6 Requirements after metric-based changes

Absolute ref	Description	Average sentence length	Average syllable per word	Reading ease	Grade level	Complexity
3	The escapology stunt shall show the usefulness of model-based systems engineering. This shall be carried out in a way which can be used by marketing	12	1.59	60.2	8	True
2	Make the audience as excited as possible before and during the stunt. The requirement to minimise risk to escapologist must still be met through the whole stunt	13.5	1.59	58.3	8.4	False

Table 6.6 shows the results of some simple textual changes to manipulate the complexity metric, and it is possible to bring the scores much closer to the pass values set by the Requirement Complexity Rule. Whether the meaning of the original stakeholder requirement has been maintained is subjective and can only be answered by a discussion between the escapologist, who set the original requirements, and the systems engineer who rewrote them. In this conversation, it was agreed to keep the new text for requirement 'Absolute ref 3'. However, for requirement 'Absolute ref 2', the escapologist didn't believe that the simplified text captured the meaning or the intent of the original. The result of this was not to revert to the original text but to provide a combination of the two statements:

> It is essential that the excitement experienced by the audience is maximised. The requirement to minimise risk to escapologist must still be met through the whole stunt.

This still does not pass the Requirement Complexity Rule, providing scores of 48.9 for the Flesch Reading Ease test and of 9.7 for the Flesch–Kincaid Grade Level test, but is an improvement on the original wording. However, there are issues associated with the level of commitment required to fully comply with the boundaries defined by a metric, not least of which is the amount of time it can take to rewrite a requirement in order to ensure compliance. It is easy to see how the use of metrics can help to identify issues and help with improvements when we look at the results shown in Tables 6.4–6.6. However, when applying these metrics in office-based or modelling tools, a number of considerations are to be taken into account including the following:

- *Can the relevant information be accessed?* In many modelling tools, the description can't be accessed directly and therefore it is difficult or impossible to automatically apply the metrics defined here. However, other tools such as standard office-based tools may provide the ability to apply such metrics to the relevant text.

- *Can the parametric constraint itself be automated within a model?* Many modelling tools will allow the presentation of parametric constraints but only a few provide the ability to apply values and execute the network. Also, those tools that allow the execution of the parametric constraints often rely on a third-party tool to achieve the mathematical calculation.
- *Are metrics such as the complexity metrics discussed here only relevant to the requirements descriptions or should they be applied to the text associated with all model elements and diagrams?* As discussed in Chapter 3, this book focuses on a model-based approach and, as such, each element and diagram should be supported by textual descriptions. Each diagram provides a window to the model, whereas the text provides the detail behind the elements. Therefore, there is a case for applying the metrics to all aspects of the model. However, it is also important to consider the impact on resources, as the time spent editing the text of a requirement to achieve a 'PASS' on a complexity metric could outweigh the benefit gained. Caution should be exercised to ensure that all the effort spent does not result in a metric-passing exercise rather than a requirement improvement exercise. It would be sensible for organisations to consider different boundary settings for specific types of text. The Requirement Complexity Rule is focused on requirements, but other text associated with diagrams or design elements may require a higher boundary given the technical nature of the requirements.
- *How should the parametric diagram be related to the descriptions being tested?* The SysML provides the concept of verification through the 'verify' dependency and it would be reasonable to say that the parametric diagram verifies the Requirement Description. However, the SysML only allows the 'verify' dependency to be used to show that sequence or activity diagrams verify requirements. Therefore, to provide a sensible relationship to show how each Requirement Description has been verified, the extension mechanisms discussed in Chapter 3 need to be used. The method of extending the language allows a graphical representation of verification which can be searchable in most tools provides the information needed to show conformance and rigour of approach to satisfy many standards and safety critical development approaches.

Parametrics used in this way can provide a strong approach to incorporating metrics into a model but must be used with care to ensure they do not monopolise the engineering effort.

6.6 The Context Definition View (identification of stakeholder contexts)

The Context Definition View identifies the points of view that are explored in the Requirement Context View. These points of view, or contexts, may take many forms including stakeholders and levels of hierarchy in a system.

Figure 6.14 shows the ACRE framework with the Context Definition View and Stakeholder Context Definition View highlighted. It is likely that the Context

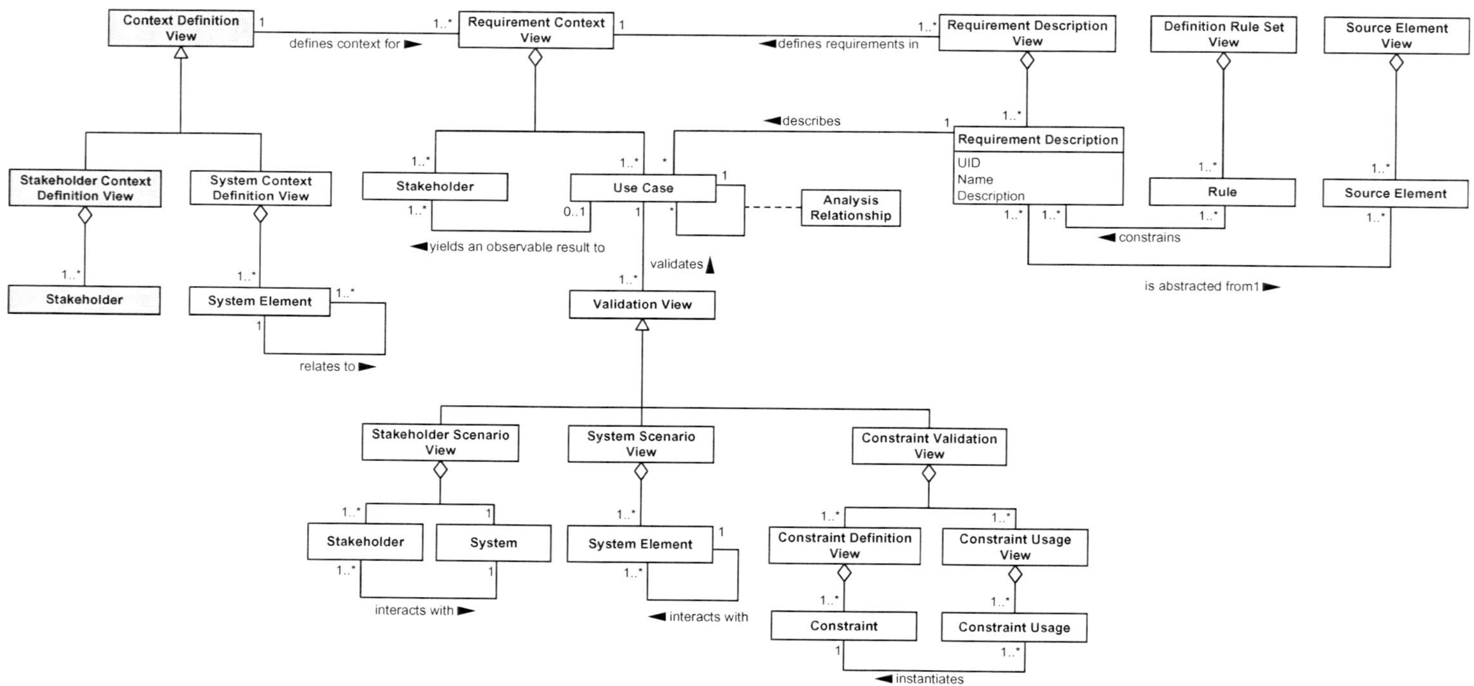

Figure 6.14 Framework – Context Definition View

Definition View, specifically the Stakeholder Context Definition View, is defined in parallel with the Requirement Descriptions in the Requirement Description View. As each requirement is captured and the related Stakeholder is defined, it may be sensible to add them to this view as it will provide a good start point for the Requirement Context Views. These Stakeholders form the focal point of the Stakeholder Context Definition View.

Figure 6.15 shows the initial Stakeholders that were identified in the form of a Stakeholder Context Definition View. No matter whether you are starting from scratch or expecting to develop Stakeholders from the Requirements Descriptions, it is always useful to have a standard starting point. Figure 6.15 shows a good starting point with some generic Stakeholder types – customer, supplier and external – as discussed in Chapter 5.

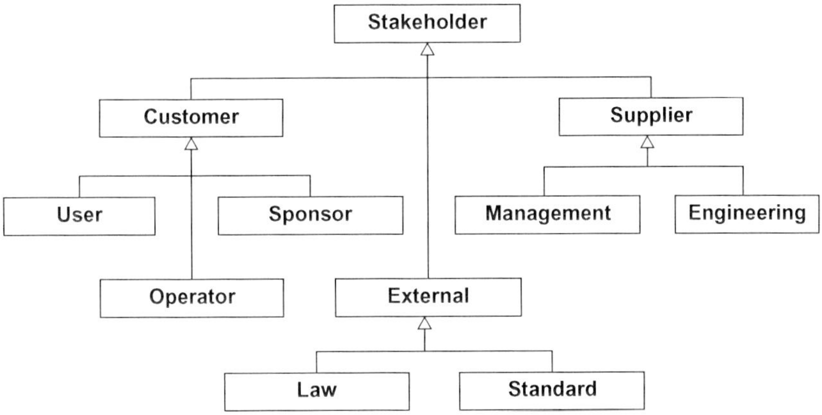

Figure 6.15 Initial Stakeholders shown in a Stakeholder Context Definition View

The 'Customer' Stakeholder is further categorised into the following:

- 'User' – those who in many cases will pay to use the system. In our example, this will be the audience – that who want to gain enjoyment from it.
- 'Operator' – those who make the stunt work. Likely to focus on the camera man and the location. It is important to remember that a Stakeholder does not have to be a person, it could be a person, place or thing which is affected by the requirements.
- 'Sponsor' – those paying to make the stunt possible.

The 'Supplier' Stakeholder is further categorised into the following:

- 'Management' – those making sure that the development of the stunt is managed and the whole exercise is delivered in a professional manner, and also responsible for ensuring that all of the modelling is complete before the stunt goes into production
- 'Engineering' – those understanding the processes and structure of the stunt and that will make the whole thing work, whether during verification, validation or delivery

The 'External' Stakeholder is further classified into the following:

- 'Law' – relates to any legislation or law which may apply, such as health and safety laws.
- 'Standards' – the documents proscribing requirements, processes or solutions which must be followed in designing and carrying out the stunt. In some ways, Houdini's book could be considered to provide the standard approach to escaping from a straight jacket.

This generic set of Stakeholder may then be populated based on these comments into a more fully formed Stakeholder Context Definition View, as shown in Figure 6.16.

Figure 6.16 shows the next iteration of the Stakeholder Context Definition View. The figure demonstrates how the final breakdown may look very different to the start point (shown in Figure 6.15). In this case, Figure 6.16 shows only the top-level classification from the generic view with the main Stakeholders being grouped under either 'Customer' or 'Supplier'. The lower level classifications have been removed to reduce the complexity of the diagram as some of the groupings would have remained empty. However, on large projects it is likely that the number and depth of the classifications will grow rather than reduce.

When classifying Stakeholders in this way, one question that often arises is concerned with where a Stakeholder should reside. Consider, e.g. the escapologist who, in this case, is considered to be a supplier but could be considered equally well to be an operator of the system. In this case, the choice has been made that the escapologist is classified as 'Supplier' as the main focus of the stunt is to entertain and inform the audience.

To understand what each Stakeholder wants from the system, or stunt in this case, we must provide more detail using the Requirement Context View.

6.7 The Requirement Context View

The Requirement Context View takes the Requirement Descriptions and gives them meaning by looking at them from specific points of view. This is known as *putting the requirements into context* and forms the basis of the approach presented in this book.

Figure 6.17 shows the ACRE framework with the Requirement Context View highlighted. It is important to remember that the main purpose of this view is to produce a single *context* based on the Context Definition Views. Each element in a Context Definition View (the Stakeholder on the Stakeholder Context Definition View or the System Element in the System Context Definition View) provides the basis for a single Requirement Context View, and this section looks at a number of these views for the escapology stunt.

Figure 6.18 shows a Requirement Context View from the point of view of the sponsor, a Stakeholder that was identified on the Stakeholder Context Definition View. It is important to consider the business when capturing requirements, whether a business unit, the whole organisation or both. Understanding the business gives rise to many of the highest level requirements, often known as capabilities, vision, mission and goals. Although, in many cases, these

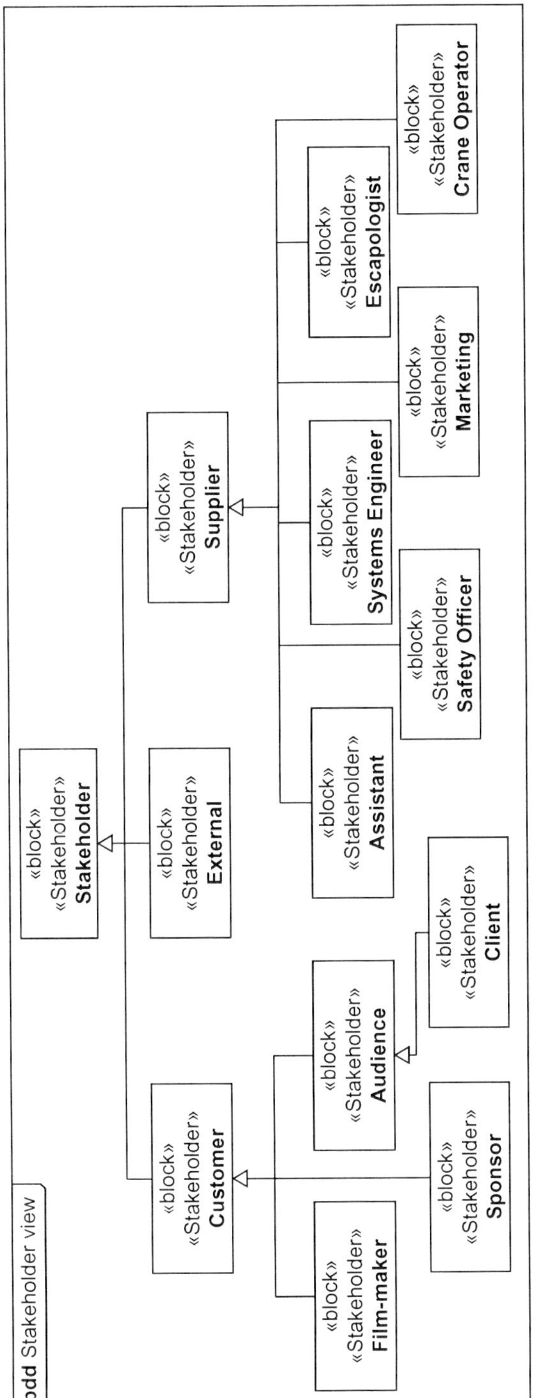

Figure 6.16 Stunt Stakeholders

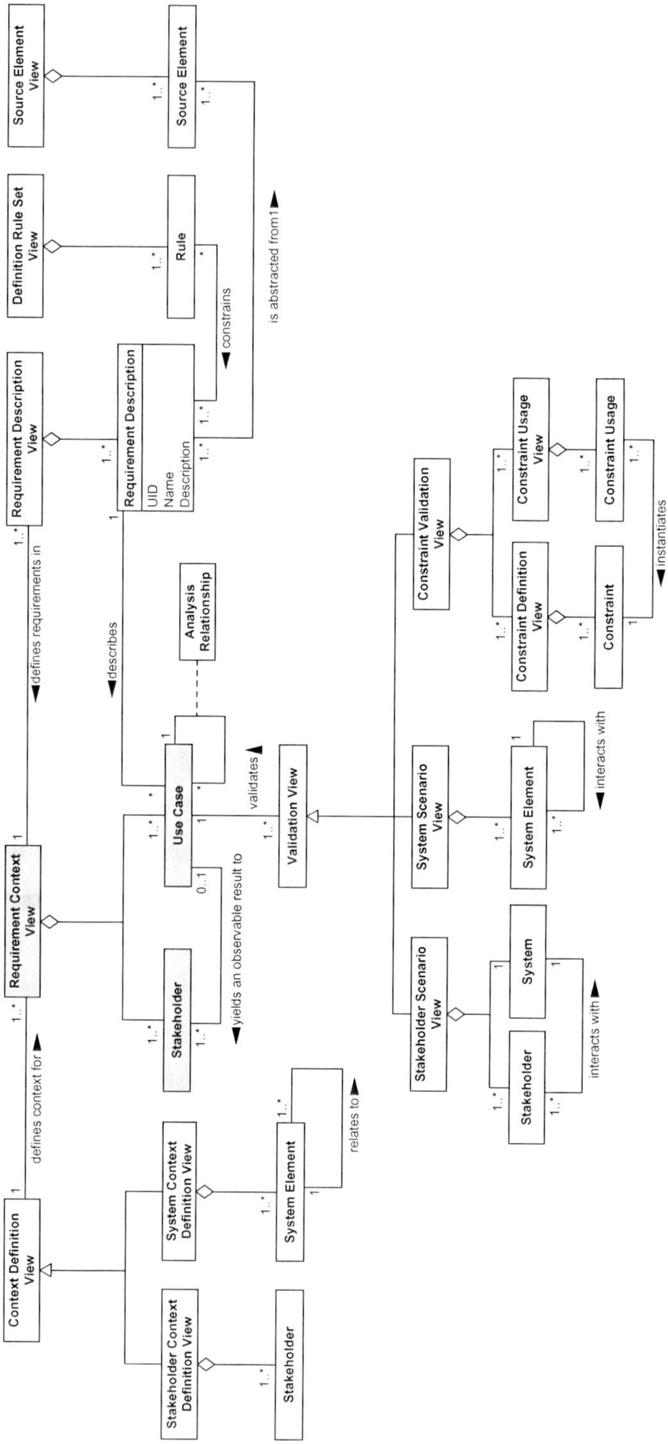

Figure 6.17 Framework – Requirement Context View

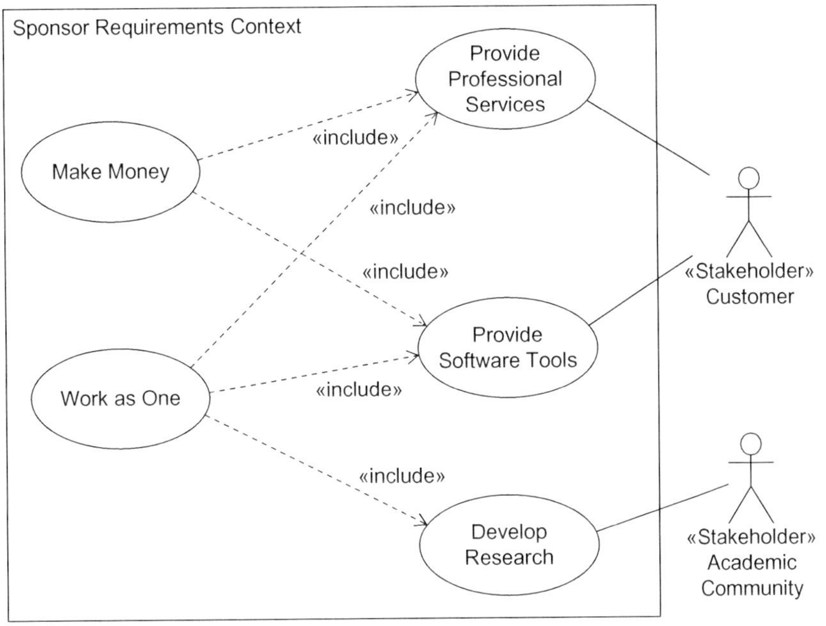

Figure 6.18 Sponsor's requirements context

are rather ill-defined, it can be essential to many groups to show how they support the overall business needs. Figure 6.18 shows that, as with many organisations commercial or otherwise, the organisation has a need to 'Make Money' through the provision of services ('Provide Professional Services') and software ('Provide Software Tools'). The business also has an underlying mission to 'Work as one', which it achieves across all of its business areas, including research ('Develop Research').

The context provided here is crucial as it represents the overall goals of the business as a whole and, therefore, has the potential to influence many other contexts. The next context considered is that of the escapologist.

Figure 6.19 shows the requirement context from the point of view of the escapologist, a Stakeholder that was identified on the Stakeholder Context Definition View. The main requirement for the escapologist is to 'Escape'. This requirement includes two other requirements, which are the following:

- 'Set up' that describes the need to set up the stunt.
- 'Get out' that describes the need to perform the actual stunt and get out of the straight jacket.

It is interesting to see that there are four high-level constraints that influence the main 'Escape' requirement, which are the following:

- 'Look Great' that describes the need to be visually impressive.
- 'Make People Sweat' that describes the need to make people excited about the whole stunt.

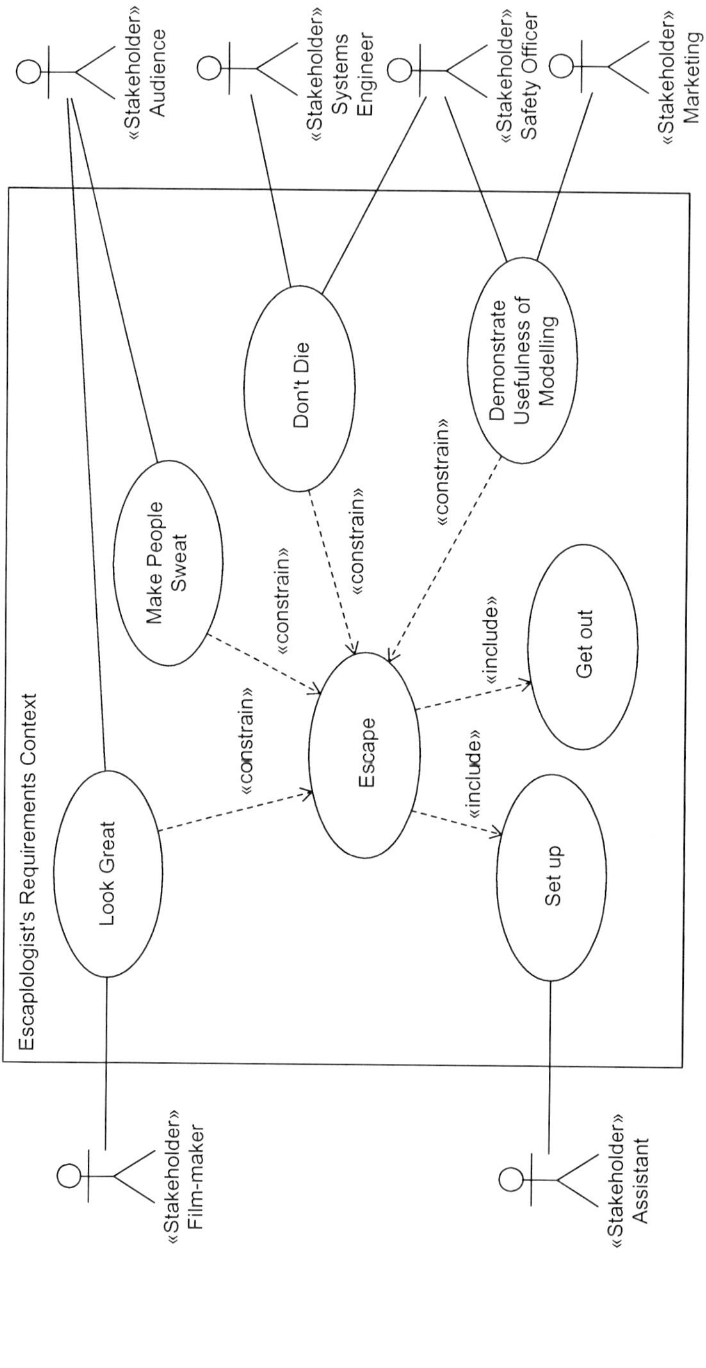

Figure 6.19 Escapologist's requirements context

- 'Don't Die' that is the highest priority requirement shown in Figure 6.19. The importance of this requirement cannot be emphasised enough and it overrides every other requirement, even that of 'Escape'.
- 'Demonstrate Usefulness of Modelling' that describes the engineering need behind the stunt.

Note that we are using the term 'requirement' here but, as they all appear on the escapologist's context, then the requirements have been given meaning and are, therefore, represented as use cases. The next context is taken from the point of view of the Safety Officer.

Figure 6.20 shows the Requirement Context View from the point of view of the Safety Officer, a Stakeholder that was identified on the Stakeholder Context Definition View. The Safety Officer's context seems to have little in common with the sponsor or escapologist context. It does, however, focus on safety – more specifically, it is concerned with ensuring that the systems have been fully analysed from a risk management and safety point of view. The safety focus of this context, and more importantly the lack of requirements or design focus, can help highlight differences between domains such as business, engineering and safety and show why some disciplines are seen as overheads. It also highlights the need to consider the relationships between the different contexts, asking such questions as 'Is it right that the Safety Officer's context doesn't highlight design or requirements? How does this relate to the stunt being performed? Why should we spend time on this?'.

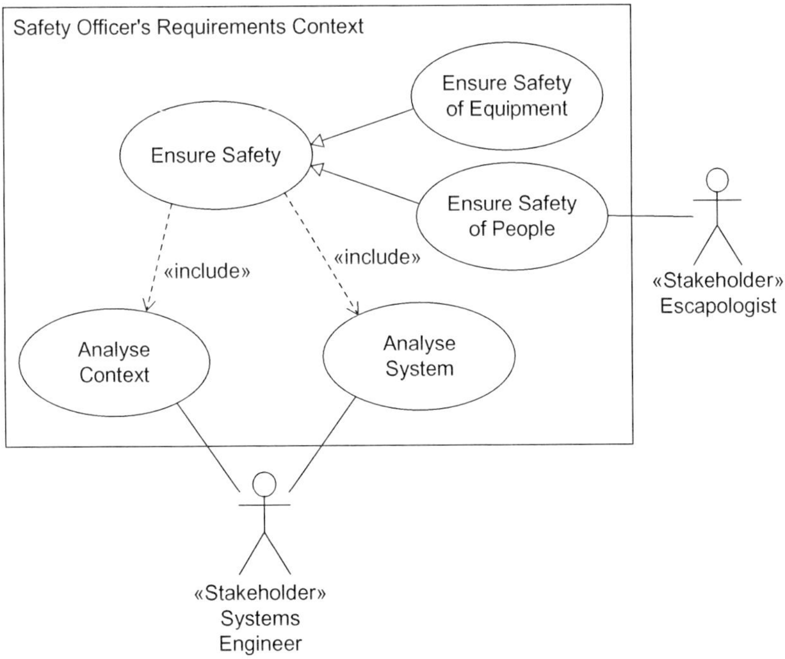

Figure 6.20 Safety Officer's Requirements Context

Creating these various contexts is very important, but it is essential that the entire set of contexts is consistent. Figure 6.21 is the starting point for considering this.

Figure 6.21 shows how the contexts that have been discussed so far relate to one another. Although this is not strictly a valid SysML use case diagram, it provides a pragmatic visualisation of the relationships between the Requirements Context Views. To further understand the relationships between the Stakeholder Requirement Context Views, the figure shows four of the Stakeholder contexts, specifically: Marketing, Escapologist, Safety Officer and Systems Engineer. It also shows how the contexts are related via the Stakeholders. There are two clear relationships between the Escapologist context and the Safety Officer's context, namely those relationships via the Safety Officer and the Systems Engineer. The link through the Safety Officer is to be expected but that through the Systems Engineer may be less obvious. However, this provides the link to the engineering and development side of the stunt. It is the Systems Engineer role that provides the research and development aspects of the work which are analysed by the Safety Officer.

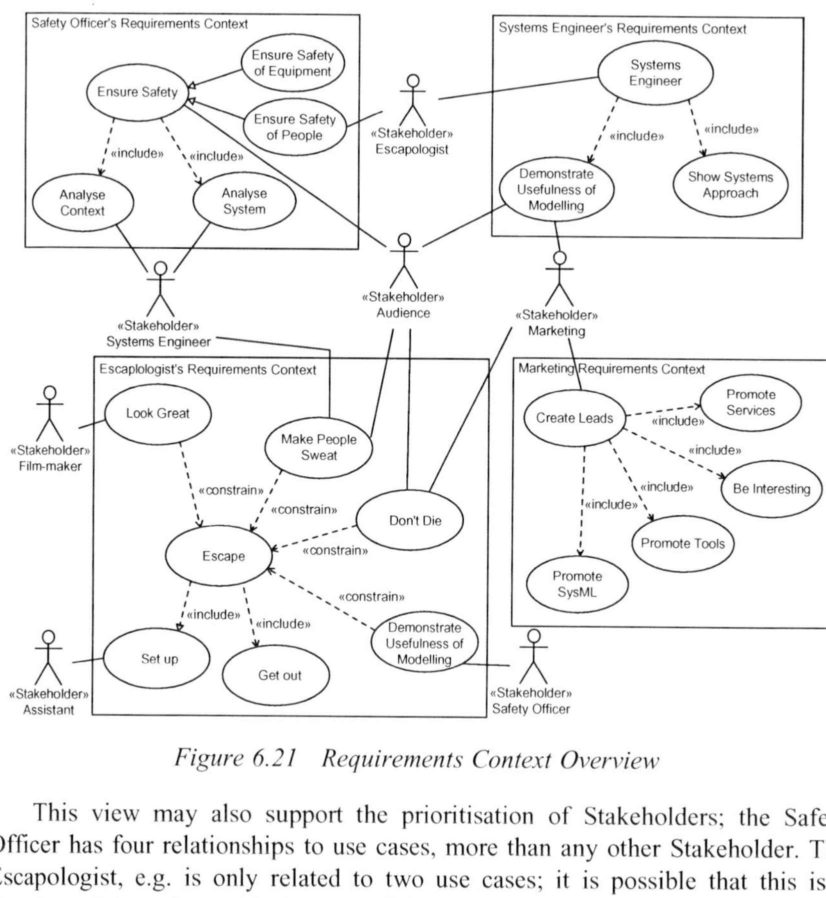

Figure 6.21 Requirements Context Overview

This view may also support the prioritisation of Stakeholders; the Safety Officer has four relationships to use cases, more than any other Stakeholder. The Escapologist, e.g. is only related to two use cases; it is possible that this is a function of the safety-critical nature of the system.

6.7.1 The Traceability View (Stakeholder Requirement Context View to Requirement Description View)

The Traceability View is used to allow traceability between different elements of the model to be explicitly shown. These traceability links may exist between views or between elements within views.

Figure 6.22 shows the ACRE framework with the Traceability View highlighted; it also highlights the Requirement Context View and the Requirement Description View which are traced to each other. As mentioned earlier, providing the Requirement Context Views based on the Stakeholder Context Definition View is crucial, but it is important that they are not developed in isolation from each other. The main aim of defining the contexts is to give meaning to the Requirement Descriptions from the Requirement Description Views by putting them into context. Therefore, each Use Case within the context must relate back to one or more Requirement Description.

Each Use Case defines the essence of a need, meaning that it can abstract multiple requirements if they carry the same inherent meaning. Although it can be difficult to establish that requirements really provide the same meaning, it is still reasonable to capture them in this way provided there is rigour behind the traceability. Having traceability in place enables checks to be carried out to ensure that any Use Case which traces to multiple Requirement Descriptions fully captures the essence of all the requirements. If this is not the case, the use cases must be reviewed and expanded, possibly meaning a change to the Requirement Context View.

When considering the traceability between use cases and Requirement Descriptions, the 'refines' relationship, discussed in Chapter 3, is used. This dependency provides a special type of traceability which allows the capture of a critical relationship between Use Case and Requirement Description.

There are several ways in which these Traceability Views may be visualised, which are described as follows.

Figure 6.23 shows a Traceability View that captures the traceability between use cases and Requirement Descriptions. The SysML requirements diagram in Figure 6.23 shows how the refinement can be achieved graphically from within a model. It can be useful to define the «refine» relationship in a diagram, such as this, as it makes it easier to see that the correct elements, in this case a Use Case and a Requirement Description, are at each end of the relationship. It also has the advantage of providing both ends of the relationship in one view, allowing multiple relationships to be made quickly and efficently as most tools will allow repeat use of the same relationship.

Once captured within the model, most tools enable the information to be output in various forms, including tabular form, as shown in Table 6.7. This shows the Use Case in the left-hand column, the type of relationship (in this case 'refines') in the grey middle column and the Requirement Description, which it refines, in the fourth column. This is often the preferred method of delivery for such a traceability or refinement system, showing different kinds of traceability links in different tables (or even all in one table) in order to avoid the overly complex digrams that would result if all the system traceability links were shown in one graphical view.

Table 6.8 also shows the refinement of Requirement Descriptions (shown on the left) by the use cases (shown across the top). It shows the same set of information as

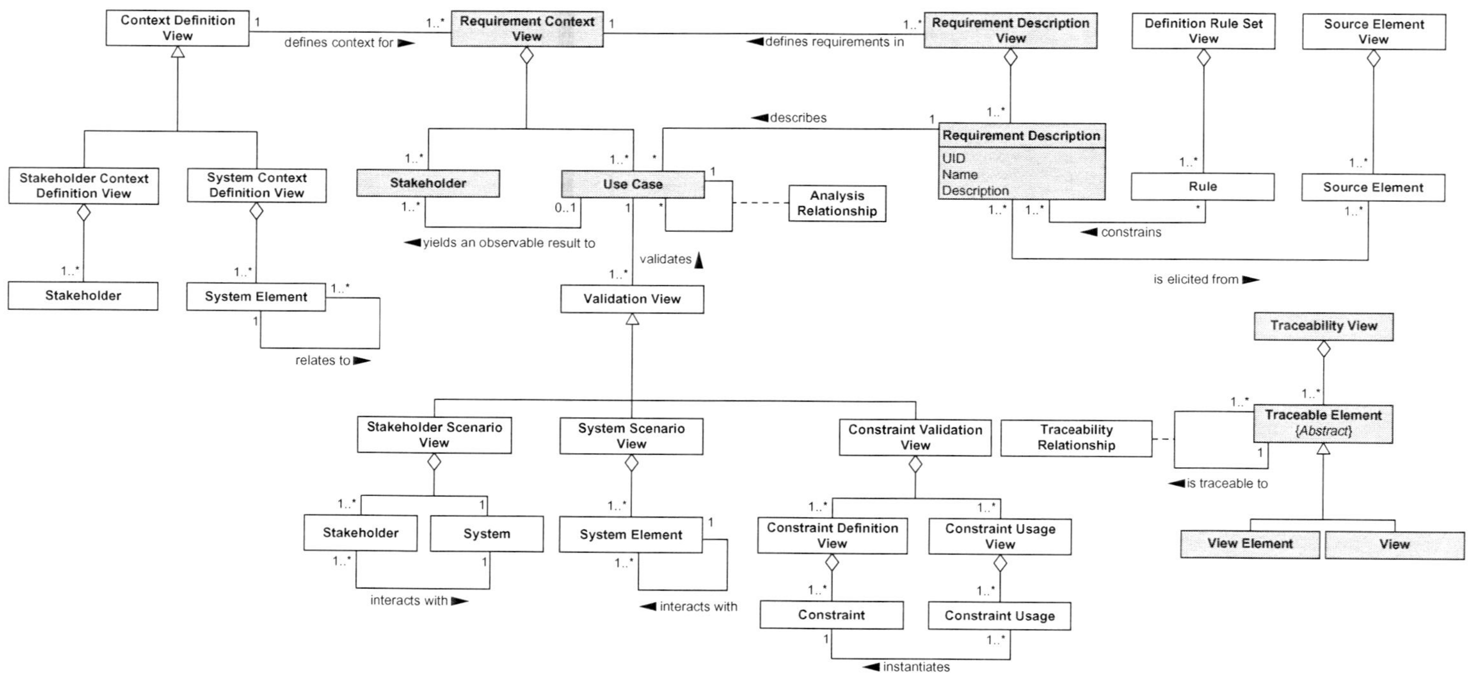

Figure 6.22 Framework – traceability

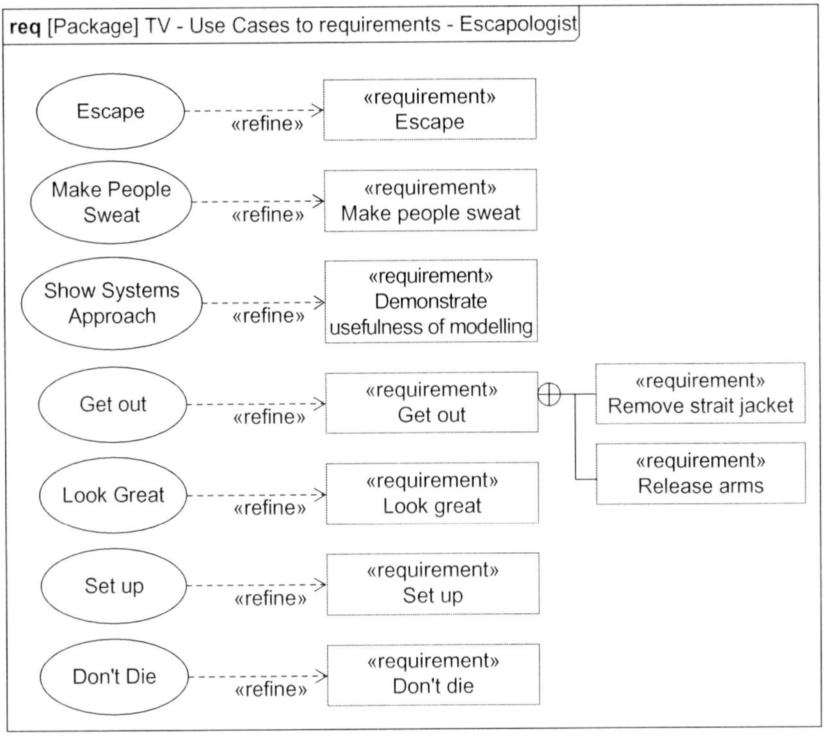

Figure 6.23 Traceability View, showing how Use Cases refine Requirement Descriptions

Table 6.7 Tracing Use Cases refine Requirements Description

Source name	Type	Relationship	Target name	Type
Don't Die	Use Case	refines	Don't die	requirement
Escape	Use Case	refines	Escape	requirement
Get out	Use Case	refines	Get out	requirement
Look Great	Use Case	refines	Look great	requirement
Make People Sweat	Use Case	refines	Make people sweat	requirement
Set up	Use Case	refines	Set up	requirement
Show Systems Approach	Use Case	refines	Demonstrate usefulness of modelling	requirement

has been shown in Figure 6.23 and Table 6.7, but this time in a matrix format. This matrix shows that for the stunt there is a simple one-to-one relationship between the Use Cases defined in the contexts and the Requirement Descriptions.

Figure 6.23, Table 6.8 and matrix have each shown the same set of information, but using different visualisations. Each technique has its own set of advantages:

- The diagram view enables the efficient viewing and definition of each refine relationship. The diagram ensures that the viewer can quickly see the element

Table 6.8 Tracing use cases refine Requirement Description

Refines \ Refined by	«Use Case» Don't Die	«Use Case» Escape	«Use Case» Get out	«Use Case» Look Great	«Use Case» Make People Sweat	«Use Case» Set up	«Use Case» Show Systems Approach
«requirement» Don't die	refine						
«requirement» Escape		refine					
«requirement» Get out			refine				
«requirement» Look great				refine			
«requirement» Make people sweat					refine		
«requirement» Set up						refine	
«requirement» Demonstrate usefulness of modelling							refine

at each end of the relationship; also in this case, most tools will allow a user to navigate based on the element to see where else the element is used, e.g. which Requirement Context View a Use Case is used in.

- The table view provides more detailed information about each Use Case and Requirement Description, if desired. This enables other information such as 'Absolute ref', 'Priority' and 'Ownership' to be displayed.
- The matrix view makes it clear that there is a one-to-one relationship and enables questions to be asked about whether this should be so. Only revisiting both the Requirements Context and the traced 'refined' Requirement Descriptions will provide the answer. However, if there is an assertion that the relationship should not be a clear one-to-one, this will require a full review and possible improvements to each of the Requirement Context Views and the relationships between them. This need to review and update should not be seen as an inhibiting overhead but a necessary useful part of the modelling process. It is the use of traceability and consistency checks in this way that promote the iterative nature of modelling, enabling confidence in a model to grow.

6.8 The Validation View (definition of stakeholder scenarios)

The Validation View provides the basis for demonstrating that the requirements can be met or complied with. These views can be informal (such as scenarios at various levels of abstraction) or may be formal (such as mathematical-based representation).

Figure 6.24 shows the ACRE framework with the Validation Views highlighted. Validation can be considered at a number of different levels with stakeholder and system being two examples.

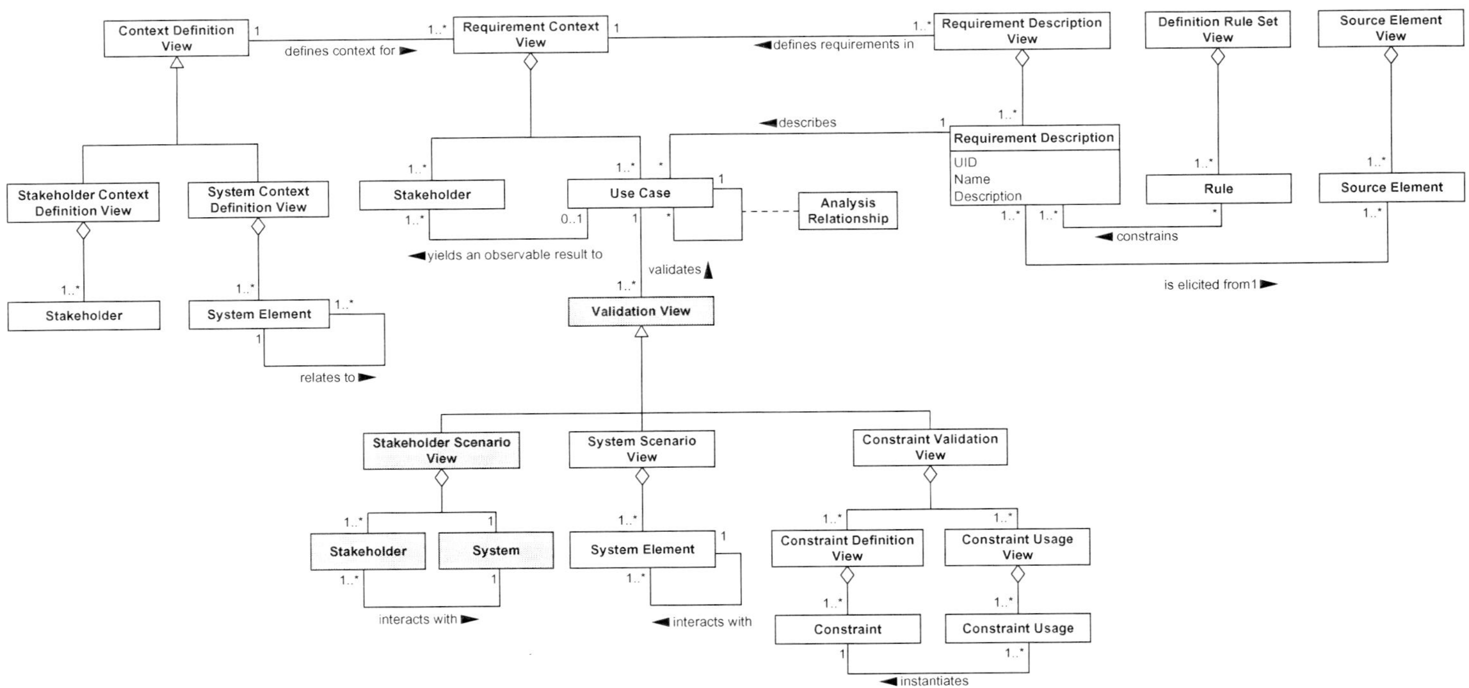

Figure 6.24 Framework Stakeholder Scenario View

Consider, initially, the Stakeholder Scenario View, which allows the interactions between stakeholders and the system as a whole, is to be explored. Each specific scenario is used to validate use cases in context.

The Validation View, shown in Figure 6.25, considers the overall stunt from the start, through lifting the escapologist, the performance of the stunt, to lowering with the escapologist having escaped successfully.

Although focused on validation, this view also provides the ability to analyse detail relating use cases and the context they are within. As the scenarios are defined, they enable, along with the consistency in the model, the number of questions to be asked, such as:

- Are these the right stakeholder roles?
- Are the interactions in the right direction?

They may seem to be trivial questions, but if we consider the case where it is not a Safety Officer but an Assistant who starts the stunt then the Validation View and, therefore, the context could be very different. It is the consideration of the impact of a change on the related areas of the model, through the consistency checks provided in the SysML and ACRE, which makes this seemingly trivial question provide a substantial answer.

Each Validation View is likely to be used for clarification and analysis before it is used for validation; however, there comes a point where a decision must be made as to whether a Validation View will be used for validation or not.

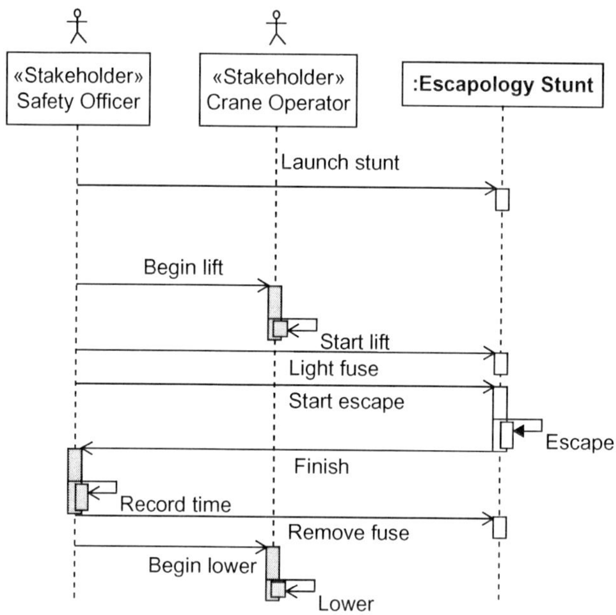

Figure 6.25 Validation View, successful escape

This exclusion from use in validation may be due to any number of engineering or management reasons, not least of which is level of rigour, availability of test facilities and coverage by other validation scenarios.

If chosen to be used for validation, this Validation View is marked with the «testCase» stereotype discussed in Chapter 3. «testCase» identifies those scenarios that are used in the validation, making it easier to find relevant Validation Views when required. To support the «testCase» definition, a «validate» stereotype, representing the relationship between the Validation View and the Use Case, has been defined and is used to relate the Validation View to the use cases. The Validation View, in Figure 6.25, validates three use cases: the main 'Escape' Use Case as well as the 'Get out' and more importantly 'Don't die'.

This Validation View has shown a successful escape, but in many cases success Validation Views are overlooked with focus given to failure cases, often due to the multiplicity of failure cases. This is not to say that failure cases should not be considered, but that they should not be considered to the detriment of the system function.

In the case of a failure such as the escapologist not getting out of the straight jacket in time so that the rope ignites, it is likely that mitigation would be defined. A placeholder for such mitigation is shown in Figure 6.26, in the reference frame called 'Rescue'. This relates to another scenario called 'Rescue', which is defined on another diagram. This reference mechanism provided by SysML allows reuse in a number of scenarios.

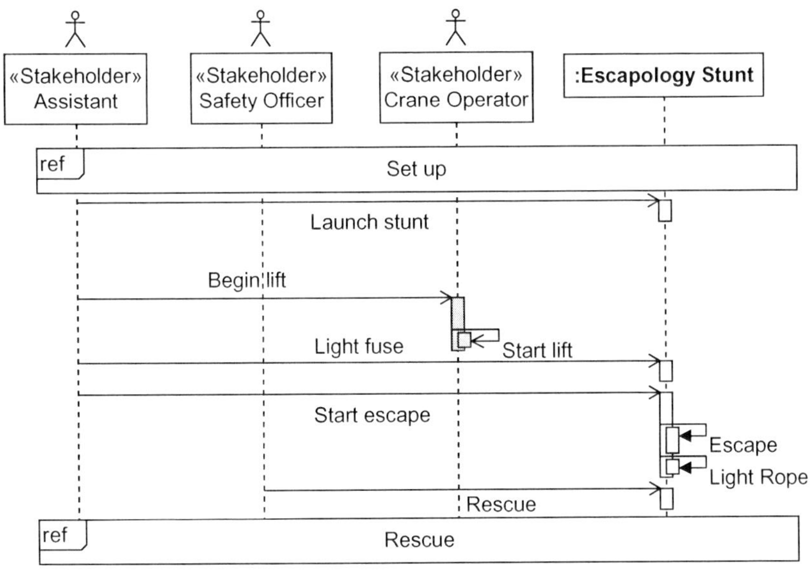

Figure 6.26 Failure to escape scenario

Other scenarios could be defined to show a rescue being carried out at earlier points in the stunt. However, these are considered based on the process-based Validation Views that are defined later in this chapter.

Having identified the need for a 'Rescue' scenario, the parallel activities within the 'Rescue' scenario itself can be seen in Figure 6.27, where it is important that the escapologist is lowered to the ground with support so that he doesn't fall. It is also important that the rope is extinguished to reduce any damage which may be caused.

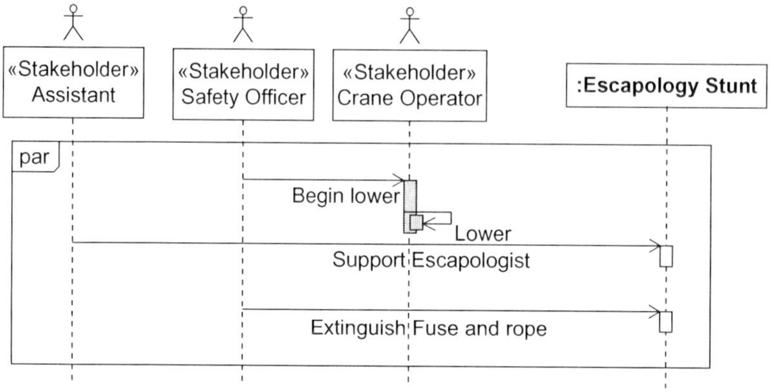

Figure 6.27 Rescue scenario

Scenarios are useful in this way as they enable an interrogation of the system before it is defined. As mentioned earlier in this section, the Assistant can also be seen interacting with the stunt during the rescue. So the consistency checks can be followed to consider the use cases validated by this scenario and the Stakeholders which interact with those use cases.

6.8.1 The Traceability View (Stakeholder Scenario View to Requirement Context View)

The Traceability View is used to allow traceability between different elements of the model to be explicitly shown. These traceability links may exist between views or between elements within views.

Figure 6.28 shows the Traceability View and also highlights the Requirement Context View and the Stakeholder Scenario View which will be traced to each other. Whenever a relationship within the ACRE framework is followed, it is also necessary to ensure that the traceability related to that relationship is in place. In previous cases of traceability, the inbuilt features of the modelling language have been used. In this case, however, the relationship to be recorded does not fit with the pre-defined SysML relationships. This means that a new modelling relationship must be defined to provide the relationship for the ACRE framework.

Table 6.9 shows which use cases are validated by each Validation View. As before, these relationships are captured within the model and automatically output

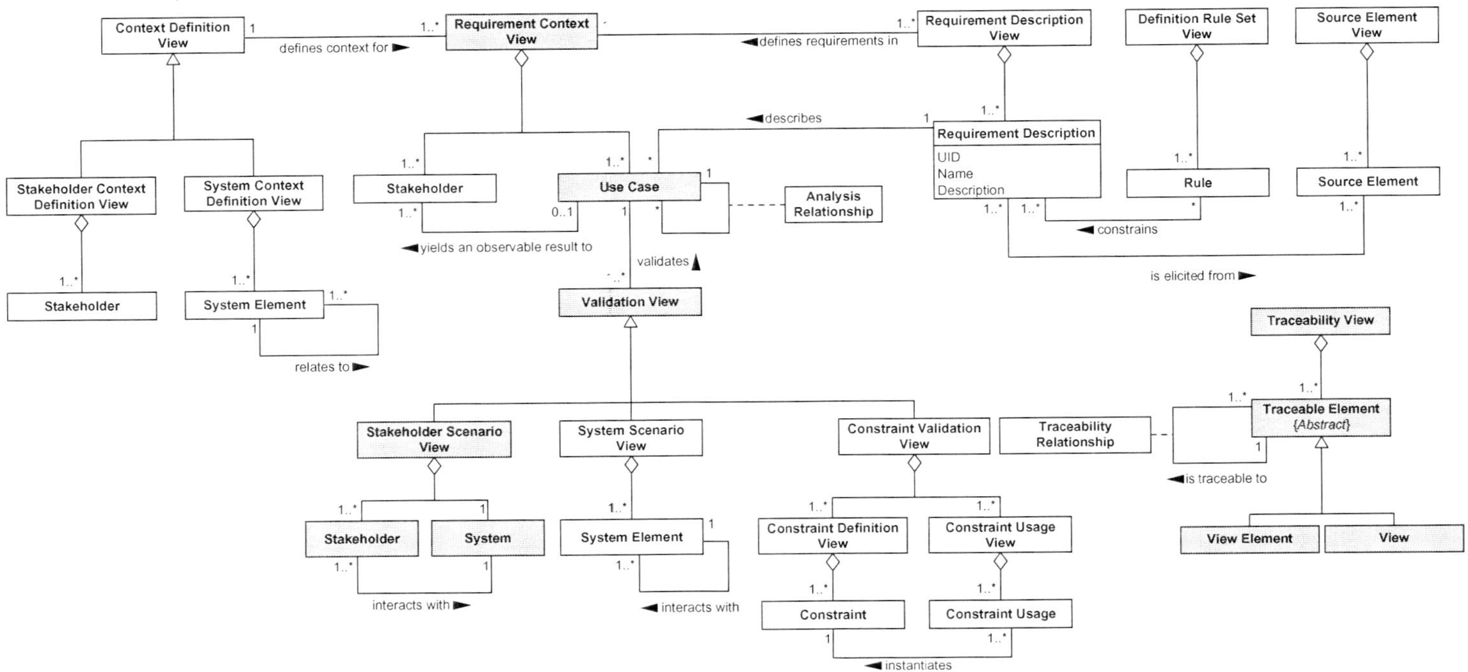

Figure 6.28 Framework – Traceability View

Table 6.9 Sequences validate use cases

Validates \ Validated by	«Stakeholder Scenario View» Failure to escape	«Stakeholder Scenario View» Rescue	«Stakeholder Scenario View» Successful escape scenario
«Use Case» Ensure safety	validate		
«Use Case» Ensure people are safe		validate	
«Use Case» Don't die			validate
«Use Case» Escape			validate
«Use Case» Get out			validate

into relevant formats. However, the addition of the «validate» stereotype brings issues of its own which may be due to the use of the SysML extension mechanism or the tools which support SysML.

Using the SysML extension mechanism, stereotypes can be added using the approach discussed in Chapter 3. In this case, the «validate» stereotype has been added to show that a Validation View specifically provides a validation of the use case it relates to. This enables compliance with the ACRE framework and ensures that the consistency it provides can be established. The main reason that the stereotype mechanism has been used is because the SysML doesn't provide the ability to validate a use case. It is not correct to say that this is verification using «verify», the next closest of the SysML relationships, as it will be used to show that the system is the right system. Although they are often referenced together, as V&V, there are major differences between verification and validation. Verification ensures that the system has been built in the right way and fulfils the design against which it is implemented. Validation, however, focuses on ensuring that the right system has been built and that it provides the functionality needed by the customers. This difference between verification and validation is often phased as 'Verification is building the product right, and Validation is building the right product' [8].

Most SysML modelling tools support the definition and use of stereotypes. However, the approach and rigour related to the application of the stereotype can vary greatly. This may mean that automated generation of tables may not be possible from non-standard stereotype relationships, or that tailoring work may need to be carried out to enable such exports.

These issues do not mean that stereotypes or tools should be avoided; on the contrary they must be used but their limitations must also be understood so as to ensure that the processes drive the tools and the tools *do not* drive the processes.

However, in many cases a pragmatic solution should be adopted to provide the best way to enable the process to drive the tool.

6.9 The Context Definition View (identification of system context)

The Context Definition View identifies the points of view that are explored in the Requirement Context View. These points of view, or contexts, may take many forms including Stakeholders and levels of hierarchy in a system.

This is the second time the Context Definition View has been used – previously it focused on defining the Stakeholders related to the system, but this time it focuses on the definition of system contexts, as shown highlighted in Figure 6.29. System contexts, in many cases, are based on historical knowledge, existing literature or initial system design ideas focusing on the major components or functions of the system. However, where none of these options exist, initial breakdowns may be defined based on the major themes running through the Requirement Context Views already defined.

In this case, the information in the books and videos has flowed through the first iteration of system design and can be used to inform the system context definition. The high-level system design tells us that there are a number of elements in the system: the rope, the height, the escapologist and fixings for the rope.

It is possible to take these initial elements within the system and record them as a list of contexts for consideration. This lists of contexts would likely start with the major components identified earlier. The simple list of contexts consists of the following:

- Fixings, which support the escapologist.
- The escapologist.
- The rope, which connects the fixings and escapologist together.

Figure 6.30 shows these high-level system elements.

When expanded into a model, the previous list provides one side of the System Context Definition View. In Figure 6.30, consideration has been given to the processes carried out during the stunt, as much of the system is focused on the activities of the escape. Therefore, the final diagram shows the two aspects of the system, physical and process, each of which is further developed.

Previously with the Stakeholders, and here with the processes, a specialisation relationship has been used showing inheritance and classification of Stakeholder roles and processes. However, the physical aspect can be seen as a SysML aggregation showing that the physical system is 'made up of' these elements. It is reasonable to use both of these relationships to define the contexts, and both of them have been used as the physical aspects of the system focus on the hierarchical breakdown of the system rather than the taxonomy of ideas.

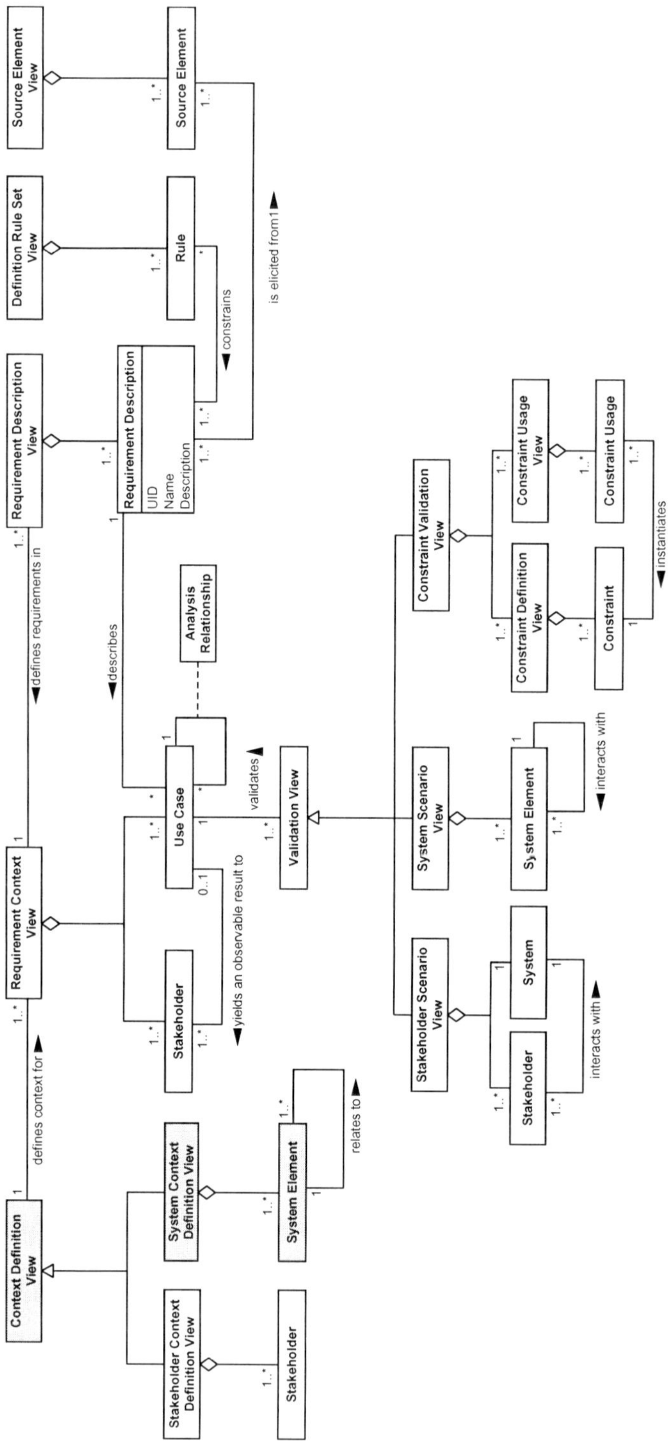

Figure 6.29 Framework – Context Definition View

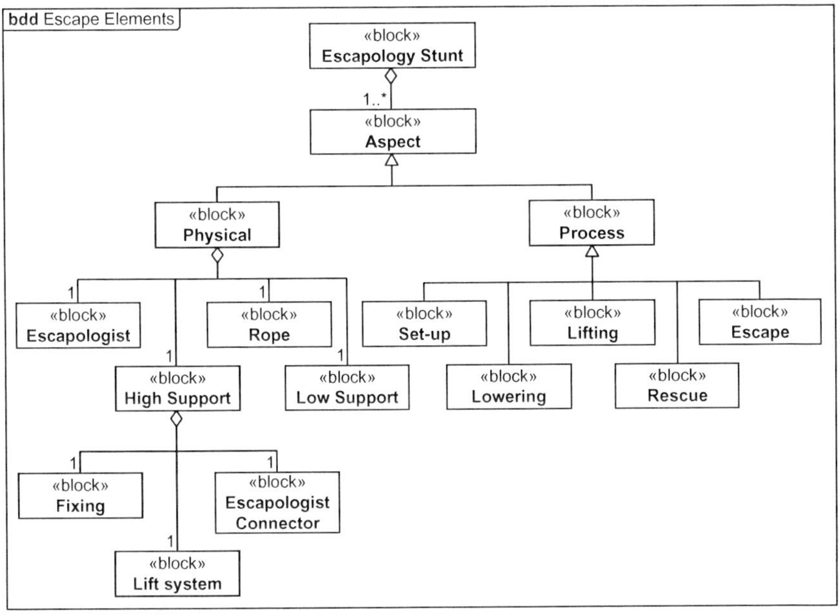

Figure 6.30 Stunt System Context Definition View

6.10 The Requirement Context View (system requirements)

The Requirement Context View takes the Requirement Descriptions and gives them meaning by looking at them from specific points of view. This is known as putting the requirements into context and it forms the basis of the approach presented in this book.

Having identified the various system contexts using the system elements in the System Context Definition View, the Requirement Context Views, highlighted in Figure 6.31, need to be defined. This means that the difference between system-level and stakeholder-level requirements need to be considered. There must be a clear separation between stakeholder and system requirements because many project issues associated with requirements engineering manifest themselves due to the lack of clarity between these requirements. The stakeholder requirements define what the stakeholder wants from the system, whereas the system requirements define what the system will provide. Confusion can occur due to the method of delivery often being the same, i.e. Requirement Context Views are developed for both stakeholder-based and system-based requirements, with requirements often directly copied from the stakeholder to the system levels.

Figure 6.32 shows the main Requirement Context View for the stunt system. The main requirement of the escapology stunt is to 'enable stunt', making sure the stunt as a whole works. This has two major constraints:

- 'Ensure safety', meaning that all aspects of the requirements, design and operation of the stunt must be safe. This will also include the definition of emergency procedures.

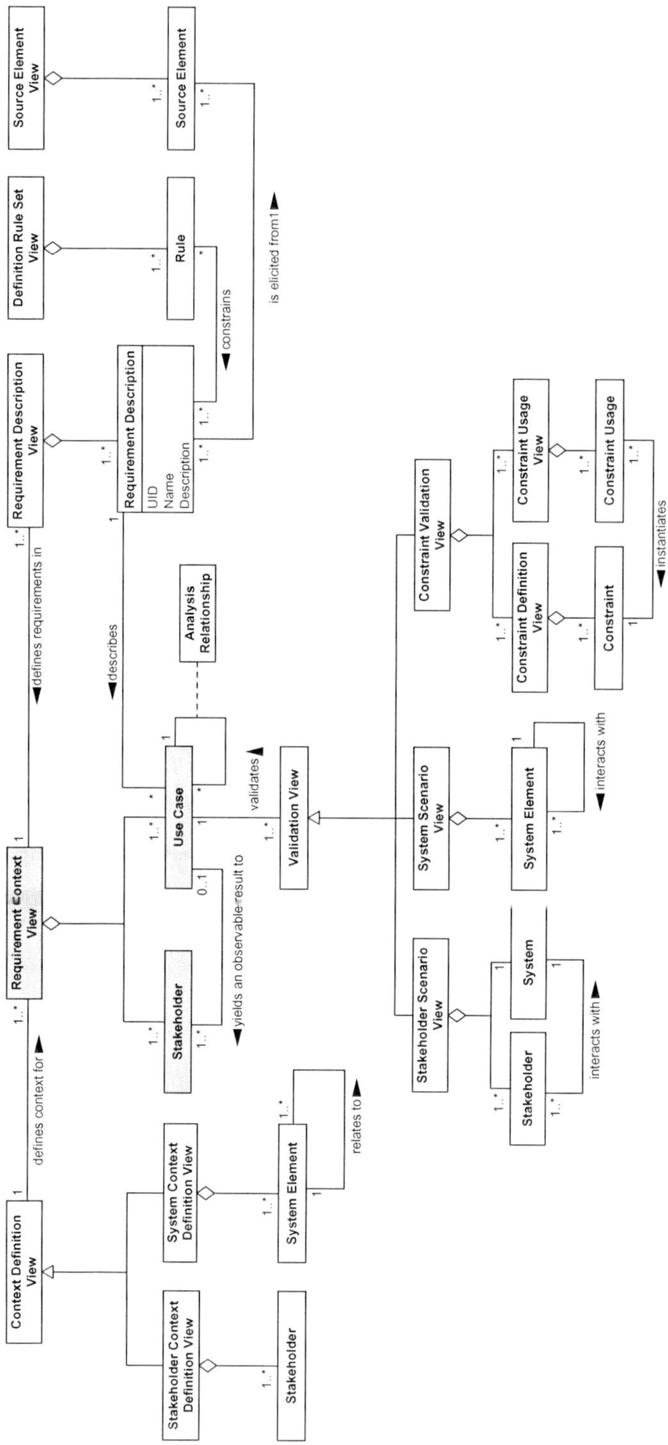

Figure 6.31 Framework – Requirement Context View

- 'Ensure competence', focusing on ensuring that all the Stakeholders involved in the definition and implementation of the stunt understand and are competent to carry out their roles.

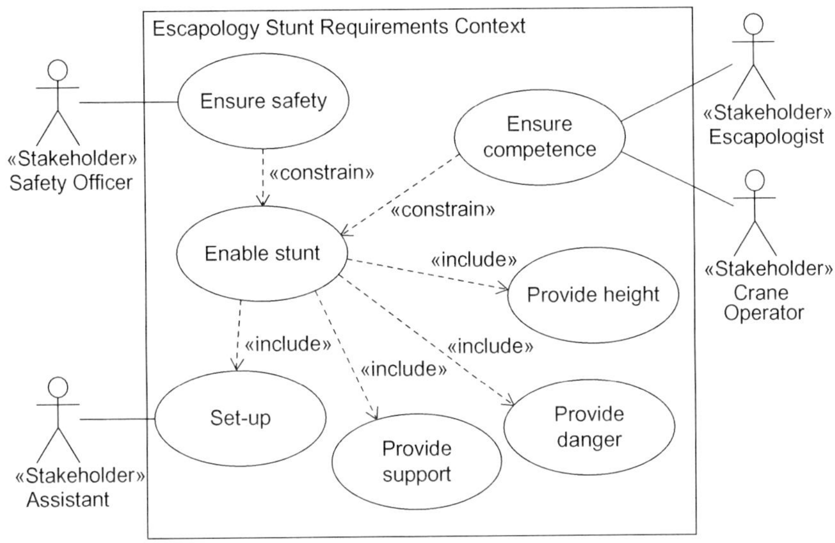

Figure 6.32 System Requirements Context View

The implementation of the stunt includes the need to 'set up', 'provide support', 'provide danger' and 'provide height', all of which relate to the process and physical contexts relating to the escapology stunt as shown on the Context Definition View in Figure 6.30.

This system Requirement Context View, including the Stakeholders and use cases, has been abstracted from the information in the Stakeholder-level Requirement Context Views. These system-level use cases can be derived by understanding the needs presented in each Stakeholder Requirement Context View, in the relationships between the Requirement Context Views and also in the priorities of requirements.

In some cases, it may be easy to see the relationships between system-level and Stakeholder-level use cases. 'Ensure safety', e.g. has the same basic name and meaning as the 'Ensure safety' Use Case on the Safety Officer's Requirement Context View. Others may be less obvious, such as 'Ensure competence' that relates to 'Ensure safety of people'. This is due to the need to show that the people supporting the stunt are not likely to cause accidents and doing everything possible to ensure that the physical equipment is safe. Finally, the relationship between 'Ensure safety' and 'Enable stunt' is abstracted from the relationship provided by the 'systems engineer' between the Safety Officer's requirements context and the escapologist's requirements context seen in Figure 6.21.

It may, at first glance, look as though many of the system-level use cases identified from Stakeholder requirements would provide constraints on the system.

This is not always the case; some Stakeholder-level use cases provide the need for included use cases in the system Requirement Context Views. An example of this is the 'Make People Sweat' Use Case in the escapologist context (see Figure 6.19), which is manifested as the included 'provide danger' Use Case at the system level.

It is useful to realise Stakeholder constraints as included use cases of the system, as it can show how the system will clearly fulfil the Stakeholder Use Case. It is likely that the constraints which are passed from Stakeholder Requirement Context Views to system Requirement Context Views are areas which need further investigation and definition.

As system-level use cases are defined, they are mapped to the Stakeholders and use cases at the Stakeholder level to ensure that they support those needs. This mapping is captured as part of the traceability, discussed later in this section.

Lower level Requirement Context Views also need to be considered. In Figure 6.33, the context of the rope is considered in more detail. Again, here we see the need to 'Ensure safety' and 'Provide danger'. However, at this level the provision of danger has changed back into a constraint. It is not unusual to see use cases provide different relationships at multiple levels in this way. In this case, the 'rope' needs to be constrained in such a way that it provides the danger. A little danger is exciting, whereas too much danger may compromise the safety of the escapologist.

The use cases included in Figure 6.33 are 'Hold escapologist' and 'Limit extension'. To investigate these use cases further, there was a need to consult specific standards relating to ropes, which meant identifying the Source Element BS EN 1891:1998 [9], and adding it to the Source Element View, extracting the requirements as Requirement Descriptions on a Requirement Description View and ensuring that traceability is maintained to the Requirement Descriptions. To further strengthen the relationship between the standards and the Requirement Context View, the standard should also be added to the Requirement Context View as a Stakeholder, which in this case would be related to the 'Provide support' Use Case.

The Source Element relating to the rope standard could have either been added to the initial Source Element View or, as in this case in Figure 6.34, kept on its own Source Element View specifically for Source Elements relating to the rope and its contexts.

The Requirement Description View for the rope is also defined as a separate diagram, as shown in Figure 6.35, enabling consideration to be given to rope-focused Requirement Descriptions so as to reduce the chance of getting sidetracked by other Requirement Descriptions. Currently, these requirements are on a diagram specifically defined for the rope Requirement Descriptions; however, it is also possible to place all Requirement Descriptions on to a single diagram. Bringing all the Requirement Descriptions into one view may show the number of requirements but is likely to make it difficult to find specific groupings, including these rope requirements. Further to this, many standards incorporate hundreds or thousands of Requirement Descriptions, each of which needs to be traced back to the source and refined by relevant use cases meaning that keeping these in their own view is beneficial.

Table 6.10 shows the refinements of the Requirement Descriptions by the use cases from the rope Requirement Context View. This is achieved in the same

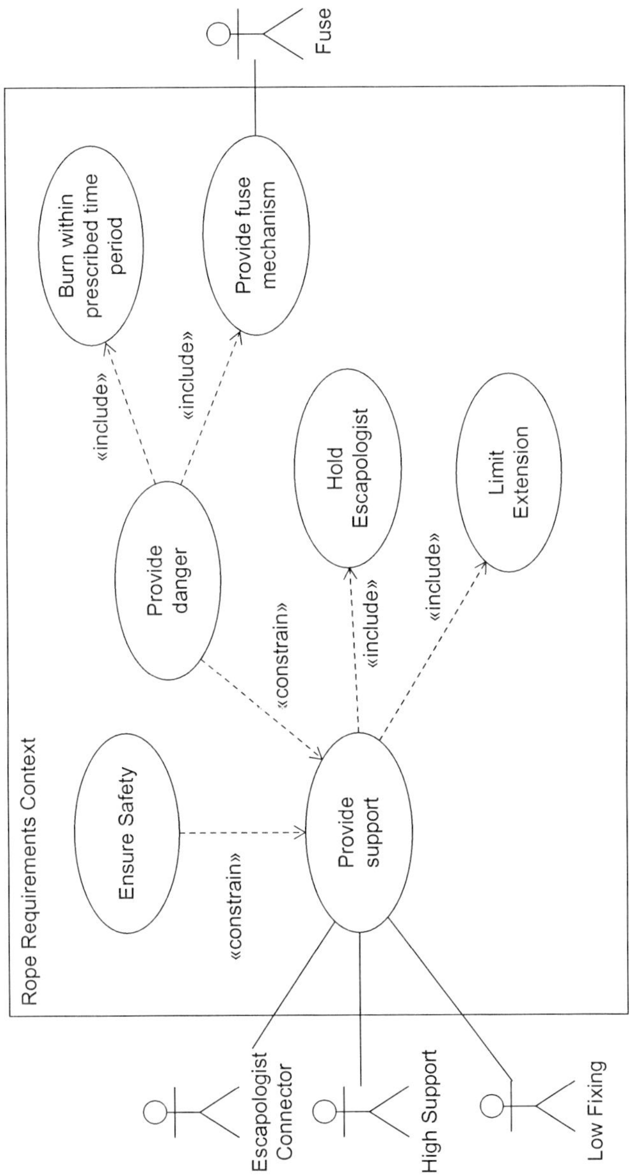

Figure 6.33 Rope Requirements Context View

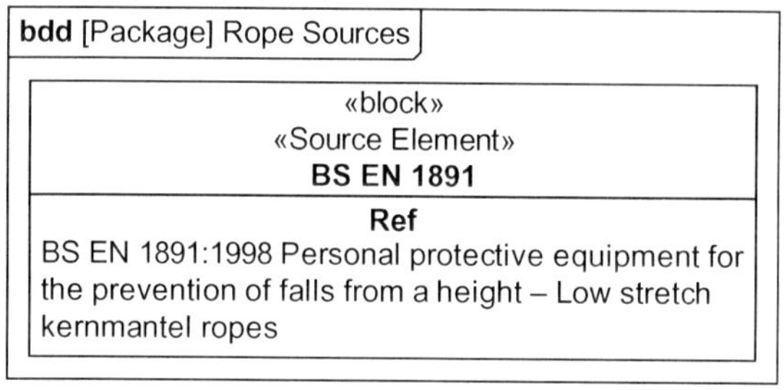

Figure 6.34 Source Elements for the rope

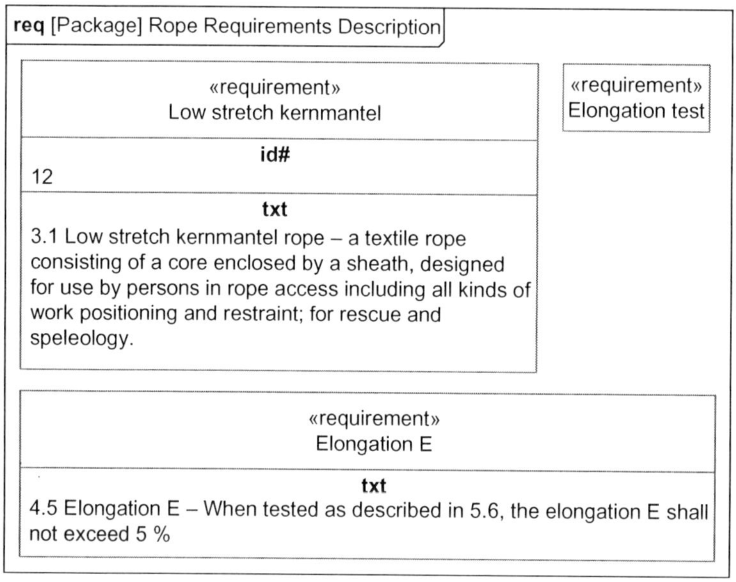

Figure 6.35 EN 1891 Requirement Description View

Table 6.10 Use cases refine Requirement Descriptions

Refines \ Refined by	«Use Case» Limit Extension	«Use Case» Hold Escapologist	«Use Case» Provide support
«requirement» Elongation E	refine		
«requirement» Elongation test	refine		
«requirement» Low stretch kernmantel		refine	refine

manner as the refinement table discussed earlier in this chapter when looking at the refinement of the Stakeholder Requirement Descriptions. The main difference here is that it is possible for the Use Case to be defined before the Requirement Description is identified, meaning that a full traceability exercise and check may need to be carried out.

Focusing on the rope Requirement Context View has meant the development of a Source Element View and Requirement Description View at a subsystem level. This highlights the scalability and positive reinforcement that modelling and the ACRE approach can provide. It has not only been used to identify and define the stakeholder and system requirements, but within a subsystem also used to identify new source information and capture the requirements at the right level to feed into the system requirements.

Figure 6.36 shows the Requirement Context View for the process aspects of the system. The rope is well within the physical aspects of the system, but consideration also had to be given to the process or activity-focused side of the stunt. The figure presents the need to 'Define process' and 'Follow process'. In this case, there are four types of process defined including 'Provide rescue', 'Set up', 'Set height' and 'Complete escape'. The 'Provide rescue' process is part of the way in which 'Ensure safety' is achieved, although the safety constraint applies to the way in which all the processes are defined.

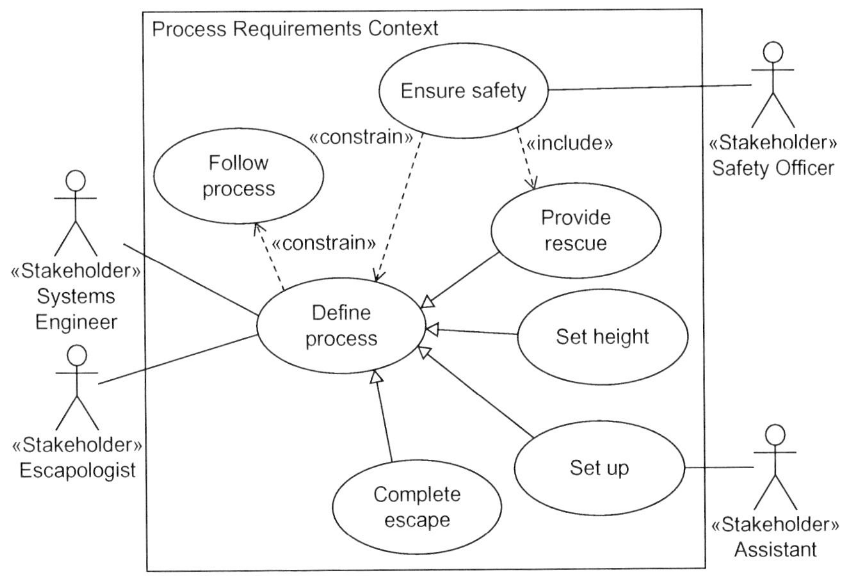

Figure 6.36 Process Requirements Context

To ensure that all processes are well defined, the seven-views approach to process modelling [10] is adopted with this figure providing the requirements view for the processes. The definition of a Requirement Context View for the processes within the system provides a basis for process definition as well as a route to validate the processes once they have been defined. Using a model-based approach for both the definition of the requirements and the processes themselves provides a strong link between the activities and structure of the system. For an example of this approach to process modelling, see Chapter 7.

Within this section the majority of the Requirement Context Views have been defined based on the stakeholder requirements contexts and knowledge of the system. Those which have been defined based on standards have Requirement Descriptions associated with them, the others do not. At this point we could have continued on and defined Requirement Descriptions for each of the use cases – for the system requirements they have been defined; however, for the lower level requirements contexts the descriptions are held within the Use Case. The reason that the descriptions have been left within the use cases is due to pragmatism, as much time could have been spent copying and pasting the text between use cases and Requirement Descriptions, but this would have added little to the model as the analysis would have already been carried out.

There are clear benefits to defining Requirement Context Views below the level of the system requirements. Requirement Context Views support a number of aspects including the following:

- The inclusion of Source Elements such as standards which apply at a detailed level
- The provision of traceability to standards without having to include detail in the system requirements
- The ability to identify stakeholders for lower level system elements
- The definition of in context requirements for each system element

Although it is advantageous to define Requirement Context Views for system elements, there must be a point at which to stop, otherwise Requirement Context Views will be defined for every system element which would not be pragmatic. There is no clear answer to how many levels within the system should have a Requirement Context Views defined; in this case, we have stopped at the level of the rope but it could be argued that this is too detailed and therefore not required. The true answer is that the number of levels of detail is dependent on the system one tries to define. Consider how many levels of detail may be needed for the dog house, house and office block discussed in Chapter 3 – only 1 or 2 for the dog house, 10–15 for the house (1 for the whole house, 1 for each room, 1 each for plumbing, heating and electrics) and maybe hundreds for the office block with heating and ventilation, lifts, fire protection, escape routes etc.

6.10.1 The Traceability View (system Requirement Context View to Stakeholder Requirement Context View)

The Traceability View highlighted in Figure 6.37 is used to allow traceability between different elements of the model to be explicitly shown. These traceability links may exist between views or between elements within views.

Figure 6.37 shows the ACRE framework with the Traceability View highlighted; it also highlights the Requirement Context View that is traced between the

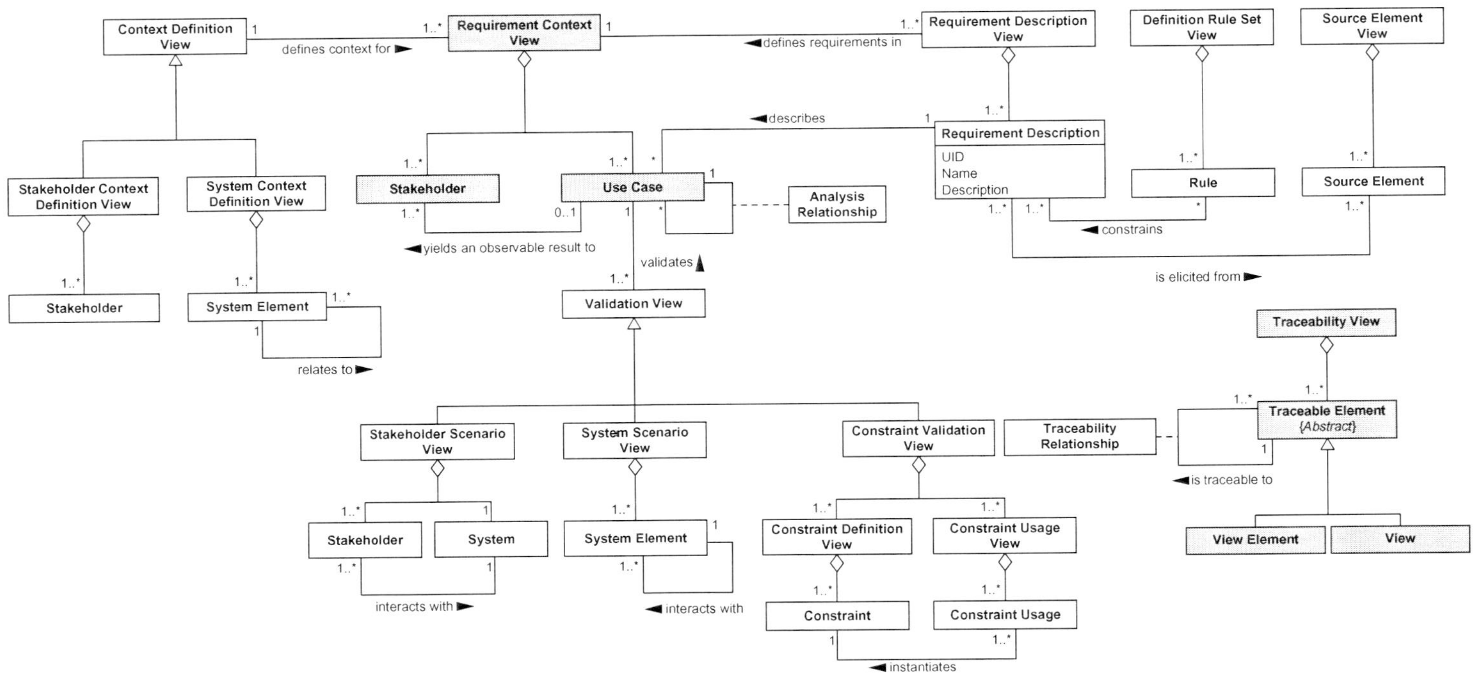

Figure 6.37 Framework – Traceability View

Stakeholder and System levels. To ensure that traceability is in place, as much traceability information as possible was captured as the project progressed. In this case, the traceability between the System Requirement Context View and the Stakeholder Requirements Context View was captured as the system level use cases were defined.

The matrix, Table 6.11, shows the «trace» relationship between the system use cases (across the top) and the Stakeholder use cases (down the left-hand side). The system Requirement Context Views have been defined based on Stakeholder Requirement Context Views, and the traceability was captured as a mapping whilst the use cases were defined. The formalisation of this mapping is the "trace" relationship between the use cases. Table 6.11 shows the system use cases across the top and the Stakeholder use cases down the left-hand side.

Table 6.11 System use cases trace to Stakeholder use cases

Traces to \ Traces from	«Use Case» Enable stunt	«Use Case» Ensure competence	«Use Case» Ensure safety	«Use Case» Provide danger	«Use Case» Provide height	«Use Case» Provide support	«Use Case» Set up
«Use Case» Engineer System	trace						trace
«Use Case» Escape	trace						
«Use Case» Get out	trace					trace	
«Use Case» Ensure Safety of People		trace					
«Use Case» Ensure Safety		trace	trace				
«Use Case» Make People Sweat				trace	trace		
«Use Case» Be interesting				trace			
«Use Case» Look Great				trace			
«Use Case» Don't Die						trace	
«Use Case» Set up							trace

If the approach discussed earlier is used to define the system Requirement Context Views, there are clear advantages to tracing directly between the system use cases and Stakeholder use cases. These advantages include the following:

- The direct correlation between Use Cases.
- Traceability for free as it is recorded during the definition of the system use cases.

However, there are also advantages to providing the traceability between the Requirement Descriptions, which include the following:

- Similarity to traditional requirements traceability approaches.
- Provision of a 'clean' requirements trace, i.e. only requirements will be seen in the trace.

Table 6.12 System requirements descriptions trace to Stakeholder Requirements Descriptions

Name	System requirement	Relationship	Name	Origin
Enable stunt	Requirement	Traces to	Develop system	Systems engineer
		Traces to	Escape	Escapologist
Ensure competence	Requirement	Traces to	Ensure people are safe	Safety officer
		Traces to	Ensure safety	Safety officer
		Traces to	Get out	
Ensure Safety	Requirement	Traces to	Ensure safety	Safety officer
Minimise risk to escapologist	Requirement	Traces to	Ensure people are safe	Safety Officer
		Traces to	Don't die	
Provide danger	Requirement	Traces to	Make people sweat	Marketing
		Traces to	Look great	
Provide support	Requirement	Traces to	Get out	
Set up	Requirement	Traces to	Develop system	Systems engineer
		Traces to	Set up	

Table 6.12 shows the traceability between the system and Stakeholder Requirement Descriptions. The first two columns show the system Requirement Descriptions; the third, grey-shaded column shows the «trace» relationship; the two columns after right of the grey-shaded column show the Stakeholder level Requirement Description and the origin of each Requirement Description.

Tables 6.11 and 6.12 provide very similar information. The differences are generally due to the naming conventions used on the Requirement Descriptions compared with those used on the Requirement Context Views, and this also suggests that a copy and paste of the text between use cases and Requirement Descriptions may not add value.

It would also be possible to mix the traces between requirement contexts and Requirement Descriptions, but this may cause confusion when you try to understand the overall traceability of the system; hence, it is not recommended.

6.11 The Validation View (System Scenario View)

The Validation View provides the basis for demonstrating that the requirements can be met or complied with. These views can be informal (such as scenarios at various levels of abstraction) or may be formal (such as mathematical-based representation).

Figure 6.38 shows the ACRE framework with the Validation Views highlighted. Validation can be considered at a number of different levels with Stakeholder and system being two examples.

Consider, now, the system level, where the emphasis is on the information exchanged between system elements. Relationships with Stakeholders may also be

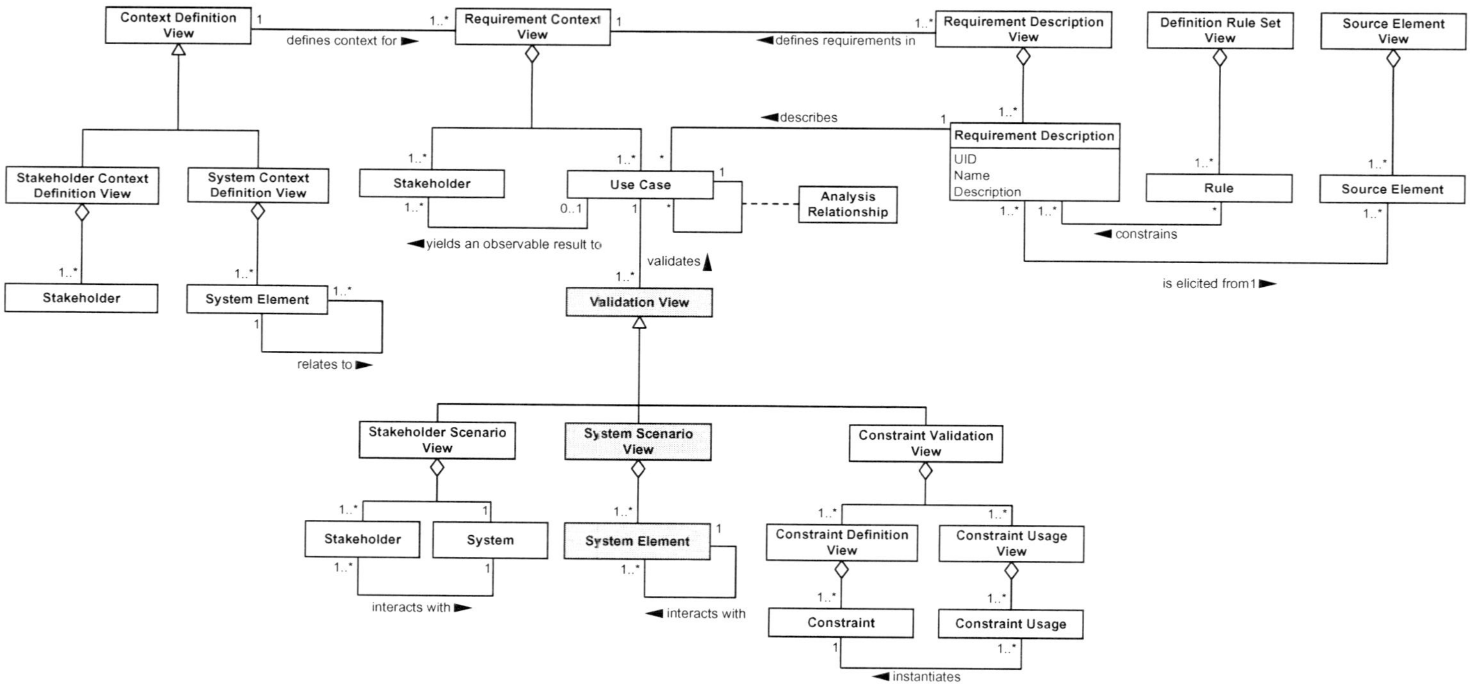

Figure 6.38   Framework – System Scenario View

shown, but these must be consistent with the relationships between the Stakeholder and the system on the Stakeholder Scenario Views.

Figure 6.39 shows the interactions between the physical components of the stunt during a successful escape. The use of white box scenarios to support the analysis and validation of the system enables inputs, outputs and functions of each system component or subsystem to be identified. The inputs, outputs and functions may be linked to the structural definition of the design as well as the behavioural aspects of the design model using state or activity diagrams.

The workings of the crane are referenced using the SysML reference frame 'Provide height' in the middle of Figure 6.39. The reference frame can be used multiple times and here it is used to show lifting and lowering of the escapologist. The way in which height is provided may be defined in another scenario within this model or may reference the design model; for example, the crane may be referenced as part of asset/component reuse. As with earlier-mentioned Validation Views, these are defined to help provide confidence in the model and the system which implement them. Therefore, areas other than the performance of the stunt may also need to be considered.

Figure 6.40 shows the physical aspects of the escapology stunt which are involved in the 'Set-up' of the escape. This Validation View shows the order in which the escapologist is attached to the equipment. First, the 'Safety rope' is secured before being attached to the 'Crane'. The 'Leg Harness' is then secured before being attached to the 'Crane', after which the main 'Rope' is secured and attached to the 'Crane'. Finally, the fuse is inserted into the 'Rope'.

Presenting the physical aspects of the system in this way provides a clear understanding of how all the component parts are related to each other. However, these views provide little insight into the process aspects of the system.

Figure 6.41 shows the relationships between four of the processes defined for the system in the case of a successful escape. These processes provide the operation of the system. 'Set-up' occurs first, concluding by sending the 'Begin lift' message which sets of the 'Lifting' process. Once 'Lifting' is complete the 'Start escape' message is sent which invokes the 'Escape' process. Finally, the 'Escape' process sends the 'Start lowering' message to begin the 'Lowering' process.

This Validation View has been used as a «testCase» to validate the processes defined for the system. It validates them by showing that the processes relate and that they can be used in sequence together to achieve the aims of the system as captured in the use cases on the system-level Requirement Context Views. The Validation View also fulfils the need for a process implementation view in the seven views process approach [10].

Figure 6.42 shows the relationships between the 'Set-up', 'Lifting', 'Escape' and 'Rescue' processes from the process aspects of the system. In this scenario an emergency situation occurs during the 'Escape' process. At this point the 'Emergency' message is sent and the 'Rescue' process is put into action.

This Validation View provides validation for some of the safety use cases of the system, ensuring that the escapologist can be rescued in the event of an

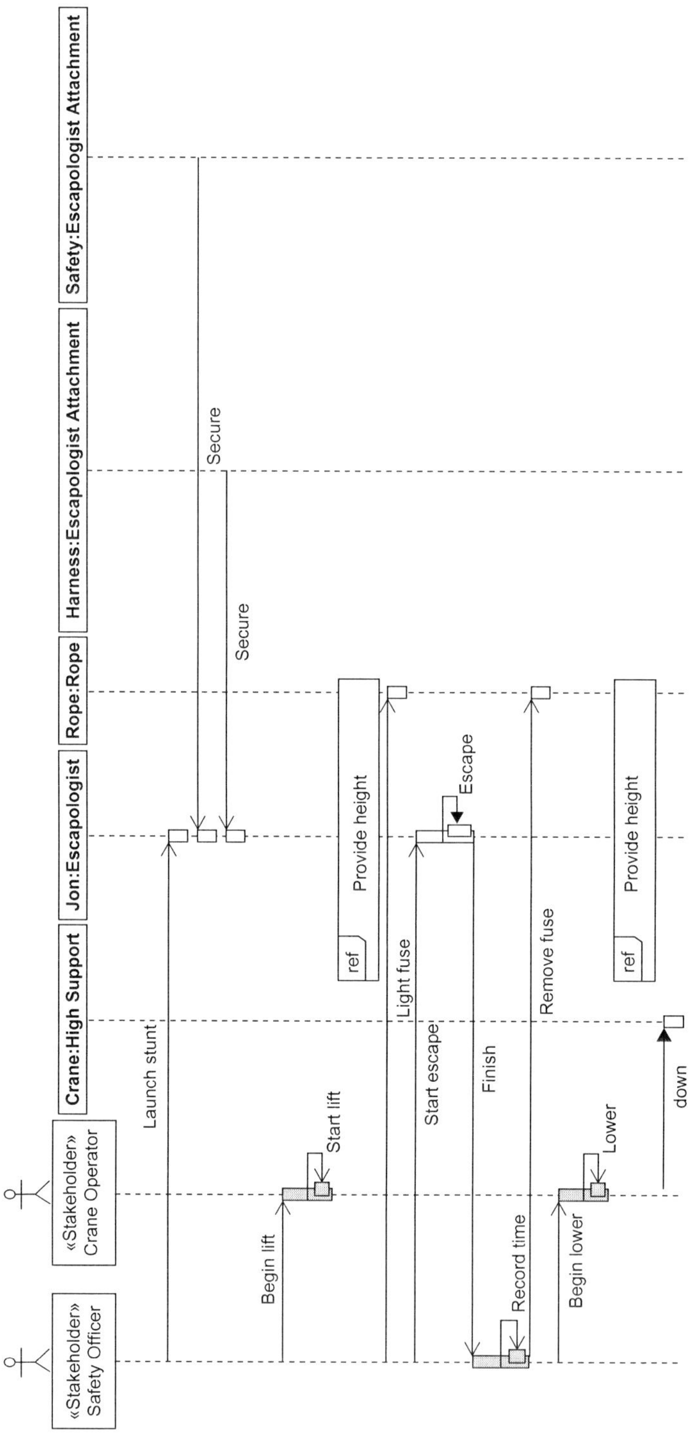

Figure 6.39 Validation View, successful escape system scenario

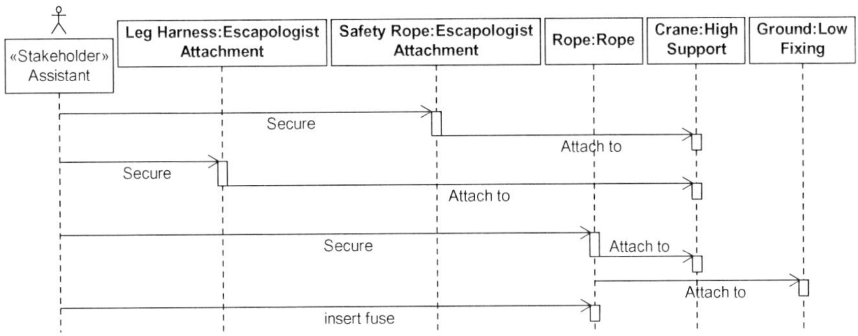

Figure 6.40 Validation View, set-up system scenario

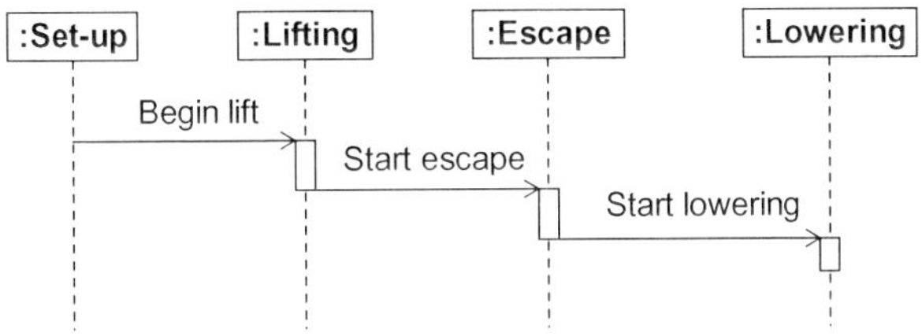

Figure 6.41 Validation View, successful escape process scenario

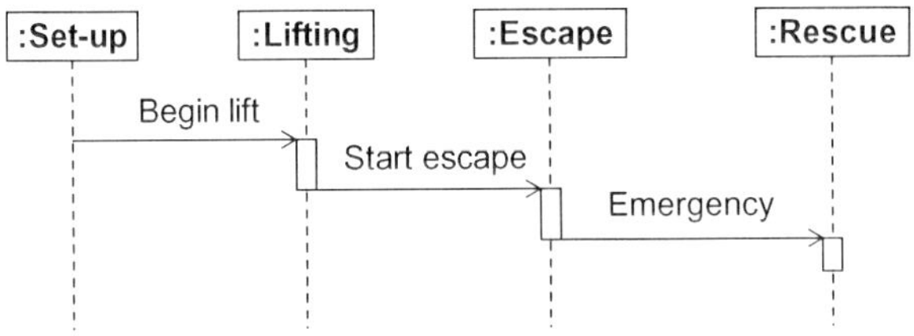

Figure 6.42 Rescue process scenario

emergency. It is also relevant to establish other Validation Views similar to this one to ensure that the rescue process can be activated from any of the system process.

These Validation Views, like the Stakeholder Validation Views, are selected and marked with «testCase» if they are to be used as part of the system validation.

6.11.1 The Traceability View (System Scenario View validates Requirement Context View)

The Traceability View is used to allow traceability between different elements of the model to be explicitly shown. These traceability links may exist between views or between elements within views.

Figure 6.43 shows the ACRE framework with the Traceability View highlighted; it also highlights the Requirement Context View and the System Scenario

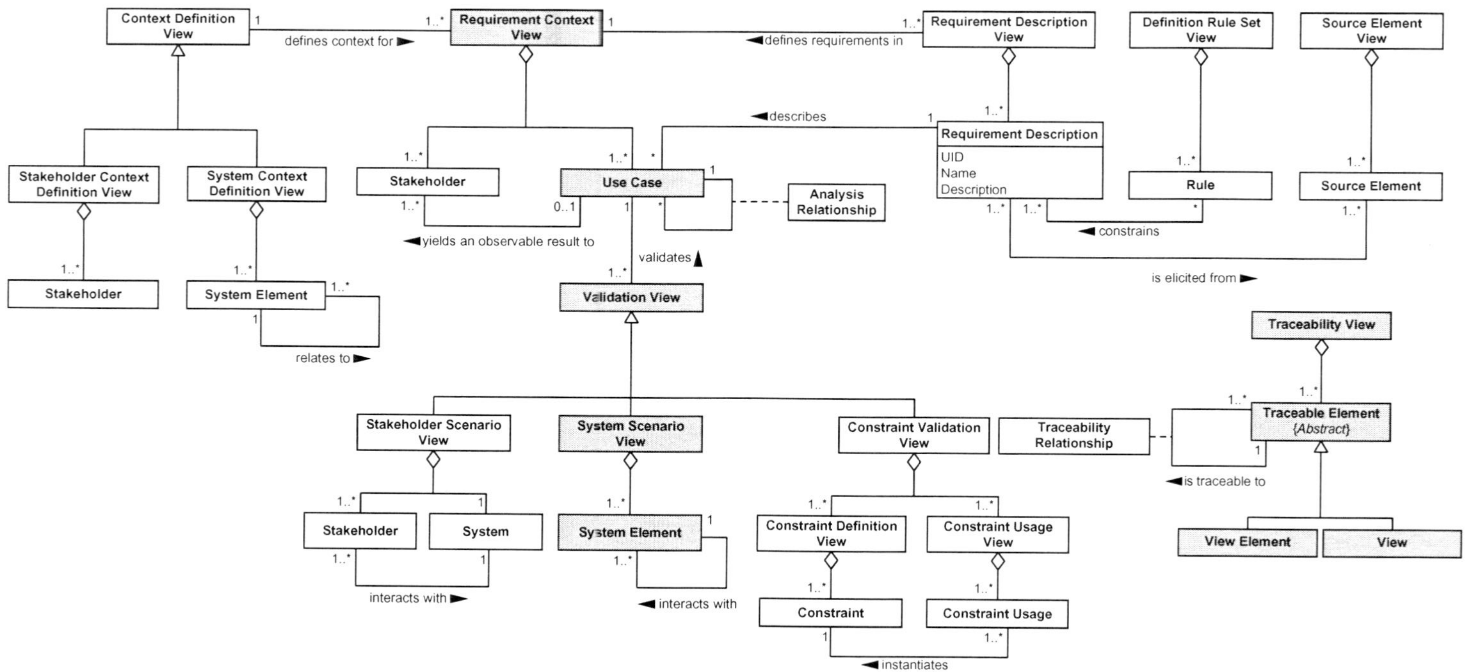

Figure 6.43 Framework – Traceability View

Table 6.13 Sequence validates Use Case

Validates \ Validated by	«System Scenario View» Successful escape processes	«System Scenario View» Normal escape	«System Scenario View» Emergency during set up	«System Scenario View» Emergency during lifting	«System Scenario View» Emergency during escape	«System Scenario View» Scenario – failed escape	«System Scenario View» Provide height	«System Scenario View» Set up
«Use Case» Enable Stunt	validate	validate						
«Use Case» Ensure Safety			validate	validate	validate	validate		
«Use Case» Provide Height							validate	
«Use Case» Set up								validate

View that are traced. As the number of Validation Views increases it becomes tempting to put traceability to one side in favour of producing the system design. However, due to the likely increase in the number of Validation Views it is imperative at this stage to maintain the traceability effort; losing control of this effort now may cause problems with the tracking of changes and evidence of validation in the future.

Table 6.13 contains a matrix showing the system use cases (in the left-hand column), which are validated by the System Scenario Views across the top. The matrix clearly shows that half of the System Scenario Views are being used to validate the Use Case 'Ensure safety'.

In many cases a «trace» relationship may be put in place between the system and Stakeholder Scenario Views. This "trace" may prove useful to verify that any Stakeholder relationships on the System Scenario Views are consistent with those on the Stakeholder Scenario Views. However, useful as this may be, it provides verification of the system and Stakeholder Scenario Views rather than validation.

6.12 The Validation View (Constraint Validation View)

The Validation View provides the basis for demonstrating that the requirements can be met or complied with. These views can be informal (such as scenarios at various levels of abstraction) or may be formal (such as mathematical-based representation).

Figure 6.44 shows the ACRE framework with the Validation Views highlighted. Validation can be considered in a number of different ways with scenarios and mathematical constraints being two examples.

In the case where calculation is needed to show that a number of system properties can come together to validate a Use Case, then the Constraint Validation View provides the ability to provide mathematically based relationships between properties within the system. The constraints used within the Constraint Validation

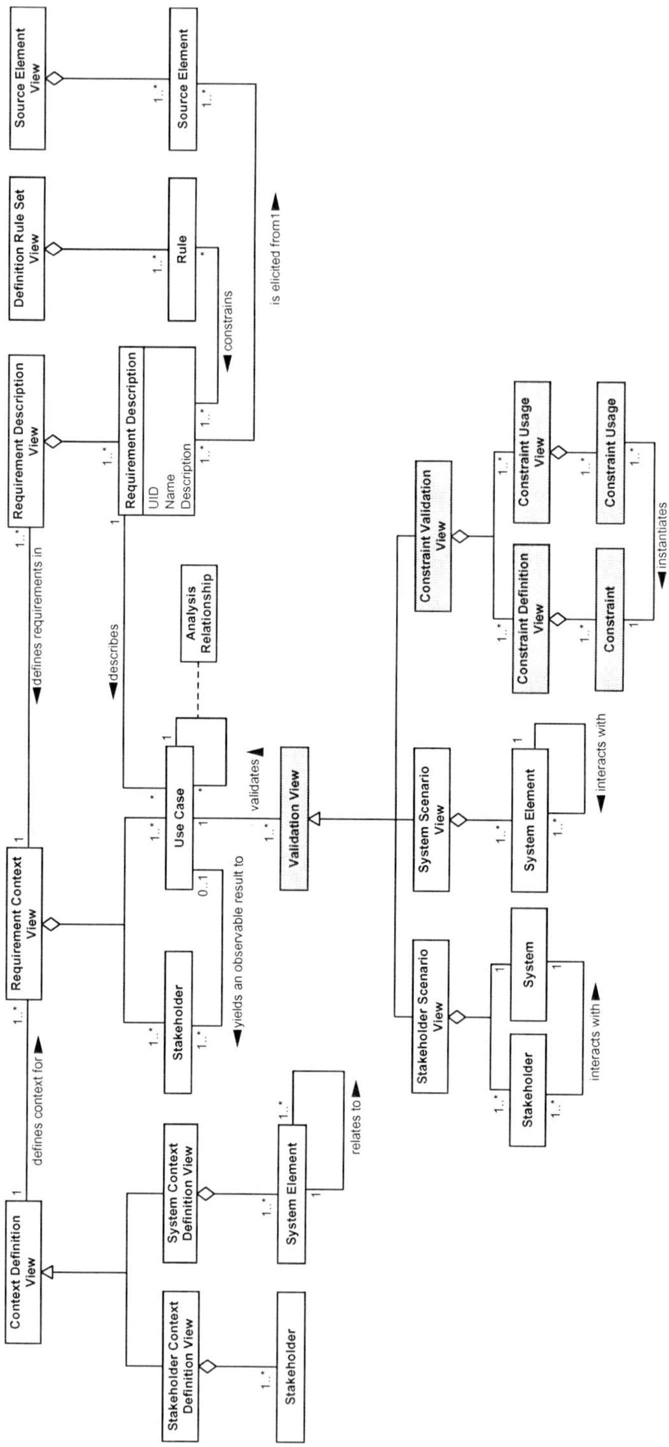

Figure 6.44 Framework – constraint Validation View

View may be mathematical operators such as plus and minus, laws such as Newton's laws or heuristics such as the boundaries provided in the Flesch–Kincaid Requirement Complexity Rule discussed earlier.

Figure 6.44 shows the definition of a number of constraints used in the case study. As discussed in Chapter 3, each includes a name, the constraint or formula and the parameters available to be used.

Constraints defined in this way can become part of a library of constraints that can be re-used on other projects. In some cases a constraint may, such as with the 'CraneHeight' constraint, be specific to a particular system or project. In these situations, it may be decided not to transfer such constraint a standard library or, if they are placed into such a library, that they be classified with specific information to ensure they are only used in relevant circumstances.

Constraints can be used by connecting instances of them together into a parametric network.

Figure 6.46 shows a parametric network made up of the constraints defined in Figure 6.45. This Validation View shows one way in which the decision as to whether the escapology stunt should be carried out may be made:

- The top left considers how long the rope should be. The calculation of the total rope length is based on the 'hypotenuse' formed by the rope and calculated from the distance between the 'Rope.AnchorDistance' and the 'crane' and the 'Crane. HookHeight'. The length must also include the distance the rope drops from the crane's hook and a calculation for the extension of the rope, as some ropes can be dangerously stretchy, e.g. climbing rope, which can extend by 30%.
- The middle of the diagram calculates the lowest point the escapologist head can reach. The 'Rope.Drop', ''Escapologist.Height' and the 'Rope.Stretch-Length' are all subtracted from the height of the hook 'Crane.HookHeight'.

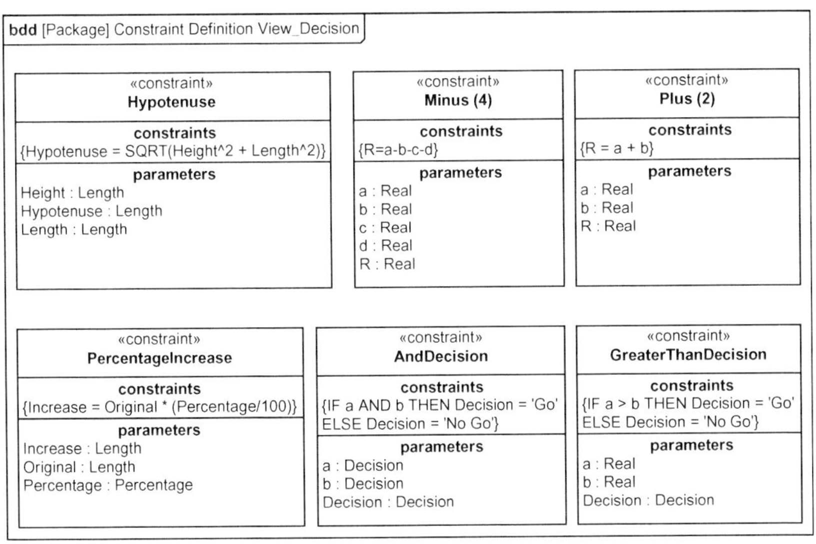

Figure 6.45 Constraint definition

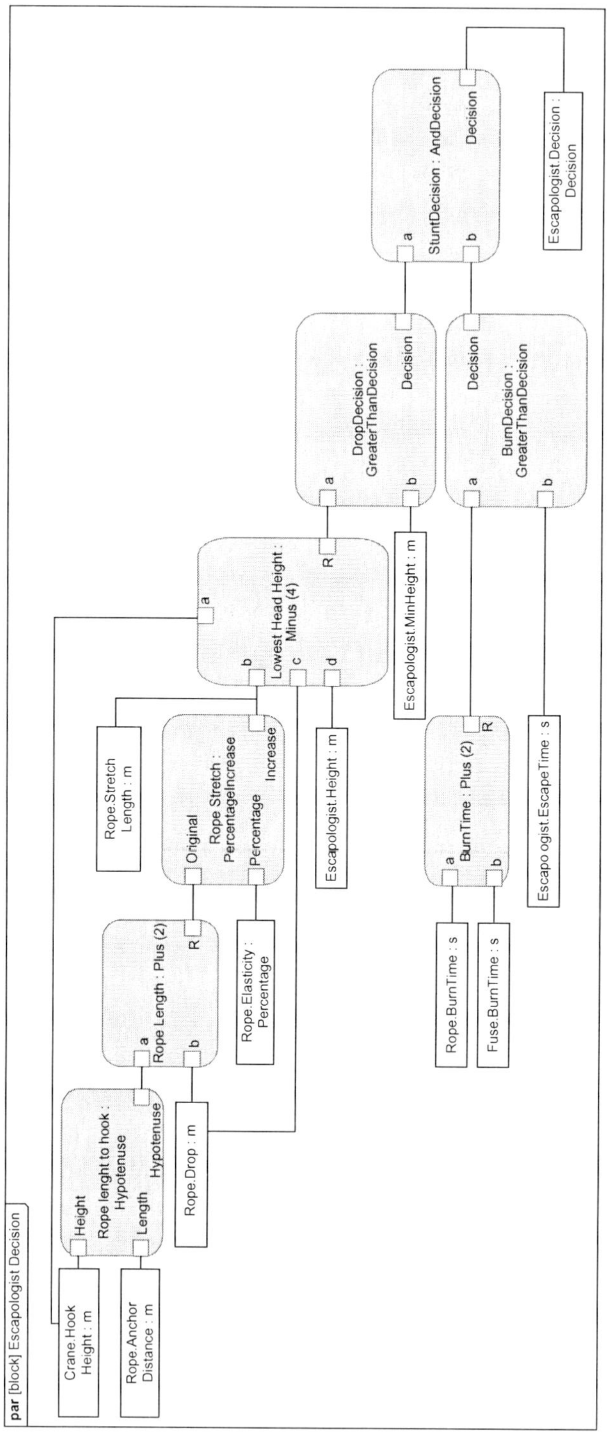

Figure 6.46 Constraint usage

- The bottom of the diagram focuses on the time taken before the rope will break. The 'Fuse.BurnTime' is measured and added to the 'Rope.BurnTime'.

The right-hand side of the diagram considers the calculations above and provides the decisions. Three decisions are made: the first two of these provide decisions on whether the burn time is long enough and whether and the rope stretches too far, meaning the escapologist will hit the floor. The final decision checks both of these and if they are both positive then the stunt can go ahead.

Table 6.14 provides initial proposed and final selected values for the system properties being analysed by the Validation View discussed in Figure 6.46. A range of values was analysed, which considered the variation of one of more system properties. Some values of these system properties preclude the running of the stunt and others enable the stunt to go ahead.

The matrix in Table 6.15 provides the calculated and decision values for a number of options shown down the left-hand side. The first row relates to the initial values from Table 6.14 and the last row relates to the final values from Table 6.14. Intermediate rows show other options considered. The columns across the top show the results of each calculation in the Validation View, Figure 6.46.

This analysis allows considerations to be made regarding the way in which the stunt is executed. For example, should it be carried out with a shorter anchor distance, a higher crane or a less elastic rope? There are two main reasons for choosing the final option presented in Table 6.15, one focuses on the ability to capture the stunt on film, the other relates to the safety of the escapologist:

- The anchor distance was reduced to ensure that the anchor and fuse were in shot during filming.
- The reduction in the elasticity of the rope means that the escapologist stays over 1 m above the ground during the stunt.

Calculations of this type are often implemented and analysed by hand using spreadsheets; however, some modelling tools automate the implementation of parametrics by using mathematical tools.

Table 6.14 System property values

Initial values		**Final values**	
Crane.HookHeight	6 m	Crane.HookHeight	6 m
Rope.AnchorDistance	8 m	Rope.AnchorDistance	4 m
Rope.Drop	1.2 m	Rope.Drop	1.2 m
Rope.Elasticity	30%	Rope.Elasticity	20%
Rope.StretchLength	3.36 m	Rope.StretchLength	1.68 m
Escapologist.Height	1.75 m	Escapologist.Height	1.75 m
Escapologist.MinHeight	0.5 m	Escapologist.MinHeight	0.5 m
Rope.BurnTime	30 s	Rope.BurnTime	30 s
Fuse.BurnTime	120 s	Fuse.BurnTime	120 s
Escapologist.EscapeTime	120 s	Escapologist.EscapeTime	120 s
Escapologist.Decision	No-go	Escapologist.Decision	Go

Table 6.15 Parametric example values

Option	Hypo-tenuse (m)	Rope length (m)	Rope stretch (m)	Lowest head height (m)	Drop decision	Burn time (s)	Burn decision	Stunt decision
Initial suggestion	10.00	11.20	3.36	−0.31	No-go	150	Go	No-go
Reduce rope stretch to 3%	10.00	11.20	0.34	2.71	Go	150	Go	Go
Increase crane height	10.63	11.83	3.55	0.50	Go	150	Go	Go
Reduce anchor distance 1 m	7.81	9.01	2.70	0.35	No-go	150	Go	No-go
Reduce anchor distance 2 m	7.21	8.41	2.52	0.53	Go	150	Go	Go
Reduce rope stretch (20%) and anchor distance (4 m)	7.21	8.41	1.68	1.37	Go	150	Go	Go

Figure 6.47 shows the output, for the parametric diagram in Figure 6.46, from a parametric solving tool. In this case, the lowest head height is less the 0.5 m resulting in a target value of '0' representing a fail or 'no go' for the decision, which is highlighted by the arrows on the diagram.

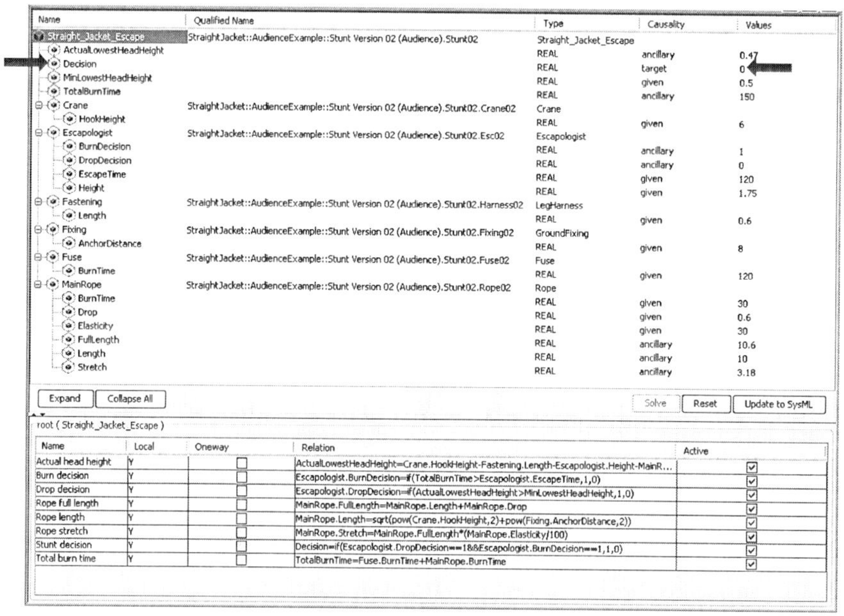

Figure 6.47 Automated parametric implementation

These calculations, whether in a spreadsheet or a parametric solving tool, provide a great deal of detail to be considered and are very useful for giving the audience confidence that the stunt poses a real danger to the escapologist. However, there is another parametric network that considers the system properties and how they relate to the safety needs of the Safety Officer.

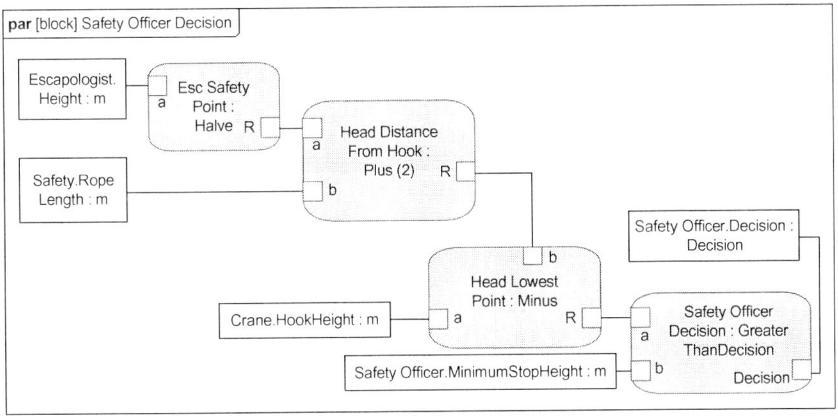

Figure 6.48 Constraint Validation View, Safety Officer's parametric network

Figure 6.48 shows the properties of the system considered by the Safety Officer. This network, rather than being used to predict the possible outcomes, was used on the day of the stunt. Each measurement was recorded and entered into the network to ensure that the escapologist would be safe no matter what the outcomes considered by the audience may be (as shown in Table 6.14).

6.12.1 The Traceability View (System Constraint Validation View to System Requirements Context View)

The Traceability View is used to allow traceability between different elements of the model to be explicitly shown. These traceability links may exist between views or between elements within views.

Figure 6.49 shows the ACRE framework with the Traceability View highlighted; it also highlights the Requirement Context View and the Constraint Validation View that are traced to each other. The use of the Constraint Validation View to validate use cases should not be underestimated. Constraints can be defined at various levels of abstraction, enabling replacement with higher or lower level concepts. This ability to abstract and provide different levels of granularity supports the consistency of the model across different levels of hierarchy, such as Stakeholder, system and subsystem, and provides a clear, repeatable validation of use cases.

The matrix in Table 6.16 shows which use cases, shown on the left, are validated by the Constraint Validation Views shown across the top. In this matrix, the use of one Constraint Validation View to validate multiple use cases can be seen.

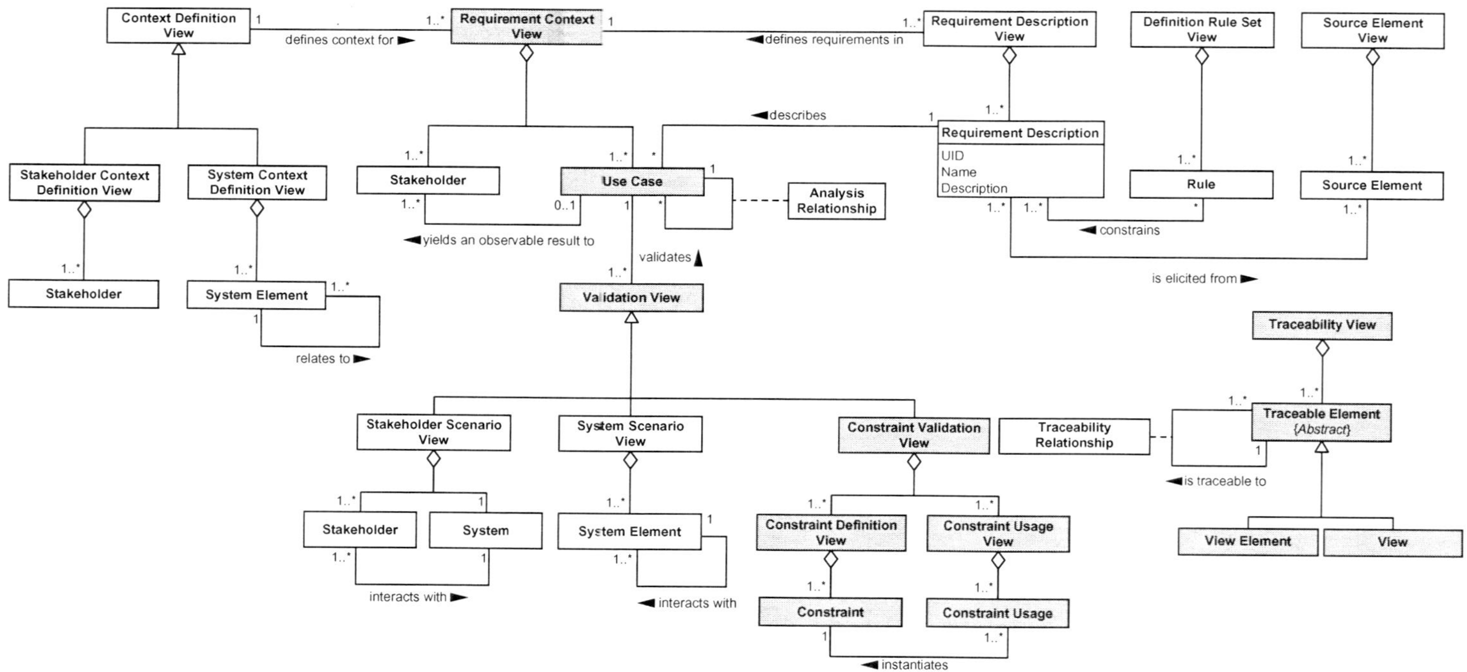

Figure 6.49 Framework – Traceability View

Table 6.16 Constraints validate use cases

Validates \ Validated by	«Constraint Validation View» [block] Escapologist	«Constraint Validation View» Initial rope length	«Constraint Validation View» [block] rope extension
«Use Case» Enable Stunt	validate		
«Use Case» Ensure Safety	validate		
«Use Case» Provide Support		validate	
«Use Case» Limit Extension			validate

Also different levels of granularity are shown by the use of the 'Initial rope length' and 'Rope extension' Constraint Validation Views to validate the 'Provide Support' and 'Limit Extension' use cases.

6.13 Conclusions

This case study exercises each of the concepts and views of the ACRE framework. The model on which this case study is based was used to implement the escapology stunt successfully in April 2010.

Figure 6.50 provides a reminder of each of the views within the ACRE framework. This case study includes many discussion points related to each of the diagrams. However, there are some which run throughout the case study, including the following:

- *The need for iteration at each system level.* To build up a consistent model, the views in the ACRE framework must be revisited and improved. Improvements will be based on developments and changes to newly developed or improved views. This may mean, e.g. changing a Requirement Context View due to the understanding gained from a Validation View.
- *The need for traceability between levels.* To ensure a complete understanding of the model, traceability must be captured between different levels, e.g. between the Stakeholder and the system. Traceability enables many types of analysis to be achieved, including impact of changes, coverage of Validation Views and validation of requirements.
- *The need for sharp tools.* Many of the views described in this chapter are shown in a number of different formats including diagrams, tables and matrices. It is important to ensure that the tools that are able to deliver views are in place in the required format from a single data source. This is discussed further in Chapter 7. However, it is just as important to ensure that the tools are driven by the process.

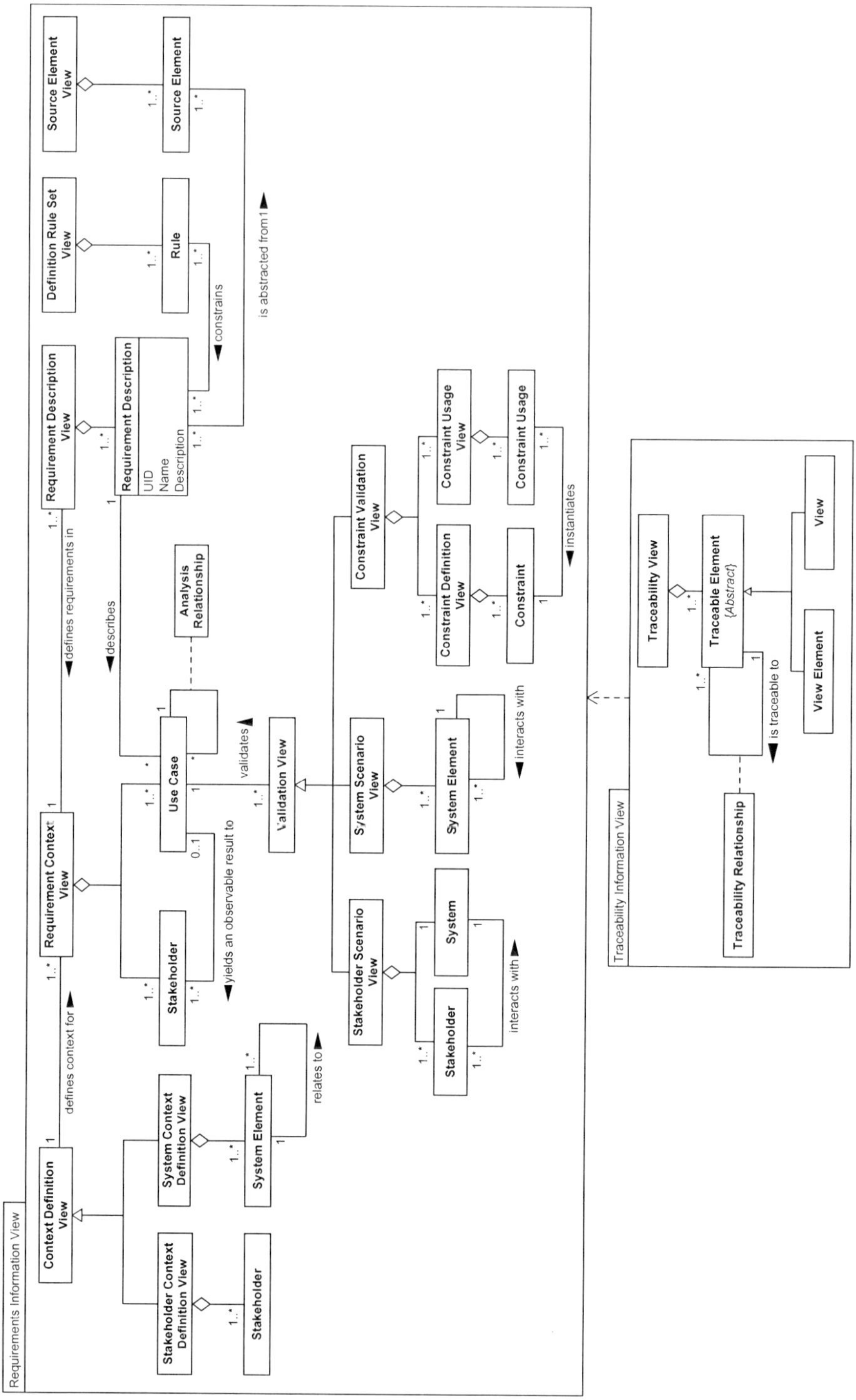

Figure 6.50 ACRE framework

One of the foundations of the ACRE approach is pragmatism. This case study has shown at least one example of each view for the escapology stunt. In many cases, it will not be relevant to develop every view from the framework and so the ACRE framework must be applied with the relevant level of rigour and pragmatism to ensure the best result is achieved. This is also discussed further in Chapter 7.

References

1. Houdini H. *Magical Rope Ties and Escapes*. Will Goldston: London; 1920
2. Houdini H. *Movie Star* [DVD]. Kino Video: New York; 2008
3. IEC 61508. *Functional Safety of Electrical/Electronic/Programmable Electronic Safety-Related Systems*. Switzerland: International Electrotechnical Commission; 1998
4. BS ISO 31000. *Risk Management – Principles and Guidelines on Implementation*. BSI: London; 2009
5. Holt J., Perry S. *SysML for Systems Engineers*. IET: London; 2008
6. Flesch R. 'A New Readability Yardstick'. *Journal of Applied Psychology*. 1948;**32**:221–33
7. Kincaid J.P., Fishburne R.P. Jr, Rogers R.L., Chissom, B.S. *Derivation of New Readability Formulas (Automated Readability Index, Fog Count and Flesch Reading Ease Formula) for Navy Enlisted Personnel*. Research Branch Report 8-75, Millington, TN: Naval Technical Training, U.S. Naval Air Station, Memphis, TN; 1975
8. Boehm B.W. 'Verifying and validating software requirements and design specifications' reprinted in Boehm B.W. (ed.). *Software Risk Management*. IEEE Computer Society Press: Washington; 1989. pp. 205–18
9. BS EN 1891:1998. *Personal Protective Equipment for the Prevention of Falls from a Height – Low Stretch Kernmantel Ropes*. BSI: London; 1998
10. Holt J. *A Pragmatic Guide to Business Process Modelling*. 2nd edn. BCS Publishing: Swindon; 2009

Chapter 7
Realising the process

Reality leaves a lot to the imagination.

John Lennon

7.1 Introduction

In the introduction of this book, and many times since, it has been stated that in order to realise any aspect of the model-based systems engineering (MBSE), it is essential to have people, processes and tools in place.

This chapter looks at the pragmatic issues involved when trying to realise the context-based requirements engineering approach in any organisation or business.

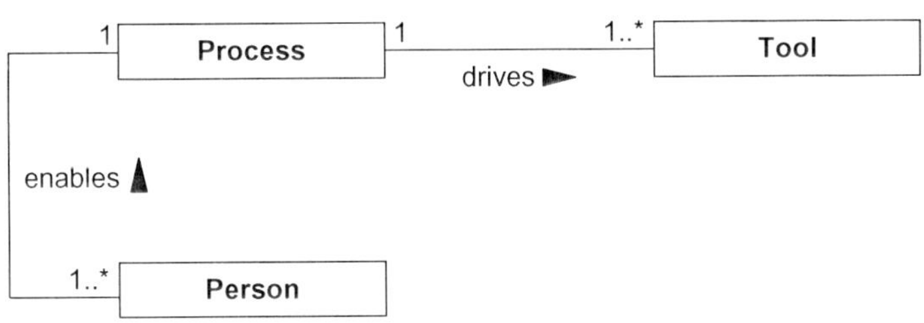

Figure 7.1 Pragmatic issues with implementing the approach

Figure 7.1 shows the three aspects that are discussed in this chapter in particular:

- The 'Person'. When talking about people, we mean *competent* people, rather than just any people. An essential element of having competent people is giving them the basic knowledge that is needed to carry out the approach. This section, therefore, contains a teaching guide that may be used to develop a bespoke teaching or training course based on the ideas in this book. To make people competent, however, involves much more than just training them to give the right knowledge – they also require skill and the right attitude. This section also contains, therefore, a competency scope that may be used as a basis for assessing the competence of people who are required to follow this approach.

- The 'Process'. It is essential that a process is in place in order to realise the approach. This involves being able to understand and define the process in order to establish a correct process. However, in order to implement the process effectively it is necessary to deploy the process in an efficient and effective manner. This section looks at how a process may be defined using modelling and how it can be applied in a number of different situations.
- The 'Tool'. To realise the full benefits of MBSE, tools are essential. In reality, we may use any tools, ranging from a whiteboard or log book, to standard office tools, to a full-blown automated toolset, to any combination of these. This section looks at a number of ways that the approach may be implemented using a variety of tools, and discusses the advantages and disadvantages associated with each.

Each of the above aspects is discussed in more detail in this chapter.

7.2 The 'Person'

This section looks at how it is possible to ensure that the right people, i.e. competent people, are in the right place, doing the right job.

The first step in ensuring that people are competent is making sure that people are trained properly, based on a well-understood analysis of the educational needs for the role that the person is to carry out. These educational needs may exist in an academic or industrial environment, depending on whether the course is being taught at a university or as part of industrial training. In this case, we are interested in context-based requirements engineering, so the scope of the education is well defined, and this is presented here in the form of a teaching guide.

To ensure that people have the appropriate competencies, it is important that they can be assessed according to the needs of their roles. To do this, therefore, it is necessary to have a good definition of the competencies for each role. To this end, two sets of competencies are presented here – one is based on an international systems engineering best-practice model, and the other bespoke set of competencies based directly on the approach adopted in this book.

7.2.1 Teaching guide

One of the main topic areas which occurs when presenting information regarding requirements modelling is that of how to teach or train people. There is no single correct way to do this, so this chapter provides a discussion on communicating the approach in this book modelling to people using teaching and training techniques. The information contained in this chapter is intended for guidance only and is based on the authors' years of experience teaching at both undergraduate and postgraduate levels as well as developing and delivering professional training courses for major industry.

This chapter suggests a course structure for teaching at university level, which is fully referenced back to this book. The main aim of this chapter is not to provide a full course that can simply be lifted out of the book and taught; therefore, no slides are provided. The intention is that this chapter, along with supporting CASE tools and models, can be used as a tool kit for someone to create a requirements modelling course based on the contents of this book. Therefore, it is intended that this book is used as the recommended course text that provides lots of additional information and many more fully worked examples.

So, please feel free to use this chapter and models as the start point to your own course – make the course your own!

7.2.1.1 Different types of teaching

There is no definitive way to teach requirements modelling, so this chapter provides a few examples of how teaching may be approached depending on the audience. One key part of any teaching or training is to know and understand the audience, and by this we mean understanding the stakeholder's requirements. The point here is that the teaching requirements differ depending on who the target audience is, and this is discussed in some detail.

Figure 7.2 shows a generic set of teaching requirements that are used as a basis for discussion. It should be borne in mind, however, that this set of requirements need to be tailored, or even started again, from scratch to fit the reader's needs. If you are interested in teaching or training, it is strongly recommended that this short requirement exercise is carried out because it will improve your own understanding of the teaching and help ensure that the course that is developed actually meets these requirements. This will clearly result in a better course and, hopefully, a better learning experience for the teaching subjects.

Figure 7.2 shows the generic system context for delivering training or teaching courses. The requirements are described in more detail as follows along with a few suggestions for each as to how the basic set of requirements may be tailored:

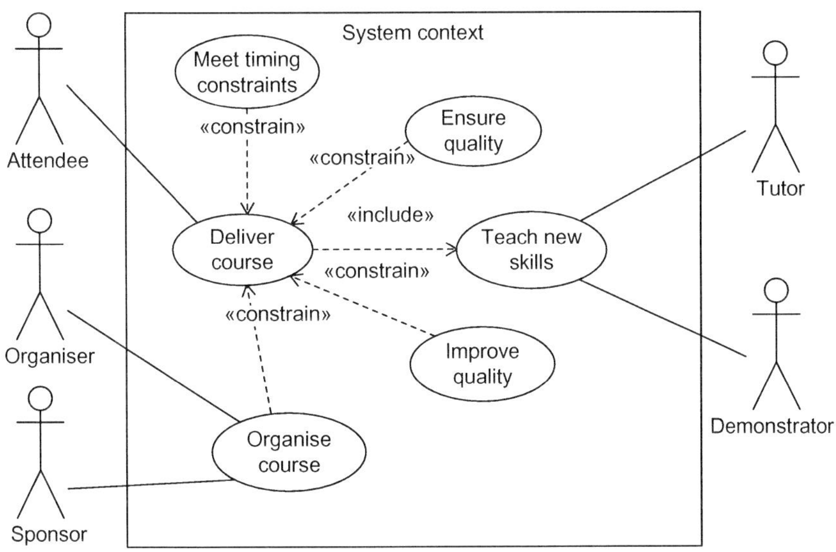

Figure 7.2 Generic teaching or training context

- 'Deliver course'. This is the overall requirement that sets the scene for the context. This could be tailored by adding different 'types of' (specialisation) relationships to the diagram to show requirements for different types of courses.
- 'Teach new skills'. Note that this is the only inclusion in the overall requirement of 'Deliver course' and hence forms the basis of the course. This could be expanded upon by adding in more included requirements. For example, it may be a requirement to provide examples or set course work – these could be added in as new requirements.
- 'Organise course'. This requirement could mean almost anything, depending on the nature of the course being taught. For example, it may be as simple as making sure that a room is booked to something as complex as making travel arrangements, renting facilities, hiring equipment and so on.
- 'Ensure quality'. This is a constraint on delivering the course and may include issues such as making sure that the course material is printed out and bound nicely, ensuring that the facilities for the course measure up. This may also be extended to include other concerns, such as making sure that the presenters wear suits and have a wash before the course or whatever else is deemed important.
- 'Improve quality'. It is always important to continuously improve everything that we do in our work and, therefore, this should be a requirement that is always present in the context. This may include collecting feedback from the course, making notes of any corrections or enhancements that could be made to the course afterwards and so on.
- 'Meet timing constraints'. This is very important as this limits what can be delivered and when. Understanding the timing constraints can often be the difference between a successful and unsuccessful course and its importance cannot be stressed strongly enough. For example, if a course is to be taught for 10 sessions, each of an hour's duration, then that course will have a different structure from a course that will be taught over 8 hours on a single day.

Due to the space and commercial constraints, the emphasis for the example provided in this chapter focuses mainly on providing a course as part of a university syllabus.

The stakeholders that are shown on the diagram differ significantly, depending on the type of teaching or training, and are discussed in more detail in sections 7.2.2 and 7.2.3.

7.2.2 Professional training

7.2.2.1 Teaching requirements

When considering a professional training course, the core requirements are as shown in Figure 7.2. With regards to the stakeholder roles identified, the following is a typical list of names that may be associated with each:

- 'Attendee'. This stakeholder role represents the actual delegates on a training course. It may be useful for the case of professional courses to record information such as name, organisation, position, and contact details.

- 'Organiser'. This may be the training company or the client company, depending on how the training is set up. This is a very important role to consider because the possible scenarios for each will differ significantly.
- 'Tutor'. This is the actual primary trainer for the course.
- 'Demonstrator'. This is the demonstrator or the secondary trainer on the course. In some cases, the role of the tutor and the demonstrator may be taken on by a single person.
- 'Sponsor'. This is the role of someone who pays the bill at the end of the day; it may be a company or a number of individuals, depending on the nature of the course.

In terms of the requirements for the course, there are some specific requirements that must be considered:

- 'Organise course'. This requirement can vary massively, depending on who is taking on the role of the organiser, as discussed previously. One of the big differences will be dependent on whether the organiser is part of the training or the client organisation. For example, if the course is being organised by the client company, the onus on the training provider may be to simply turn up and deliver the course. If the organisation of the courses, however, is being managed by the training organisation, a number of logistical processes will then be necessary, such as arranging the event venue, refreshments and meals, and accommodation. This is a good example of where a project will vary enormously depending on the nature of the people or organisations that map onto the generic stakeholder roles from the requirements view.
- 'Teach new skills'. This is the main core requirement for any training or teaching. In the case of a professional course, this may be related directly to staff assessments, competency profiles or standards of some description.
- 'Ensure quality'. When part of a professional training organisation, the quality of the course may be driven by an external source, such as an independent or industry-driven endorsement from a recognised body. Another aspect of quality here relates to mapping the course content to recognised competency frameworks.
- 'Improve quality'. This entails capturing any problems or mistakes on the course notes, capturing and addressing any comments that are made by the attendees of the course, updating course materials, ensuring that best practice is being adhered to with regards to the course content, and so on.
- 'Meet timing constraints'. The timing constraints for a professional training course usually are concerned with making sure that the course is delivered over the duration of perhaps two or three working days. There may also be some client-specific constraints that come into play here, e.g. some organisations only allow training on particular days of the week or it may be desirable to avoid school holidays.

There are a lot of considerations to bear in mind with regards to professional training. Interestingly, depending on which of the above requirements and stakeholders apply to your organisation, the diagram itself will change. For example, new stakeholders may be introduced that represent, say, a professional body that accredits trainers.

7.2.3 Teaching as part of an undergraduate/postgraduate course

This section considers the situation where process modelling needs to be taught as part of a university or college course. The generic requirements are revisited and discussed in more detail within the context of an educational establishment.

7.2.3.1 Teaching requirements and stakeholders

The generic stakeholders remain the same as discussed previously, but the following points need to be borne in mind:

- 'Attendee'. This stakeholder role represents the actual students who are enrolled in the course.
- 'Organiser'. This is the department that offers the course.
- 'Tutor'. This is the actual lecturer for the course.
- 'Demonstrator'. This may be a lecturer or an Assistant who may supervise example classes and laboratory sessions.
- 'Sponsor'. This is someone who pays the university fees for the students.

In terms of the requirements for the course, these again remain the same, but the way in which they are realized will change. These are discussed in the following list:

- 'Organise course'. This involves ensuring that the rooms are booked and available and that any necessary resources are also available. In the case of a college or a university, however, this also includes ensuring that the information regarding the course is disseminated to students, such as the time and location of the course.
- 'Teach new skills'. In the case of a university environment, there may be a specific set of skills that is required to be taught.
- 'Ensure quality'. This involves making sure that the course maps onto any generic teaching requirement, such as Bloom's taxonomy that is often used in the United Kingdom [1,7].
- 'Improve quality'. This entails capturing any problems or mistakes on the course notes, capturing and addressing any comments that are made by the students of the course, updating course materials, ensuring that best practice is being adhered to with regards to the course content and so on. In fact, most universities have an established means of student feedback that applies to all courses.
- 'Meet timing constraints'. The timing constraints for a university course are very strict and rely on the number of teaching and access hours with students, the structure of the timetable, holidays and so on. For example, some courses may be taught in intensive 2-week modules, whereas another course may be for 1 hour per week over a 20-week duration.

7.2.3.2 A generic course structure

It is possible to identify several key elements that should be considered when defining a course structure. The structure provided here is intended as a guide only and should be used as a starting point for developing a full course and its associated resources. The structure presented here is based primarily on the experience of the author in presenting material to students in a university environment.

Figure 7.3 shows a generic structure for a university-type course and each of the main elements of this course is explained in more detail later.

Introduction

The introduction section of the course contains three main elements, as detailed below:

- *Background.* It is important to put the course into context and get the students to understand where the course has come from and why it is necessary. For example, the course may form part of or a whole module in a larger course. This may be, e.g. part of a larger systems engineering course, or an engineering or IT course.
- *Aims and objectives.* It is important that the aims of teaching the course are identified early on. A good way to think about this is to generate a use case diagram that has the aims and objectives of teaching represented as the use cases and the main stakeholders represented as the actors. An important consideration here is to identify any constraints that may come into play with regards to source standards or information. For example, it may be that the course needs to map onto the teaching objectives of Bloom, in which case the standard (realised by the actual Bloom taxonomy) would be represented as an actor and there would be an associated use case, named something along the lines of 'meet source standards' or similar. Of course, the use cases also reflect the more functional aims and objectives. In fact, there is an almost endless set of aims for a course such as this, but it is crucial to identify what they are and then to ensure that course content addresses these aims.
- *Course structure.* This section is relatively simple as it simply states the major elements of the course and the relationships between them. For using the information in this chapter, the course structure is simply the diagram in Figure 7.3 along with some explanatory notes.

It is important to use this section to make the course your own and to make sure that it meets all the stakeholder requirements.

Concepts

There are two main conceptual areas that need to be introduced here, the ontology and the framework:

- The ontology provides a clear indication of exactly what concepts and definitions will be used in the world of model-based requirements engineering. These form the basis for the rest of the course.

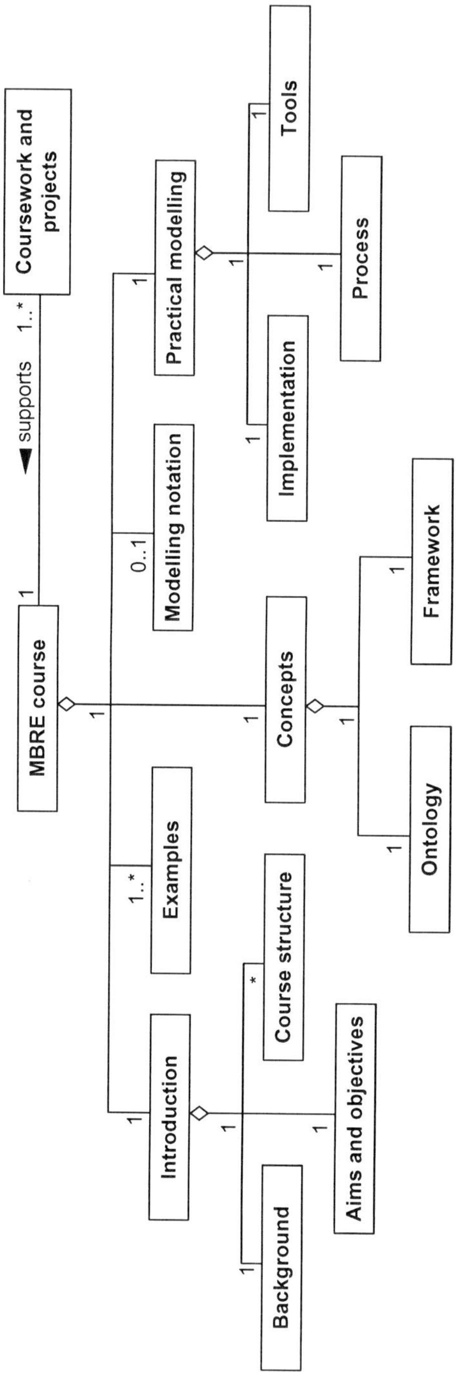

Figure 7.3 Generic course structure for a university-type course

- The framework elaborates on all the concepts introduced in the ontology. Simple examples of each view should be provided along with a clear definition of why the view is required.

These concepts relate directly to the information in Chapters 4 and 5.

Modelling notation

When introducing the modelling notation, it should be stressed that any notation deemed suitable for systems modelling may be used here. In the context of this book, the notation that is chosen is the SysML [8] for reasons that have been discussed previously. It is also worth considering that the notation chosen should not rely on any specific tool or application, and that students should be able to work out as much as possible, in the first instance, using a PAPS (pen-and-paper system) tool. This is for very pragmatic reasons. When attending a course, students will potentially be learning about a number of new ideas and concepts simultaneously. For example, they will be learning about requirements for the first time, modelling for the first time and also the SysML (in this case) for the first time. It is important to try to isolate each of these initially when communicating the information to the students and then bring them together to form a complete knowledge. If any tool is introduced too early, students will immediately dive into trying to use the tool which adds another layer of complexity and shifts the students' focus away from understanding what they should try to do.

It is suggested that the actual notation is underplayed and to concentrate on examples and emphasise on the consistency checks that are contained in the ACRE framework, rather than on the individual parts of the notation. The notation should be correct but should also be kept to a minimum.

The use of summary sheets is also highly recommended, such as the ones found in the Appendix A of this book. One approach is to have the two conceptual and realisation views on one side of a sheet of paper, with the four notation guides on the reverse. This forms an excellent quick-reference guide for all students. See Chapters 2 and 3 for a full description and the appendices for notation summary information.

Examples

Examples are best worked out as a group, rather than just providing detailed case studies. Another approach is to provide partial models and then get the students to fill in the gaps in the model. This is a good way to emphasis the consistency between the views and, if used sensibly, can be an excellent way to show how the SysML model may be navigated by asking the right questions at the right time.

The best types of examples are ones that are based on either real-life situations or situations that most people would have some knowledge of, such as films and books. An excellent example that should be considered is the 'coffin escape' application that is discussed in the book. Students should be encouraged to look how they can make the model their own, and see how small changes to any aspect of the model may results in quite large changes to the system as a whole. See Chapter 4 for the coffin escape and Chapter 6 for the main case study.

The subject of specific examples most appropriate to the course being taught is left to the readers to decide.

Practical modelling

This is a section that may or may not be the part of the course, depending on the teaching constraints. This covers the areas of process, implementation and tools and is described in detail in this chapter.

Coursework and projects

It is suggested that any course work that is given out is phrased using the terminology of the ontology. An example of a generic project description is provided in the following box.

Project description

Choose any example project, such as the development of a robot, and produce the following information, in line with the ACRE process:

- *Requirement Description Views*. These should use SysML requirement diagrams to describe a set of individual requirement.
- *Context Definition Views*. These should take the form of a SysML block definition diagram to show a taxonomy of stakeholders.
- *Requirement Context View*. These should be SysML use case diagrams that relate back to the stakeholders that are identified in the Context Definition View.
- *Validation Views*. These should be generated using SysML sequence diagrams and relate directly to the use cases from the Requirement Context Views.
- *Traceability Views*. These should show the traceability relationships that can be derived from the ACRE ontology.

Each diagram should be accompanied with a short textual description, no more than half a page.

Please note that marks will only be awarded for the information requested above. Any missing views will lose marks and any additional diagrams will not warrant extra marks. Many of the marks will be awarded for consistency of the diagrams as discussed in the lectures and shown on the summary sheet.

Also, do not choose a cash point machine (ATM), any library system (or variations thereof) or a petrol pump.

This project description should be treated purely as a guide. For example, this example asks for some of the ACRE views but not others. It is also worth putting some constraints on the solutions, such as not allowing ATMs or library systems as these are standard examples used in many, many text books. If you don't want 30 copies of a petrol pump submitted, then please consider these constraints seriously.

Due to the rigorous nature of the project description, it is possible to have an equally rigorous marking schedule. The bulk of the marks should be awarded based directly on the views asked for in the project description, which include both the views and the relationships between the views and their various elements.

In fact, if a true MBSE approach is applied to the teaching of the course, it would be possible to automate many of the mundane aspects of the marks, such as consistency checking, by using appropriate tools.

7.2.3.3 Conclusions

This chapter has provided a start point for developing teaching courses and material, no matter if it is for professional training of university-based teaching. Much of the teaching is subjective and depends on the nature of the person who is teaching, the format of the courses, the type of attendees or students and so on. The information contained in this chapter is based on many years of experience in teaching and training at many levels and is offered to promote thought rather than to be proscriptive.

7.2.4 Competence

When considering individuals for a specific role or set of roles, it is essential that they have the right knowledge, skills and attitude to perform the activities required for the role. Before their suitability can be assessed, it is important that the processes they will be involved in, in particular the activities that they will be responsible for and contributing towards, have been identified and are well-understood. Once the processes are identified, it is then possible to consider which competencies are necessary for the role.

Before progressing any further, it is worth defining and differentiating between a few terms that are used when discussing competence. These are shown graphically in Figure 7.4.

Figure 7.4 shows that a 'Competence' describes the ability of a 'Person'. A 'Person' takes on one or more 'Role'. A 'Competency Scope' describes the required ability for each 'Role', whereas the 'Competency Profile' describes the actual ability of a 'Person'.

In particular, the following observations should be noted:

- The term 'Competence' is used to refer to the total ability of an individual, rather than a single element. Therefore, each person in an organisation has an overall competence.
- The term 'Competency' refers to a single element of competence that can be measured and, hence, assessed. The totality of a person's competencies forms their competence.
- The term 'Competency Scope' refers to the defined set of competencies that are required for a specific 'Role'. This competency scope forms one of the major inputs to a competency assessment exercise.

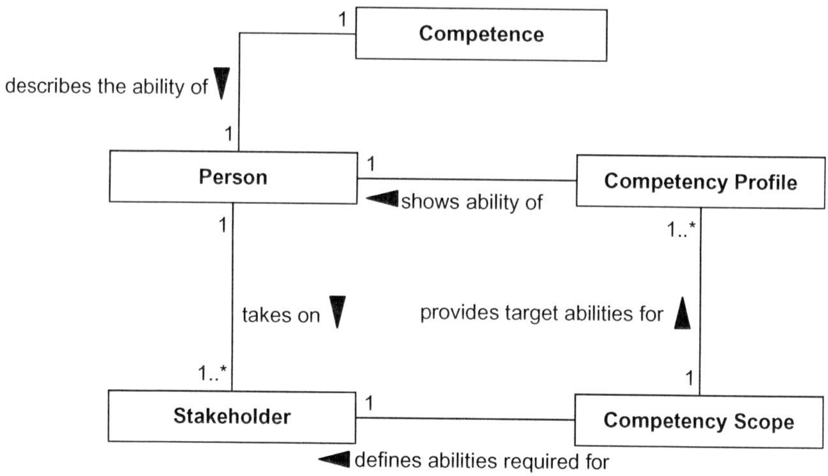

Figure 7.4 Terms relating to competence

- The term 'Competency Profile' refers to a defined set of competencies that is held by a person and that relates directly to the competencies required for a role, as defined in the 'Competency Scope'. The competency profile forms the main outcome of a competency assessment exercise.

To illustrate how competencies can be used to assess the suitability of a person, some examples of competency scopes are presented in the following sections. It should be stressed that these are shown here for illustrative purposes and, although they may suit your assessment needs exactly, in many cases they need to be tailored to meet your specific needs.

Competencies may exist at different levels. For example, it is possible to define a set of competencies that are based on an accepted industrial best practice of standard which can be used at a generic level. Competencies may also be defined at a very specific level. The next section discusses the former, generic, competencies, whilst the section following discusses the latter, more specific competencies. Each approach has its own set of advantages and disadvantages, which are presented as part of the discussion.

7.2.4.1 Generic competencies

This section looks at generic competencies that are based on industrial best practice. There are many sources available when it comes to defining competencies, most of which may be described as 'competency frameworks'. A competency framework defines a set of competencies that is usually specific to a particular industry or technology. In the example presented here, the framework chosen is a systems engineering best-practice model that is known as the 'INCOSE competencies framework' [2]. The International Council on Systems Engineering (INCOSE) is an international body that is committed to furthering the discipline of systems engineering. INCOSE has produced a competency framework that maps back to the UK-SPEC and covers various cross-cutting concepts associated with systems engineering. This has been chosen because INCOSE has defined a set of competencies that is associated with the world of systems engineering. As has been discussed earlier in this book, the whole field of requirements engineering is an important part of systems engineering, so this is a good choice for a generic framework to be used as a basis for competency assessment.

The focus of the framework is concerned with the concept of 'Systems Engineering Ability', which is described in Figure 7.5.

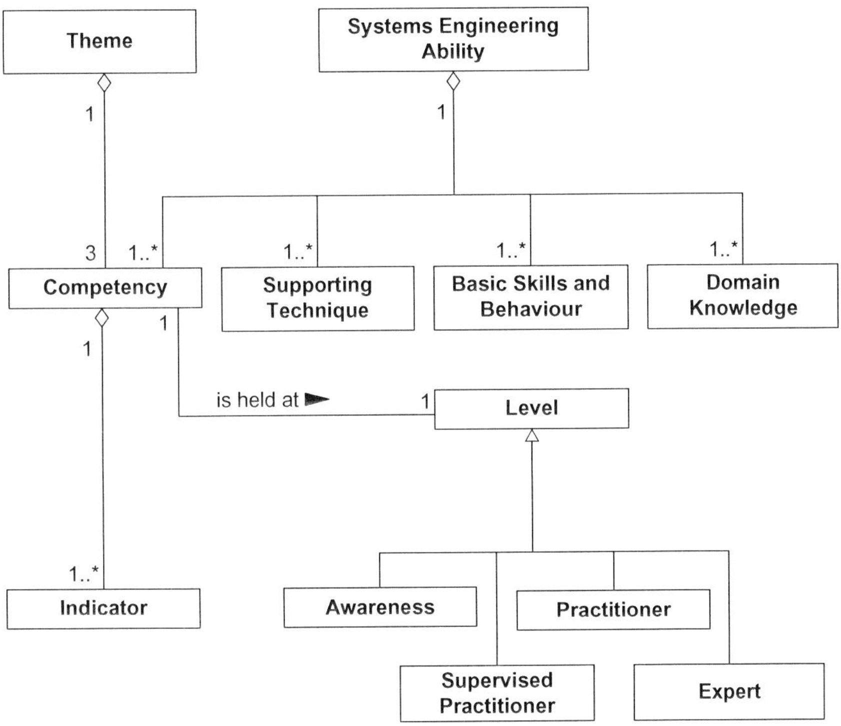

Figure 7.5 High-level view of the INCOSE competencies framework

The INCOSE competencies framework has a concept of 'Systems Engineering Ability' that may be categorised into four main areas:

- 'Supporting Technique'. A supporting technique is a specific technique that is used to support the main competencies, e.g. failure analysis, decision analysis, the use of specific notations and languages and so on. The supporting technique is very important but is not of much value by itself because it starts to add true benefits only when it is used to support and enable competencies. These supporting techniques tend to be of a more technical nature and, therefore, easier to teach and measure. Due to the sheer number of these different techniques, the INCOSE framework does not go into any detail in this area, but simply provides a checklist that one may want to refer to when considering such techniques.

- 'Basic Skills and Behaviour'. These represent the soft skills that are required in order to be a systems engineer. This includes skills such as abstract thinking and communication (verbal/non-verbal, listening, writing etc.). These softer skills tend to be less easy to teach or to measure and can often rely on the objectivity of an assessor. Again, the INCOSE framework does not enter into much detail in this area and provides only a simple list of suggested areas that may be considered.
- 'Domain Knowledge'. This knowledge is related directly to the domain in which the person is working. As systems engineering is a multi-disciplinary subject it can cover, potentially, any domain. As the scope of 'any domain' is so wide, it is not covered in any detail in this framework.
- 'Competency'. So far, the INCOSE framework has managed to sidestep all three areas covered so far, but the focus of the framework is very much on what is referred to as 'Competency' that refers to the core skills required for a systems engineer. These are discussed in more detail in the remainder of this section.

Each of the competencies identified in the previous list may be held at a particular level of competency. The INCOSE framework identifies four levels of competency:

- Level 1: 'Awareness'. The awareness level indicates that the person is able to understand basic concepts, understand how the system fits into their enterprise and ask relevant questions associated with each competency. It may be that the person has no actual experience of the competency but does display some theoretical knowledge and understanding of it.
- Level 2: 'Supervised Practitioner'. An individual who has competencies held at the supervised practitioner level has some real experience of the competency. They are able to display true understanding through the application of systems techniques and concepts as part of their work.
- Level 3: 'Practitioner'. An individual who has competencies held at the practitioner level provides guidance and leads activity in this area. They are able to supervise people at lower levels of competency and may very well lead teams or groups of people.
- Level 4: 'Expert'. The expert level represents the rare individuals who truly lead the field in a particular area. They are able to display their experience by defining best-practice policy or process within an organisation or industry.

Each of the competencies identified may be held at any of these four levels. At each of the levels and for each competency, there are a number of indicators defined, and it is these indicators that are actually assessed. Each indicator is a simple statement of what must be demonstrated to contribute towards meeting a competency. The indicator should be measurable in some accepted form.

Examples of competency scopes for two roles are now discussed to illustrate how the INCOSE competencies framework may be used pragmatically.

Requirements engineer

Figure 7.6 shows the scope that is defined for the role of 'requirements engineer' in the organisation.

The chart in Figure 7.6 shows the competency scope for the requirements engineer role. The relevant levels for each of the competencies are shown by shading the relevant cells.

There are some interesting features to this scope when the shape itself is considered. First of all, note that it is not a 'flat' shape, but has highs and lows. The highest level on this scope is 'Level 3: Practitioner', which is typical for most engineers. The areas in which the requirement for Level 3 is present are related to the role name. Anyone who is involved in requirements engineering would be expected to have a good appreciation of systems engineering generally (the 'Systems thinking' themed competencies) and would be expected to also be at the same level for requirements-related life cycle competencies. This includes 'Determining and managing stakeholder requirements', which is the obvious competency, and also two other competencies that require this high level are closely related: 'Functional analysis' and 'Modelling and simulation'.

Looking at the 'Systems engineering management' theme, there is an interesting pattern there also. Both 'Life cycle process definition' and 'Planning, monitoring and controlling' require competencies, but only at 'Level 1: Awareness'. This is quite typical because the scope is asking that the individual understands management (Level 1), but is not expecting any relevant experience in this area.

Requirements manager

Figure 7.7 shows the scope that is defined for the role of 'requirements manager' in the organisation.

The chart in Figure 7.7 shows the competency scope for the requirements manager role. The relevant levels for each of the competencies are shown by shading the relevant cells.

There are some interesting features to this scope when the shape itself is considered. First of all, note that it is not a 'flat' shape, but has highs and lows – as was seen in the scope for the requirements engineer.

The requirements manager, unlike the requirements engineer, does not need a high level of competence in many of the competencies because they will not be carrying out the activities, merely managing them. It is essential that the manager does possess 'Level 1: Awareness' in each of these competencies as it is important to understand the principles and concepts of any subject in order to manage it. In a similar way, the requirements manager requires a higher level of competence in other areas that the requirements engineer does not – clearly the 'systems engineering management' competencies are relevant here.

The two scopes show that different roles have different areas in which they need to hold higher and lower levels of competence. Of course, it is possible for a person to hold both of these roles, in which case the person would need to hold competencies that combine the two scopes.

Scope – requirements engineer

	Systems thinking		Holistic life cycle view					Systems engineering management	
	Systems concepts	**Super-system capability issues**	**Determining and managing stakeholder requirements**	**Integration and verification**	**Validation**	**Functional analysis**	**Modelling and simulation**	**Life cycle process definition**	**Planning, monitoring and controlling**
Expert	Professional qualification, publication, activity definition	Professional qualification, publication, activity definition	Professional qualification, publication, activity definition	Professional qualification, publication, activity definition	Professional qualification, publication, activity definition	Professional qualification, publication, activity definition	Professional qualification, publication, activity definition	Professional qualification, publication, activity definition	Professional qualification, publication, activity definition
Practitioner	Educational qualification, lead activity	Educational qualification, lead activity	Educational qualification, lead activity	Educational qualification, lead activity	Educational qualification, lead activity	Educational qualification, lead activity	Educational qualification, lead activity	Educational qualification, lead activity	Educational qualification, lead activity
Supervised practitioner	Formal course, activity	Formal course, activity	Formal course, activity	Formal course, activity	Formal course, activity	Formal course, activity	Formal course, activity	Formal course, activity	Formal course, activity
Awareness	Informal course, tacit knowledge	Informal course, tacit knowledge	Informal course, tacit knowledge	Informal course, tacit knowledge	Informal course, tacit knowledge	Informal course, tacit knowledge	Informal course, tacit knowledge	Informal course, tacit knowledge	Informal course, tacit knowledge

Figure 7.6 Competency scope for the 'Requirements Engineer' role

Scope – requirements engineer

	Systems concepts	Super-system capability issues	Determining and managing stakeholder requirements	Integration and verification	Validation	Functional analysis	Modelling and simulation	Life cycle process definition	Planning, monitoring and controlling
Expert	Professional qualification, publication, activity definition	Professional qualification, publication, activity definition	Professional qualification, publication, activity definition	Professional qualification, publication, activity definition	Professional qualification, publication, activity definition	Professional qualification, publication, activity definition	Professional qualification, publication, activity definition	Professional qualification, publication, activity definition	Professional qualification, publication, activity definition
Practitioner	Educational qualification, lead activity	Educational qualification, lead activity	Educational qualification, lead activity	Educational qualification, lead activity	Educational qualification, lead activity	Educational qualification, lead activity	Educational qualification, lead activity	Educational qualification, lead activity	Educational qualification, lead activity
Supervised practitioner	Formal course, activity	Formal course, activity	Formal course activity	Formal course, activity	Formal course, activity	Formal course, activity	Formal course, activity	Formal course, activity	Formal course, activity
Awareness	Informal course, tacit knowledge	Informal course, tacit knowledge	Informal course, tacit knowledge	Informal course, tacit knowledge	Informal course, tacit knowledge	Informal course, tacit knowledge	Informal course, tacit knowledge	Informal course, tacit knowledge	Informal course, tacit knowledge
	Systems thinking		**Holistic life cycle view**					**Systems engineering management**	

Figure 7.7 Example competency scope for the 'Requirements Manager' Role

7.2.4.2 Bespoke competencies

This section also looks at some examples of competencies, but this time bespoke competencies have been generated based on the requirements of the ACRE process. The structure of the competencies is similar to the INCOSE framework that has been discussed in the previous section but with some subtle yet important differences, that will be discussed once the competencies have been introduced.

Table 7.1 shows a set of three competencies that should be considered when assessing competency against the ACRE process. This table is included here for

Table 7.1 Bespoke competencies for the ACRE process

Competency	Description	Indicators
Modelling	This competency relates to generic modelling	• Understands the need for modelling • Can provide an appropriate definition of modelling • Understands the concept of abstraction • Understands the concept of connection to reality • Understands the concept of different approaches to modelling • Understands the concept of consistency • Can define what a view is • Understands the difference between modelling and drawing pictures • Understands structural modelling • Understands behavioural modelling
SysML Modelling	This competency reflects the ability to use the Systems modelling language	• Must hold the 'Modelling' competency • Is familiar with the background to SysML • Is familiar with the ownership of SysML • Can name the five structural diagrams • Can name the four behavioural diagrams • Understands the use of each of the structural diagrams • Understands the use of each of the behavioural diagrams • Understands the relationships between the diagrams • Understands the language extension mechanisms
ACRE Process	This competency relates specifically to the ACRE process	• Must hold the 'Modelling' competency • Must hold the 'SysML Modelling' competency • Understands the concepts and terms used in the ACRE ontology • Understands the purpose of each view in the ACRE framework • Understands the relationship between the framework and the ontology • Is able to identify and define each of the views • Is able to define the relationships between the views • Is aware of the level, or levels, of rigour required • Is aware of the level, or levels, of scalability required • Understands how the use of tools applies to the process

completeness and may be either used as is or considered food for thought for anyone considering generating their own competencies.

Table 7.1 shows three competencies that are required in order to be able to use the ACRE process properly. Like the INCOSE framework that has been used in the previous section, these competencies are also held at one of four levels. However, the way these competency levels are defined is somewhat different. Although the INCOSE framework introduces new indicators at each level, the bespoke competencies here only specify unique indicators at 'Level 1: Awareness' and then use the same set of generic indicators for each of the other levels. This is more in keeping with common approaches to capability determinations (such as ISO 15504 and CMMI). For a full explanation of this approach to defining competencies, see Reference 3.

The four levels used in the bespoke approach are described below.

Level 1: Awareness

This level identifies a number of indicators that are required to be understood and that are used as the basis for assessment of all the other levels. These indicators are specific to the competency at Level 1, whereas at higher levels, the indicators are the same across *all* competencies.

The main aim of this level is for the assessee to demonstrate that they possess the ability to '*speak knowledgeably about a particular aspect of the competency. The main aim is for the assessee to demonstrate that they understand each indicator fully, and back this up with examples – either theoretical or real-life*'. In terms of the topics that the assessee must share their knowledge about, this is defined by the key indicators shown in the Table 7.1.

For example, to obtain 'Level 1: Awareness' in the 'Modelling' competencies, the assessee must demonstrate that they understand the need for modelling (first indicator), can provide an appropriate definition of modelling (second indicator) and so on (all the indicators).

By understanding all the concepts discussed in this book, Level 1 can be easily achieved.

Level 2: Support

The indicators defined at 'Level 2' are the same across all of the competencies, whereas the ones at 'Level 1' are different for each competency.

The main goal of this level is for the assessee to demonstrate that they can '*reflect the ability to implement the concepts that have been discussed at Level 1 for this competency*'. In this example, the indicators that are defined *apply* to the indicators that have been *identified* in Level 1. These indicators are defined as follows:

- *Has achieved Level 1, 'Awareness', for this competency.* Therefore, for the 'Modelling' competency, the assessee must have met the criteria – they are able to demonstrate their knowledge – for the indicators specified.
- *Has implemented the concepts discussed at Level 1.* Therefore, for the 'Modelling' competency, the assessee must have actually worked on a project

where they have been able to understand the need for modelling (first indicator), provide an appropriate definition of modelling (second indicator) and so on (all the indicators).

- *Has been trained in some way*. This is usually by a course or in some cases by on-the-job experience, in the areas described by the indicators at Level 1. Therefore, for the 'Modelling' competency, the assessee must have actually been trained to understand the need for modelling (first indicator), provide an appropriate definition of modelling (second indicator) and so on (all the indicators).
- *Has supported other people in the implementation of work activities that use the indicators in Level 1*. Therefore, for the 'Modelling' competency, the assessee must have supported people on a project where they have been able to understand the need for modelling (first indicator), provide an appropriate definition of modelling (second indicator) and so on (all the indicators). Examples of this on a real project include contributing to creation and generation of artefacts, participation in workshops etc.
- *Has created artefacts related to the competency as characterised by the indicators for Level 1*. Therefore, for the 'Modelling' competency, the assessee must have produced or contributed to the production of artefacts to demonstrate that they can understand the need for modelling (first indicator), provide an appropriate definition of modelling (second indicator) and so on (all the indicators).
- *Has controlled artefacts (e.g. applied version control) related to the competency as characterised by the indicators for Level 1*. Therefore, for the 'Modelling' competency, the assessee must have applied version control to artefacts that they have produced to demonstrate that they can understand the need for modelling (first indicator), provide an appropriate definition of modelling (second indicator) and so on (all the indicators).
- *Has had artefacts reviewed and has been able to address any issues that have arisen as a result of the review*. Therefore, for the 'Modelling' competency, the assessee must have actually worked and had their artefacts reviewed by others to demonstrate that they understand the need for modelling (first indicator), provide an appropriate definition of modelling (second indicator) and so on (all the indicators).
- *Can identify best practice in the competencies, such as standards, books, and methodologies*. Therefore, for the 'Modelling' competency, the assessee must be able to reference best practice, techniques, approaches, standards etc. to demonstrate that they can understand the need for modelling (first indicator), provide an appropriate definition of modelling (second indicator) and so on (all the indicators).

This list of Level 2 indicators is then applied to all competencies in the competency assessment in exactly the same way. Therefore, when assessing against Level 2 using this approach, there are never 'new' indicators that are introduced, just the same ones applied to all the competencies.

Level 3: Lead

The indicators defined at 'Level 3' are the same across all of the competencies, whereas that at 'Level 1' are different for each competency.

The aim of this level is for the assessee to demonstrate that they can *'reflect the ability to be able to lead the activity that was described at Level 1 and implemented at Level 2'*.

The company is heavily involved in consultancy work and it wanted to ensure that all of its consultants were suitably qualified and experienced; therefore, 'level 3 – lead' is the minimum requirement for a member of staff to be able to be considered as a consultant for the related competency.

In the same way that generic indicators were defined at Level 2 that applies to all competencies, the same is done for Level 3. These indicators are described as follows:

- *Has achieved Level 2: Support*. The assessee must have achieved Level 2 and, therefore, Level 1 (which was one of the Level 2 indicators).
- *Has led activity at a project level*. Therefore, for the 'Modelling' competency, the assessee must be able to demonstrate that they have led a team or group of people where they have been able to understand the need for modelling (first indicator), provide an appropriate definition of modelling (second indicator) and so on (all the indicators). Typically, the group that was led by the assessee would be made up of primarily Level 2 people in the relevant competencies.
- *Has supervised Level 2 activity*. Therefore, for the 'Modelling' competency, the assessee must have supervised people who are at Level 2, where they can understand the need for modelling (first indicator), provide an appropriate definition of modelling (second indicator) and so on (all the indicators). This supervision may be management supervision in the same group or may also include mentoring of Level 2 people, perhaps from other groups in the business.
- *Has managed Level 2 activity (version control, release, setting work, assessing review responses etc.)*. Therefore, for the 'Modelling' competency, the assessee must be able to demonstrate that they have been involved with assessing work, setting work etc. where they can understand the need for modelling (first indicator), provide an appropriate definition of modelling (second indicator) and so on (all the indicators). Again, note how the level of responsibility is increasing – at Level 2, the assessee was required to have their work set and managed; at Level 3, the assessee sits on the other side of the table and performs the setting of the work.
- *Has formally reviewed artefacts*. Therefore, for the 'Modelling' competency, the assessee must be able to demonstrate that they have reviewed artefacts on real projects where they understand the need for modelling (first indicator), provide an appropriate definition of modelling (second indicator) and so on (all the indicators). Yet again, the Level 3 assessee is now sitting across the table from the Level 2 assessee and is performing the reviews.

- *Has experience facing clients.* Therefore, for the 'Modelling' competency, the assessee must be able to demonstrate that they can represent the organisation where they can understand the need for modelling (first indicator), provide an appropriate definition of modelling (second indicator) and so on (all the indicators). This is the first indicator that reflects an outgoing image to the outside world, where the company's reputation may be at stake.
- *Has some formal affiliation to a professional body, such as associate or full membership.* Therefore, for the 'Modelling' competency, the assessee must be able to demonstrate that they have found the relevant professional body that relates to modelling and that shows that they can understand the need for modelling (first indicator), provide an appropriate definition of modelling (second indicator) and so on (all the indicators).

Again, these Level 3 indicators, similar to the Level 2 indicators, are applied to all the competencies in the competency scope. Therefore, when assessing against Level 3 using this approach, there are never 'new' indicators that are introduced, just the same ones applied to all the competencies.

Level 4: Expert

The indicators defined at 'Level 4' are the same across all of the competencies, whereas the ones at 'Level 1' are different for each competency.

The aim of Level 4 is for the assessee to demonstrate that they can *'reflect the ability to be a true, recognised expert in the field that is described by this competency'.* In terms of a small company, such as the one in this case study, this will mean having an established reputation in the general field. This is for two reasons, the first would be that as the company is small, to be an expert in the company does not necessarily indicate general expertise in the field. If the company was larger and had, say 1000 or more employees, to be the company guru may qualify them for expert status. The second reason is that, due to the nature of the main work of the company, being a recognised expert in the field is beneficial to the organisation.

The indicators for Level 3 are as follows:

- *Has achieved Level 3: Lead.* This is similar to the criteria of both Level 2 and Level 3, each of which requires attainment of the previous level for qualification.
- *Holds formal chartered status from a recognised professional body.* Therefore, for the 'Modelling' competency, the assessee must be able to demonstrate that they have found the relevant professional body that relates to modelling and shows that they can understand the need for modelling (first indicator), provide an appropriate definition of modelling (second indicator) and so on (all the indicators). The assessee must hold the chartered status qualification or equivalent.
- *Has published in the field.* This includes books, first or second author on papers and first author on published public presentations. Therefore, for the 'Modelling' competency, the assessee must be able to demonstrate that they have published work that shows that they can understand the need

for modelling (first indicator), provide an appropriate definition of modelling (second indicator) and so on (all the indicators). Due to the size of the organisation, this will again mean publications in the public domain, whereas for a large organisation, internal publications may or may not be considered.

- *Has external recognition*. This includes speaking at public events, invited presentations, awards, panels etc. This is similar to the previous indicator, but this time it relates to oral communication rather than written. Therefore, for the 'Modelling' competency, the assessee must be able to demonstrate that they have presented papers, spoken at events etc. that show that they can understand the need for modelling (first indicator), provide an appropriate definition of modelling (second indicator) and so on (all the indicators).
- *Has led activity at the strategic or programme level*. Therefore, for the 'Modelling' competency, the assessee must be able to demonstrate that they defined process, policy etc. that relate to modelling and that show that they can understand the need for modelling (first indicator), provide an appropriate definition of modelling (second indicator) and so on (all the indicators). At the expert level, the assessee must be seen to be driving the relevant disciplines forward both within and without the organisation.
- *Has mentored Level 2 and Level 3 staff*. Therefore, for the 'Modelling' competency, the assessee must be able to demonstrate that they have mentored staff in relation to modelling and that shows that they can understand the need for modelling (first indicator), provide an appropriate definition of modelling (second indicator) and so on (all the indicators). Note again that the level of responsibility is increasing all the time and that the mentoring for this indicator applies to the lower levels 1, 2 and 3.
- *Has contributed to best practice*. This includes development of recognised methods, methodologies, tools etc. Therefore, for the 'Modelling' competency, the assessee must be able to demonstrate that they have contributed to the knowledge pool and shows that they can understand the need for modelling (first indicator), provide an appropriate definition of modelling (second indicator) and so on (all the indicators).
- *Is currently active in recognised professional bodies*. Therefore, for the 'Modelling' competency, the assessee must be able to demonstrate that they have found the relevant professional body that relates to modelling and that they are actively involved in activities that show that they can understand the need for modelling (first indicator), provide an appropriate definition of modelling (second indicator) and so on (all the indicators).

These indicators apply to all the indicators identified at Level 1. Therefore, when assessing against Level 4 using this approach, there are never 'new' indicators that are introduced, just the same ones applied to all the competencies.

Examples of competency scopes using these bespoke competencies are defined and discussed later in this chapter, when the stakeholder roles for the ACRE process are discussed.

7.2.4.3 Generic versus specific competencies

Each of the two approaches that was described in the previous sections has its own set of advantages and disadvantages. One of the main advantages of using the generic framework is acceptance. One of the main reasons to use an industry best-practice model or standard is that it is recognised in the industry and, in the case of some, at an international level. Of course, this can be very advantageous as it provides a common way that people's competence can be assessed and recognised in more than one organisation. This has a lot of attraction from an individual's point of view because it provides a mechanism to demonstrate one's own competence that may be used in a number of organisations and, hence, it will make moving between organisations simpler. From a company's point of view, this is also good because it makes the whole area of recruitment far simpler. When recruitment is based (either wholly or partly) on competency assessments, it is possible to define a set of competency scopes that represent the roles that make up the post and issue them to potential employees.

The downside of using the generic approach is that, because the competencies are defined at a high level, the definitions may neither map onto the way that an organisation does their business, nor will they reflect any specific techniques, methodologies, or tools that may be required by the company. This is one of the main strengths of the specific approach. When defining specific competencies, it is possible to get an exact match on the skills required by the organisation.

Therefore, in summary:

- The generic approach is good for establishing a common base for competencies that are recognised by different organisations.
- The specific approach is weak in terms of recognition, as it is tailored to a particular organisation.
- The generic approach is weak when it comes to specific tools, techniques and methodologies, as it is, by its nature, aimed at the high level.
- The specific approach is strong in meeting the exact requirements of an organisation.

The use of competencies is very important when it comes to getting the right people for a role. For a far more detailed discussion, see Reference 3.

7.2.5 Summary of 'people'

One of the common themes of this book is that 'people, process and tools' are essential for realising any systems engineering capability. This section looks at how the 'people' element of this may be addressed by discussing, in particular, the following:

- *The importance of teaching, training education.* This was achieved by discussing the requirements for education and providing a high-level teaching guide for anyone involved with model-based requirements engineering education.
- *The importance of competence and competency assessment.* This was achieved by discussing why competency assessment is so important when deciding the

suitability of a person for a role or set of roles. Two approaches to defining competencies were discussed – one generic and one specific – and an example set of competency scopes provided that may be used as a start point for competency assessment.

The issue of 'people' is just one of the three main enablers, and 'process' and tools' are discussed in the next two sections.

7.3 The 'process'

The backbone of any approach is the process itself. This section provides an overview of a simple, yet effective process for model-based requirements engineering, known as the ACRE process. The emphasis here is on simply describing the process at a high level.

Any process that is defined needs to be flexible in a number of ways:

- The process needs to be able to be realised using a number of different techniques. A good process should be independent of any specific techniques. Of course, when considering model-based requirements engineering, there is an immediate constraint that modelling techniques must be used, but the process should be flexible enough to allow the use of different modelling notations.
- The process must be able to be realised using a number of different tools. The process should drive the tools and not the other way around. With this in mind, it is essential that the process may be implemented using any number of tools.
- The process must be scalable in terms of size and duration of project. The process must be able to be applied to very small project lasting only a few days or weeks, right up to long-term projects lasting many years.
- The process must be able to be applied at different levels of rigour. For example, it must be able to be applied to a system where there are no safety concerns right up to a safety-critical or mission-critical system.

This chapter describes how the ACRE process can be implemented for real-life projects and be used in a flexible way to meet the requirements already described. This involves looking at the following three areas in more detail: process definition, process use and process deployment.

7.3.1 Defining the process

The generic process for the approach in this book has been defined by modelling the process. The approach taken for this modelling is an industrial best-practice model, the so-called 'seven views' approach. This approach uses the same modelling notation that is used in this book, SysML, but applies it to process modelling. In fact, the 'seven views' approach may be realised using any number of notations, such as UML, SysML, BPMN, and flowcharts, but it is the SysML implementation that is being used here. For more information on the 'seven views' approach, see Appendix C and Reference 4.

7.3.1.1 The ACRE process: the Requirement View

The requirements view for the process describes why the process is needed and how the key stakeholders benefit from the process. In this example, a SysML use case diagram is used to describe the requirements from the point of view of an organisation.

Of course, there is ample room for confusion here, but just to make it clear, this section is looking at why we need a requirements engineering process in the first place. At the risk of appearing a little chicken and egg, this is achieved by applying the requirements modelling techniques that form the basis of this book.

Figure 7.8 shows the use cases for the ACRE process that are defined from the point of view of an organisation (the 'Organisation' context). Hence, the main use case here is to 'Define official opinion for requirements engineering'.

There are three main constraints on this high-level use case, which are as follows:

- 'Must be realisable by existing tools'. This constraint reflects the requirement that the approach must be able to be used by using the tools that are already owned by the organisation. From the point of view of the ACRE, this means that the process must be able to be realised using anything from a PAPS right up to the most sophisticated CASE tools. This is discussed in detail later in this chapter.
- 'Must be realisable using other tools'. This constraint reflects the fact that the tools being used by the organisation changes as time goes on. This may be because the maturity of the modelling used by the organisation increases, because customers use a specific tool, because a tool may no longer be fit for purpose and so on. There are many reasons why the toolset may change and, in real life, this is almost a certainty.
- 'Must be model based'. This constraint reflects the whole philosophy of this book, that the whole approach must be model based. If you find that you are disagreeing with this constraint, please close the book, put it down and try to re-gift it to someone more worthy.

There are three main lower level use cases that make up the highest level:

- 'Define the process'. This is what this section is doing for ACRE. This has three main constraints in that it 'Must be scalable' both in size and in rigour, it must 'Meet best practice' and must 'Comply with standards'. All of these points have been discussed in the previous section.
- 'Demonstrate process'. This is an essential requirement as you want to obtain buy-in to the process once it has been defined. It is always useful to have case studies and examples of how the process may be implemented.
- 'Disseminate process'. People need to be told about the process, whether they are internal to the company as part of the organisation's capability or they are clients of the company, to inspire confidence on the approach to requirements engineering.

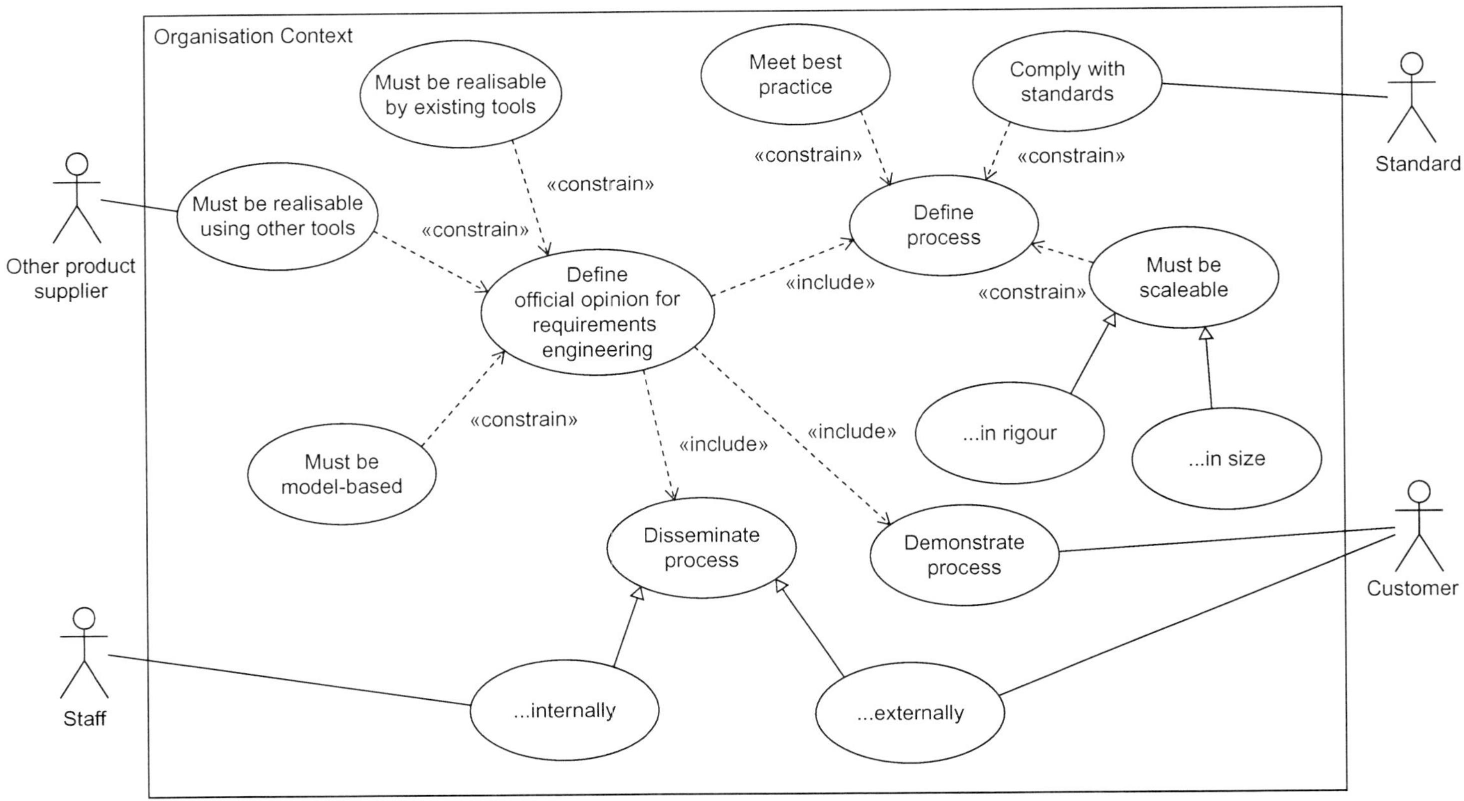

Figure 7.8 Requirements view for the ACRE process

The key stakeholders that have been identified at this point are the following:

- 'Staff'. This stakeholder represents the role of all the employees who work for the organisation.
- 'Supplier'. This stakeholder represents the role of the suppliers to the organisation. Examples of these include tool providers and service providers.
- 'Standard'. This stakeholder represents the role of any standard information provider. Examples of these include standards, best-practice models, books, papers and industry-specific processes.
- 'Customer'. This stakeholder represents any customer or client of the organisation who may be exposed to the process in any way.

The context here represents the 'Organisation', hence there is no organisational stakeholder shown on the diagram.

7.3.1.2 The ACRE process: the Process Structure View

The 'seven views' approach to process modelling that is used here calls for the generation of a process structure view (PSV) that defines the key concepts and terminology that is to be used in the process model. This PSV is a sobriquet for the ontology, so most of this has already been discussed so far in this book. The basic ontology that has been developed so far is shown in Figure 7.9.

Figure 7.9 shows the ACRE ontology that has already been discussed in great detail in this book; therefore, no further description is provided at this point.

The only part of the ontology that has not been defined yet is the area that is concerned with the definition of the concepts and terminology associated with the process itself, which is shown in Figure 7.10.

Figure 7.10 shows the main concepts and terms that are associated with the ACRE process. These concepts are discussed in three main areas: the life cycle, the process and the competency.

Any process must be executed within the context of a 'Life Cycle'. The type of life cycle depends on the type of project and there are many different types, such as acquisition life cycles, development life cycles, product life cycles, and capability life cycles. The ACRE process that is described here may be used in any of these life cycles. Figure 7.10 shows that a 'Life Cycle' is made up of one or more 'Stage'. A 'Stage' is defined as a discrete period of activity during the overall life cycle, where a number of processes are executed. Each 'Stage' will have a 'Gate' where the outputs of the stage are assessed and a decision is made to decide whether to move on to the next stage, continue with this stage, go back a stage, pause the project or to cancel the project.

The relationship between the 'Process' and the 'Stage' is not as straightforward as simply executing processes directly within a stage, and they are usually grouped into one or more 'Process Execution Group'. This term is often referred to as an 'iteration' but, experience has shown, the use of this term is massively misleading to most people, hence the use of the term 'Process Execution Group'.

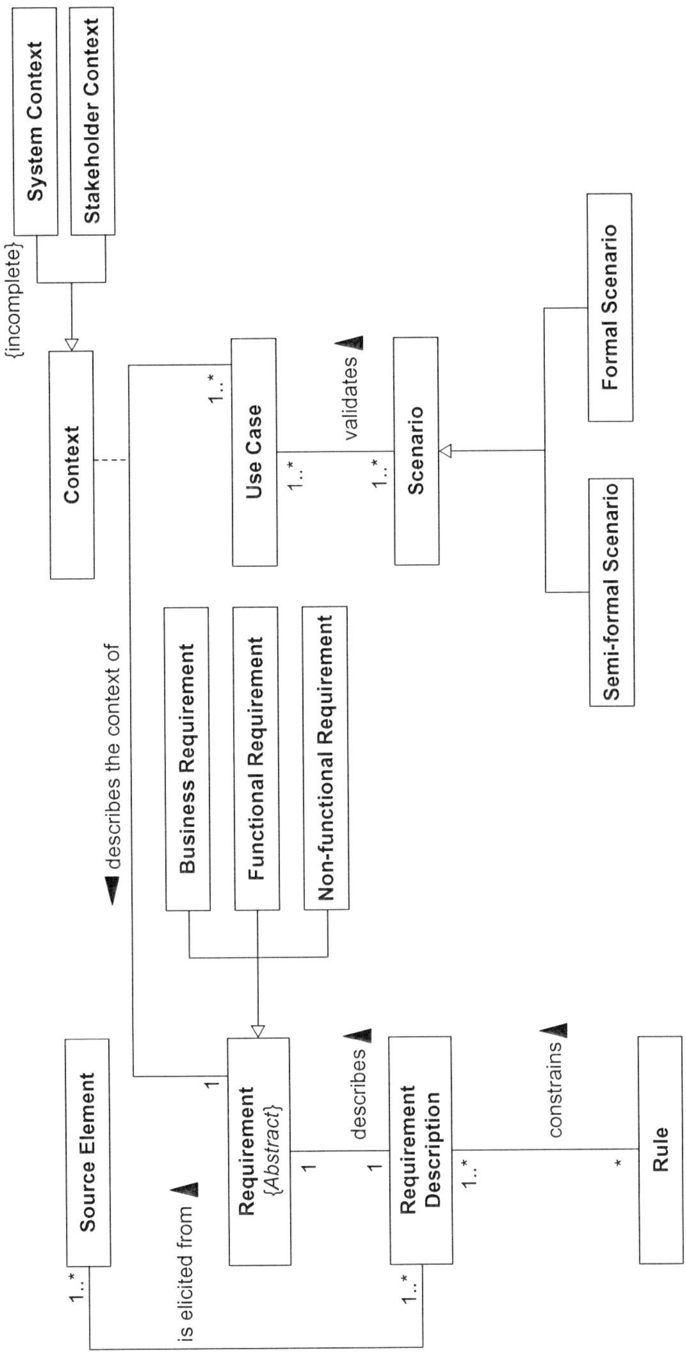

Figure 7.9 Ontology masquerading as a process structure view for ACRE

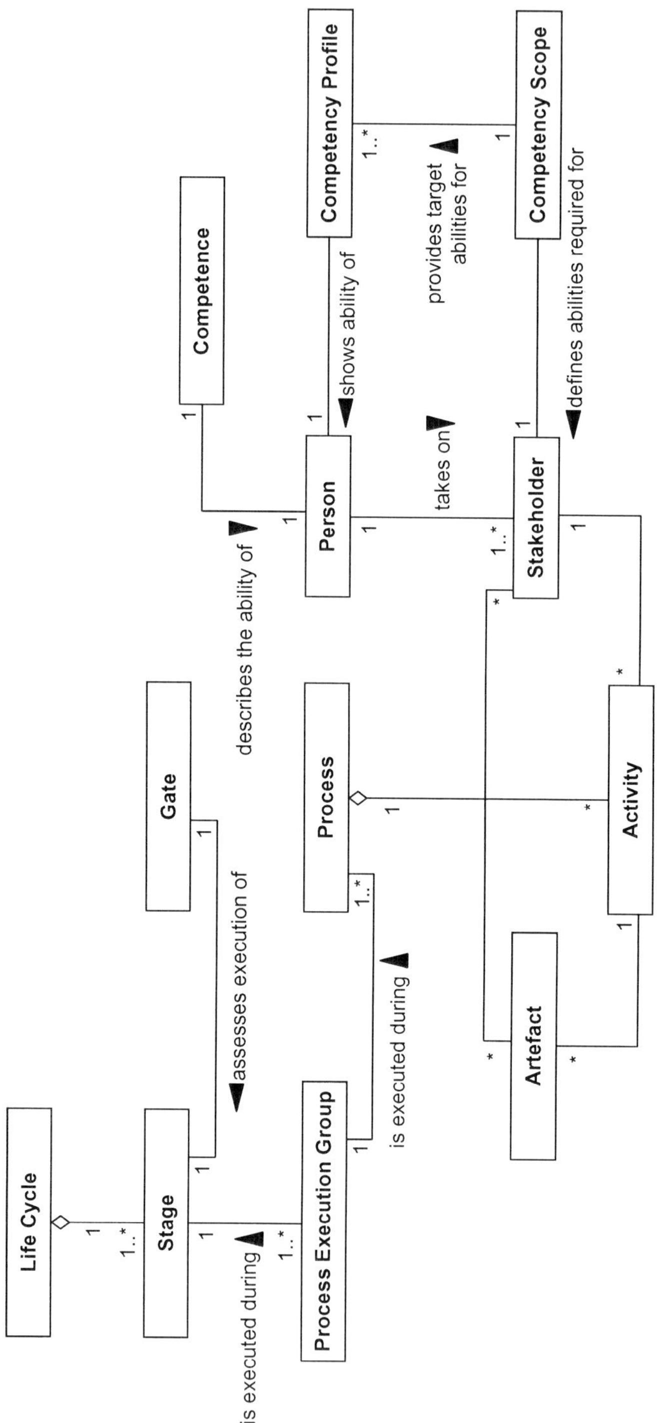

Figure 7.10 Ontology showing process concepts and terminology used for the ACRE process

Each process is made up of the following three main types of element:

- *One or more 'Activity' that describes something that is done during the execution of a single process.* This term is sometimes referred to under different terms, such as task, step, operation, and action. An 'Activity' produces and consumes one or more 'Artefact'.
- *One or more 'Artefact' that describes an input or output to an activity.* Examples of artefacts include models, model elements, documents, specifications, system elements, systems, services, material and so on.
- *One or more 'Stakeholder' that describes the role of a person, organisation or thing that is responsible for one or more 'Activity'.* The 'Stakeholder' also provides the basis for defining competencies that relate to the process activities.

Each 'Person' in the organisation takes on one or more 'Stakeholder', and it is these roles, rather than the people themselves, that have competencies defined for them. These competencies are expressed in several ways:

- The 'Competence' describes the overall ability of a 'Person'. The 'Competence' of a person may be assessed by looking at a set of individual competencies.
- The 'Competency Scope' defines a set of competencies that is *required* by the 'Stakeholder' and that form one of the main inputs to the competency assessment exercise.
- The 'Competency Profile' defines the set of competencies that is *held* by the person and forms the main output of a competency assessment exercise.

The world of competency assessment can be quite confusing because the terminology adopted contains many similar terms, so it is important that they can be clearly differentiated.

This PSV (ontology) defines the concepts and terminology that is used throughout this book whenever processes are referred to.

7.3.1.3 The ACRE process: the Process Content View

The process content view (PCV) defines the processes that are available in an organisation and may be thought of as a 'process library'. In the case of the ACRE, only a single process is described here, but this is intended to be a start point for people to use when defining their own, bespoke processes for model-based requirements engineering.

Figure 7.11 shows a single process represented as a single SysML block, which is a rather neat way to show process graphically. In the description of the PSV, we defined a process as being made up of three elements: artefacts, activities and stakeholder roles. This view allows the artefacts and activities to be shown for a single process in the following way:

- An 'Activity' describes something that has to be done in the process. In SysML modelling terms, if we represent the process as a block, then the activity maps directly onto the SysML concept of an 'operation'. Therefore, we can show all the activities for the process as operations on the block.

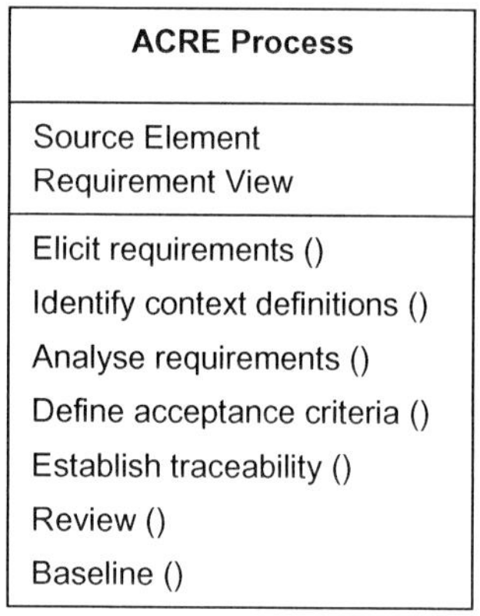

Figure 7.11 Process content view for the ACRE process

- An 'Artefact' describes a feature of the process. In SysML modelling terms, if we represent the process as a block, then the artefact maps directly onto the SysML concept of a 'property'. Therefore, we can show all the artefacts for the process as properties on the block.

The artefacts in the process must all be consistent with the ACRE framework. In this case, the artefacts have been simplified and are defined as follows:

- 'Source Element'. This is taken directly from the ontology and represents anything that is used as a source for the requirements engineering exercise.
- 'Requirement View'. This represents the set of all the views that are used as part of ACRE, which are the Source Element View, the Requirement Description View, the Definition Rule Set View, the Requirement Context View, the Context Definition View and the Validation View. For a full description of each view, see Chapter 5.

The activities in the process describe what actually needs to be done to execute the ACRE process. These activities are described as follows:

- 'Elicit requirements'. The main aim of this activity is to identify, gather and define the Source Elements that are used as a basis for the requirements. In terms of the framework, this activity consumes the Source Element View (containing Source Elements) as an input and produces the Requirement Description View (containing Requirement descriptions).
- 'Identify context definitions'. The main aim of this activity is to identify and define the stakeholders and system elements that are used as a basis for the Context Definition Views. In terms of the framework, this activity consumes the Source Element View (containing Source Elements) as an input and produces the Context Definition Views (the Stakeholder Context Definition View and/or the System Context Definition View).
- 'Analyse requirements'. The main aim of this activity is to understand the requirements of the system by looking at the use cases for them. Note that a use

case is a requirement in context, and this activity is mainly concerned with generating a number of contexts, based on the context definitions in order to understand the requirements. In terms of the framework, this activity consumes the Context Definition Views (the Stakeholder Context Definition View and/or the System Context Definition View) and the Requirement Description View as inputs and produces the Requirement Context Views (containing stakeholder and use cases along with all their inter-relationships).

- 'Define acceptance criteria'. The main aim of this activity is to consider how each use case is validated. In terms of the framework, this activity consumes the Requirement Context Views as an input and produces the Validation Views (a combination of Stakeholder Scenario Views, System Scenario Views and Constraint Validation Views).
- 'Establish traceability'. The main aim of this activity is to ensure that traceability between all the views has been defined. In terms of the framework, this activity consumes, potentially, all of the view and produces the Traceability Views.
- 'Review'. The main aim of this activity is to assess, consider and provide an indication of how fit for purpose the process artefacts are. In terms of the framework, this activity consumes all of the views that have been generated and produces commented forms of the views.
- 'Baseline'. The main aim of this activity is to apply configuration management and version control to the model. In terms of the framework, this activity consumes all of the process artefacts and produces a configured baseline of the model.

The way these activities are executed depends on the way the process is being used, according to the type of project it is being applied to. The basic activities remain the same, but the techniques and tools that are used to realise each activity may differ enormously – this is discussed later in this chapter along with the consideration of the use of the process and the use of tools.

7.3.1.4 The ACRE process: the stakeholder view

The stakeholder view identifies all the stakeholder roles that are needed to implement the ACRE process. These stakeholders are shown in Figure 7.12 for the ACRE process.

Figure 7.12 shows that there are three types of 'Stakeholder':

- 'Customer' that describes all roles that may make use of the system or service or that the supplier needs to keep satisfied. In many cases, the supplier may have some influence over the customer and be in a position to discuss issues and agree on compromises with them.
- 'External' that describes the roles that must be satisfied by the supplier, but that cannot be negotiated with in any way. In other words, the supplier has no influence on external stakeholder roles.
- 'Supplier' that describe all the roles that are involved with the development and delivery of the product or service.

Each of the stakeholder roles has one or more names associated with it that describe who or what realises this role.

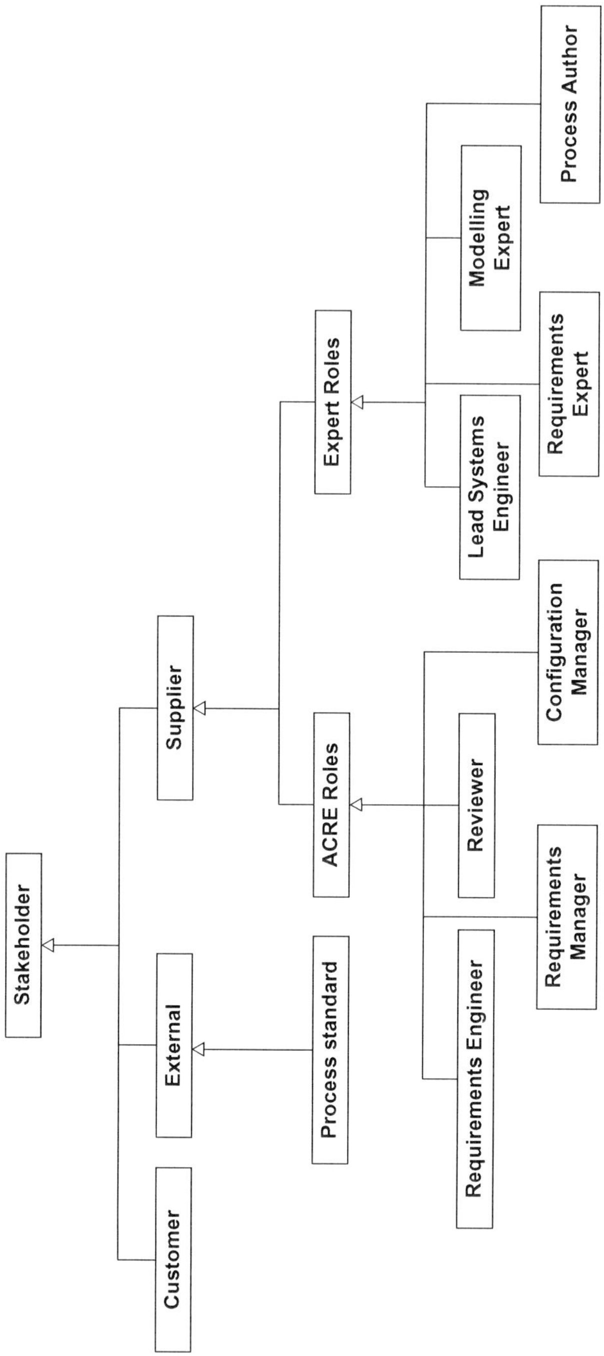

Figure 7.12 Stakeholders for the ACRE process

The stakeholder roles that are specific to the ACRE process are described below, along with some additional roles that may be required when setting up the process. For two of the roles, a competency scope is presented so that it can be compared with the generic competency scopes described previously in this chapter.

Requirements Engineer
'Requirements Engineer' is responsible for much of the work that goes on in the ACRE process, including the 'Elicit requirements', 'Identify stakeholders', 'Analyse requirements' and 'Define acceptance criteria' activities.

An example competency scope for the 'Requirements Engineer' is shown later, based on the bespoke competencies described previously in this chapter (Figure 7.13).

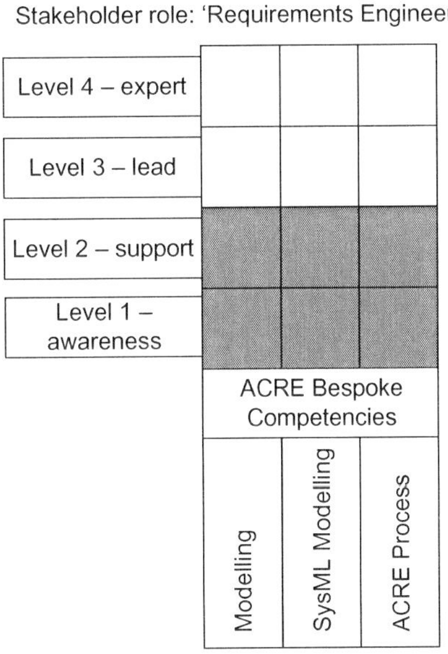

Figure 7.13 Example competency scope for the 'Requirements Engineer' stakeholder role using bespoke competencies

Requirements Engineer needs a strong, practical background in all three of the bespoke competencies. In an ideal world, these should all be held at 'Level 2 – lead', as shown here. However, if there is a team of requirements engineers, it may be possible to have one role held at 'Level 3 – lead' supported by a number of people who hold 'Level 2 – support'.

This scope may be compared with the one described in Figure 7.6, which shows the generic competencies based on the INCOSE competencies framework.

Requirements Manager
'Requirements Manager' is responsible for the management of the work that goes on in the ACRE process. Despite the fact that the Requirements Manager is only

responsible for the 'Establish traceability' activity in the ACRE process, it should be noted that the scope shown here is only the bespoke competencies and not the management-specific competencies.

An example competency scope for the 'Requirements Manager' is shown in Figure 7.14, based on the bespoke competencies described previously in this chapter.

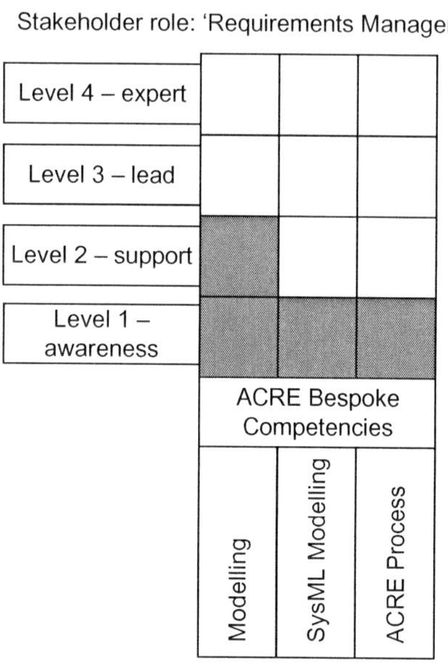

Figure 7.14 Example competency scope for the 'Requirements Manager' stakeholder role using bespoke competencies

Requirements Manager does not need to hold these bespoke competencies at a high level, but does need to be aware of all the concepts that are being used. Therefore, both the 'SysML Modelling' and 'ACRE Process' competencies are required to be held only at 'Level 1 – awareness' whereas the 'Modelling' competency is seen to be fundamental to everything that is being carried out in the process; therefore, it is held at 'Level 2 – support'.

It should be remembered that there are other competencies that are required by the requirements managers as discussed previously in this chapter.

Other ACRE stakeholder roles

There are a number of other stakeholder roles that are required by the ACRE process, which are as follows:

- 'Reviewer'. The Reviewer is responsible for the 'Review' activity and, as such, requires a high level of competence in all three of the bespoke competencies. It is suggested that, as a general rule, the level of competence of any reviewer in any competency must be *at least* as high as the people carrying out the work to be reviewed. Therefore, all three competencies here should be held at either 'Level 3 – support' or even 'Level 4 – expert'.

- 'Configuration Manager'. The Configuration Manager is responsible for all the configuration management and version control activities in the process. Although the three bespoke competencies do not immediately scream out as being directly relevant, the whole subject of configuration management is essential to any successful modelling activity. Therefore, the Configuration Manager should have a good understanding of the Modelling competency which should be held at 'Level 1 – awareness' or, even better, 'Level 2 – support'.

This completes the set of roles that have been identified as being used directly in the ACRE process, but there are some other roles that may be required to enable this process.

Other important roles

There are a number of other roles that are necessary when setting up a process, which are as follows:

- 'Process Author'. The Process Author is a person who models and then describes the process (in this case, the requirements process) and as such must have a high degree of competence in the 'Modelling' and (if SysML is being used for the process modelling) the 'SysML Modelling' competencies. Both of these should be held at least at 'Level 2 – support'. Other competencies that would be required for this role would be ones concerned with life cycle and process definition.
- 'Requirements Expert'. The Requirements Expert provides the source knowledge for the process that is described by the Process Author. The person filling this role does not necessarily need to be skilled on the process definition (this is the role of the Process Author) but must hold the 'Acre Process' competency absolutely at 'Level 4 – expert'.
- 'Modelling Expert'. The Modelling Expert also has an input to the process definition for the requirements process because an MBSE approach to requirements engineering has an essential element of modelling within it. Therefore, there is a need for a person who holds both the 'Modelling' and 'SysML Modelling' competencies at 'Level 4 – expert'.
- 'Lead Systems Engineer'. The Lead Systems Engineer is the person who deploys and executes the requirements process, so it is vital that they are involved in the process definition as an added assurance to the practicality and feasibility of the process.

These are some of the supplier roles that are necessary in order to define an effective MBRE process. Of course, it may be possible that a single person takes on several of these roles (such as Requirements Expert and Modelling Expert) or multiple people may take on the same role.

7.3.1.5 The ACRE process: the Information View

The information view defines all of the process artefacts and the relationships and dependencies between them. Fortunately, this has already been done in great detail in this book in the form of the framework. This is because the ACRE process is an information-driven process.

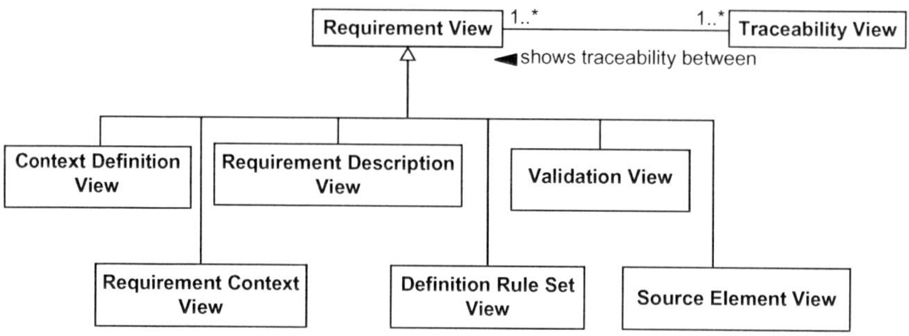

Figure 7.15 The different types of view in the ACRE process

Figure 7.15 shows that there are six different types of views in the ACRE process: the 'Source Element View', the 'Requirement Description View', the 'Definition Rule Set View', the 'Requirement Context View', the 'Context Definition View' and the 'Validation View'. One or more 'Traceability View' shows the traceability between one or more 'Requirement View'.

Each of these views is shown in more detail in Figure 7.16.

Figure 7.16 shows the main information view for the ACRE process. As the ACRE process is an information-driven process, this view is the framework that has been discussed in detail in Chapter 5.

7.3.1.6 The ACRE process: the Process Behaviour View

The process behaviour view shows how each process is executed. In other words, it shows us the order that the activities are executed in, the conditions under which they are executed, the information flow and which stakeholder is responsible for each activity. When it comes to process modelling, this is the view that most people are familiar with it because it shows a 'flow-chart-type' view that is sometimes used for showing how a process works.

Figure 7.17 shows how the ACRE process is executed. The activities are shown in terms of their order of execution and the conditions under which they are executed. The basic activity flow can be seen by following the arrowed lines through the process. The information that is produced and consumed by each activity is also shown by using objects between either activities or external inputs to activities. The responsibility for each activity is indicated by placing each activity inside a swim lane and then allocating a stakeholder to it.

This view may vary depending on the type of project and is deliberately kept simple for this reason.

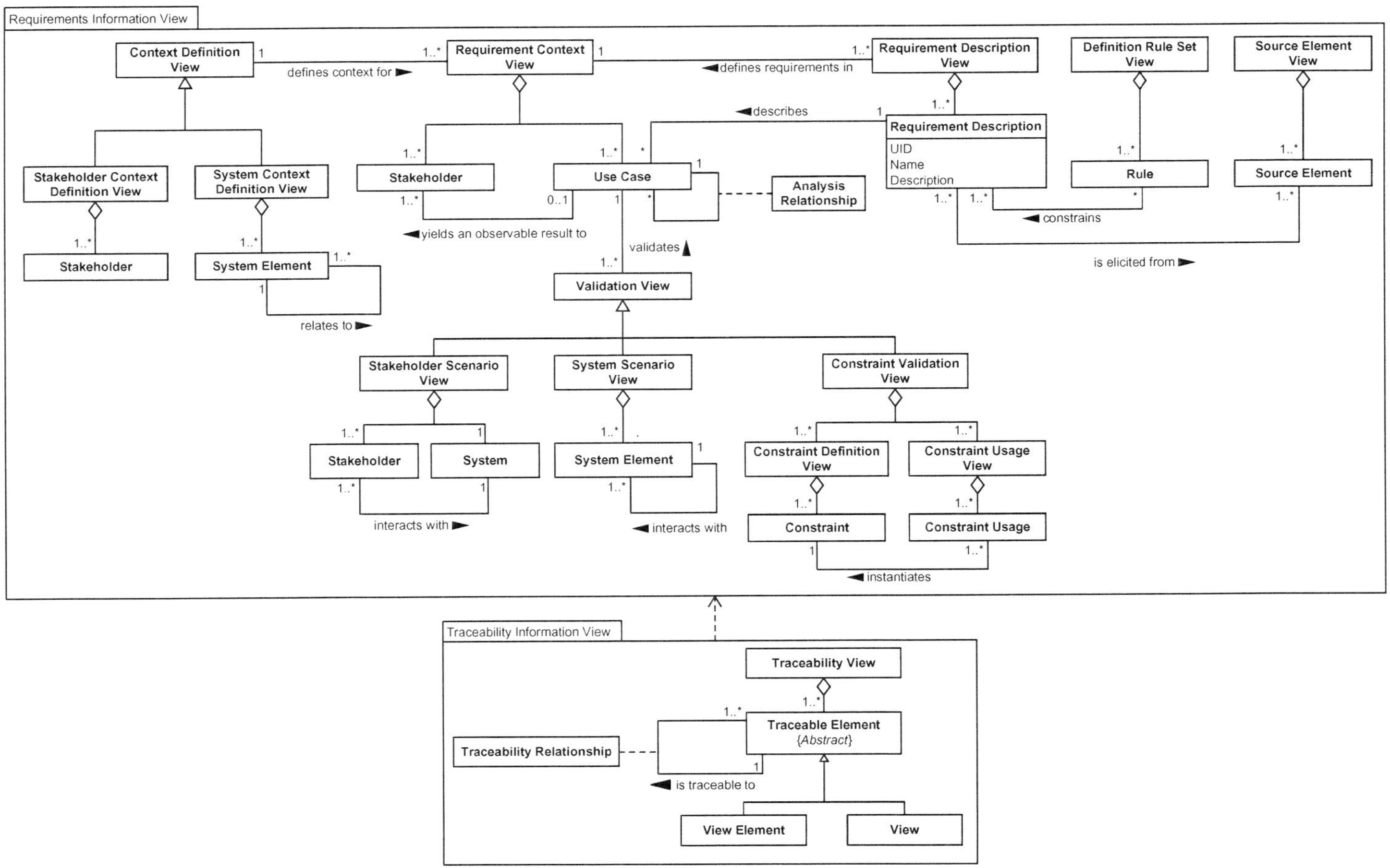

Figure 7.16 Information view for the ACRE process

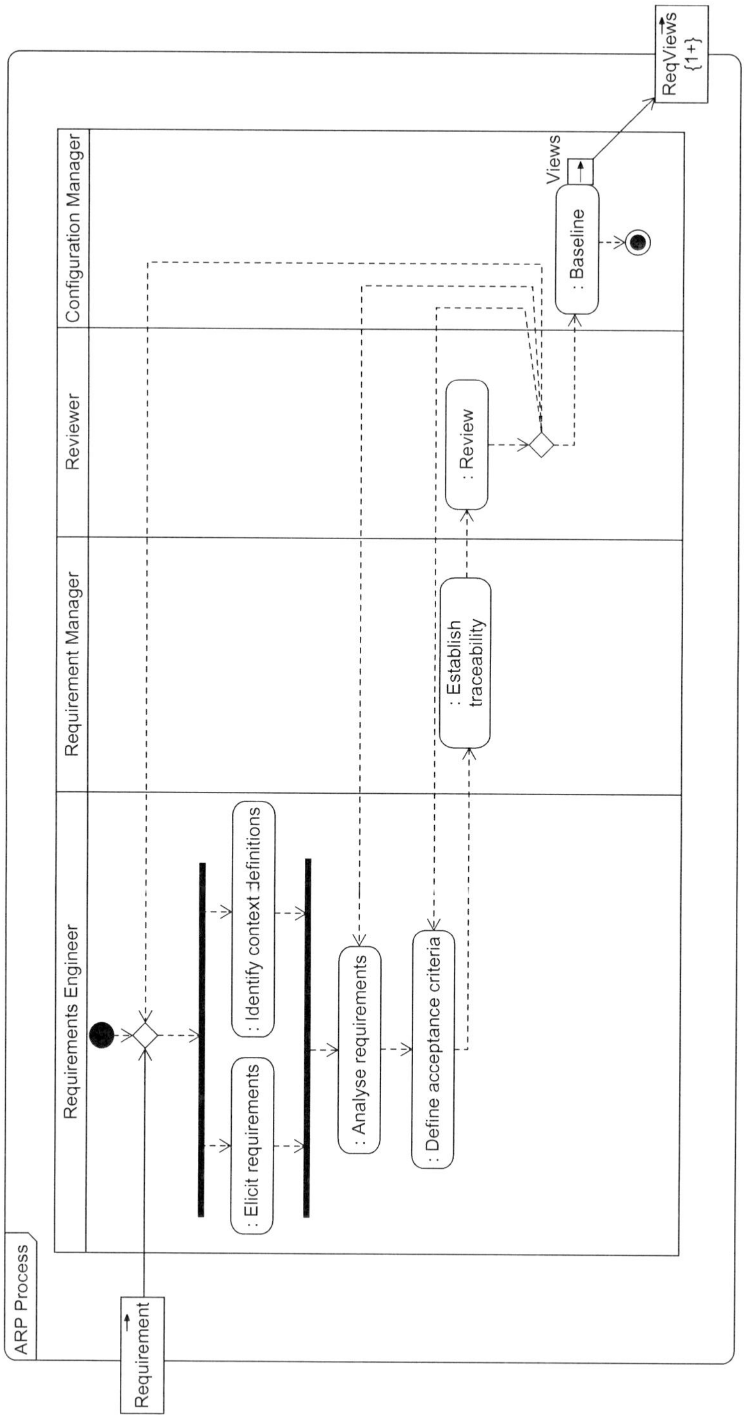

Figure 7.17 Process behaviour view for the ACRE process

7.3.1.7 The ACRE process: the Process Instance View

The process instance view shows how processes are executed in sequences to satisfy original requirements. This view uses sequence diagrams to satisfy use cases for process validation in exactly the same way as they are used in ACRE for requirements validation. The sequence diagram-to-use case pattern is a common one that is used throughout modelling.

Figure 7.18 shows an example of how the ACRE process may be executed in conjunction with a number of other processes. The sequence diagram shows how the 'ACRE Process' is executed before the 'Analysis Process' and the 'Design process'. This is, of course, just one simple scenario, but this is an important exercise as it allows us not only to satisfy the use cases but also to verify that the processes can be executed with one another in terms of their inputs and outputs. See Chapter 3 for a more in-depth discussion on the use of sequence diagrams.

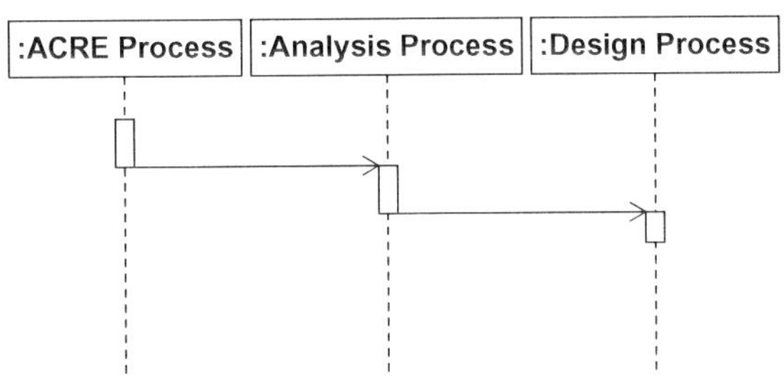

Figure 7.18 Example process instance view

7.3.2 Mapping the process to source standards

One of the constraints that was identified as part of the approach was to ensure that the process complies with best practice in the form of standards, best-practice models and so on. This section provides a simple example of how traceability back to source standards, best practice models and other processes may be defined. The standard that is used as a basis for this example is ISO 15288, 'Systems engineering – system life cycle processes'. The approach taken to establish this mapping is the one used in the 'seven views' approach to process modelling as described in Reference 4.

Before any of the processes or activities in the process can be compared, it is essential that we ensure that the two processes can 'speak' to one another. This requires making sure that the concepts and terms used in both the source standard [5] and the target process (ACRE) can be mapped together. When using the 'seven views' approach, this entails comparing the ontologies in the PSVs for each process. This can be seen in Figure 7.19.

Figure 7.19 shows how the concepts and terms used in both processes compare with one another. The specific mapping of the terms is shown in Table 7.2.

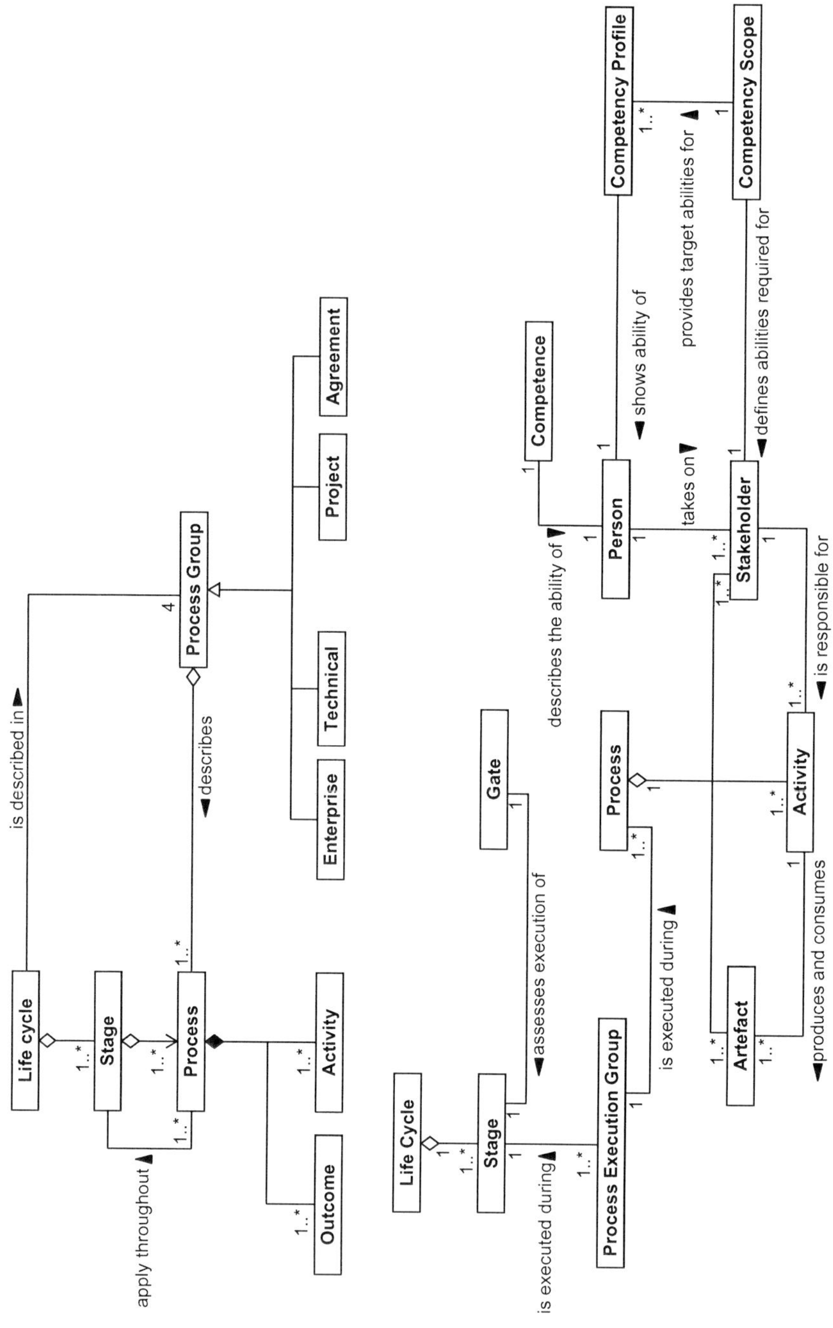

Figure 7.19 Mapping between two process ontologies

Table 7.2 shows the mapping between the terms and concepts used in both processes. The mapping is really quite straightforward, with most of the terms and their associated meaning being the same. One slight discrepancy is in the mapping between 'Artefact' in ACRE and 'Outcome' in ISO 15288. In ISO 15288, as in most standards, specific artefacts are not defined, but the general aim of each activity is described at a high level. In the case of ISO 15288, these are the outcomes. Another discrepancy is the lack of the term 'Stakeholder' in ISO 15288. Although this standard does use the term, it is defined as '*a party having the right, share or claim in a system or in its possession of characteristics that meets the party's needs and expectations*.' This definition does not relate directly to responsibility of processes in the standard, and hence is left off the table here.

Table 7.2 Mapping of concepts and terms between ACRE and ISO 15288

ACRE Concept/Term	ISO 15288 Concept/Term
Life cycle	Life cycle
Stage	Stage
Process	Process
Activity	Activity
Artefact	Outcome
Stakeholder role	*No direct mapping*

The basis of mapping between any two processes (using the term 'process' to refer to any standard or best-practice model) is the PCV from the process model which is then enhanced by a number of tables. The mapping between the PCVs is shown in Figure 7.20.

Figure 7.20 shows the PCVs for the processes. The mapping for the activities in these views is shown in Table 7.3.

Again, the mappings here show a good match between all the activities in the two processes. In reality, this may be used as a basis for process audits or assessments either internally to the organisation or by third parties. These tables may also be used as part of the process automation so that initial compliance with any process may be shown automatically as part of the process deployment.

7.3.3 Using the process

The process presented here may be used in a flexible way in terms of the size and rigour of the project. The process may be used at any level of abstraction of the system and in a number of different ways. This section looks at three examples of how the process may be implemented, although it can be implemented in many more ways.

When using the process for different levels of rigour or for different scale projects, the fundamental process stays the same, but it is the number of views produced that changes and the way in which they are realised. The way in which each view is realised is discussed in more detail in section 7.5.

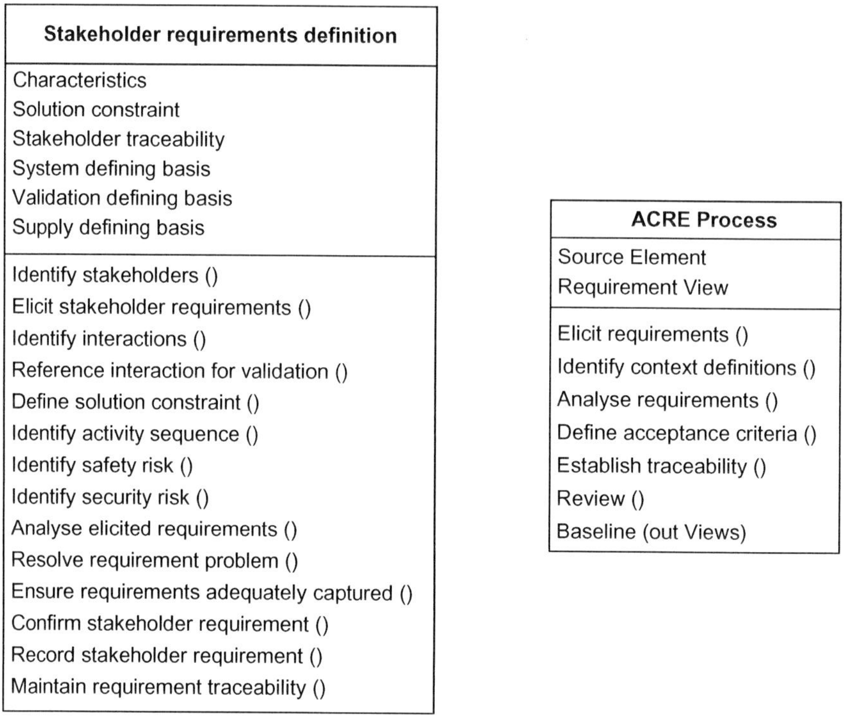

Figure 7.20 Comparison of processes

Table 7.3 Mapping between activities of ACRE and ISO 15288

ACRE Concept/Term	ISO 15288 Concept/Term
Identify context definition	Identify stakeholder
Elicit requirements	Elicit stakeholder requirement
Define acceptance criteria	Identify interaction
Define acceptance criteria	Reference interaction for validation
Analyse requirements	Define solution constraint
Define acceptance criteria	Identify activity sequence
Analyse requirements	Identify safety risk
Analyse requirements	Identify security risk
Analyse requirements	Analyse elicited requirement
Analyse requirements	Resolve requirement problem
Review	Ensure requirement adequately captured
Review	Confirm stakeholder requirement
Baseline	Record stakeholder requirement
Establish traceability	Maintain requirement traceability

The complete set of views needed to perform a complete requirements modelling exercise has been introduced previously in this book, but the number of views produced may vary depending on the scale or rigour of the project in the following manner:

- *Scale of the project.* When referring to the 'scale' of a project, we talk about the size of the project in terms of the resources. For example, a small project may only last for a single week, which may only leave a morning, or even a few hours, to carry out all the requirements activities. At the other end of the spectrum are projects that may last many years and the requirements activities themselves may initially take months or even years, and then they will also be reiterated throughout the project.
- *The level of the project.* When referring to the level of 'rigour' of the project, we talk about the criticality of the project or system. For example, the project may develop a system where, if it fails, lives are put at risk. Such 'safety-critical' systems have a high level of rigour compared with, say, a system that allows someone to edit documents. Other systems that have a high level of rigour include mission critical systems (where the failure of the system may result in some critical activity such as the navigation system on a missile or spaceship not), real-time systems (where the correctness of a result depends not only on the logical value, but also on the timing characteristics of the system) and business-critical systems (where the failure of the system may result in the organisation going out of business).

For the purposes of the discussion in this chapter, three levels of scale (small, medium and large) and three levels of criticality (non-critical, semi-critical and critical) are considered.

This section discusses which views are produced for three different types of application of the process, but the emphasis is on the description of the activities, rather than how each activity is implemented. This is covered later in this chapter in the section on process deployment.

7.3.3.1 Example use: quick and dirty process

The quick and dirty process is often used on projects that have a small scale and that have a non-critical level of rigour.

For example, consider a project that has a timescale of a single week for a proof-of-concept project.

Figure 7.21 shows the subset views that are realised when applying a quick and dirty process. The activities in the process may be executed as follows:

- 'Elicit requirements'. In this case, the requirements would come from informal sources, indicated here by the lack of a Source Element View.

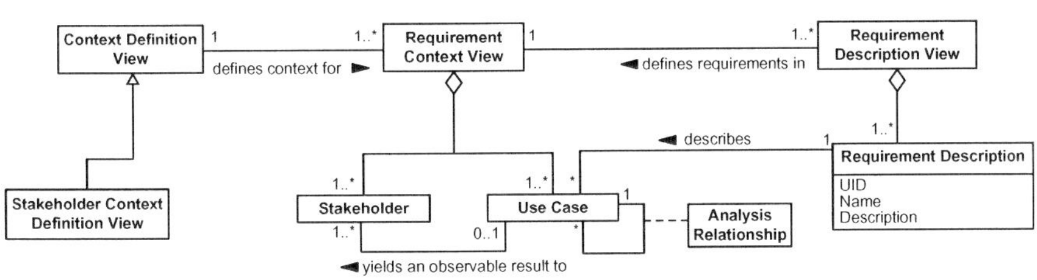

Figure 7.21 Example of a quick and dirty process

- ‘Identify context definition’. In this case, only the stakeholders are being considered, so the ‘System Context Definition View’ is missing.
- ‘Analyse requirements’. This forms the main activity in this process and involves creating a context for each of the stakeholders identified.
- ‘Define acceptance criteria’. This is not performed formally, indicated by the lack of Validation Views. In reality, however, many people may consider validation as they define the use cases even if they are not documented.
- ‘Establish traceability’. This is not performed formally, indicated by the lack of Traceability Views.
- ‘Review’. This may take the form of a simple, non-documented review where the model may be annotated with review comments.
- ‘Baseline’. Regardless of the scale or level of rigour, it is essential to perform this activity, even at a basic level. It is far better to simply have a version number for the model with no formal configuration management processes applied rather than nothing at all.

As can be seen, we are still following the bones of the ACRE process here, albeit in an informal way. Despite this informality, the type of tools used to implement the process has a bearing on how rigorous these artefacts are.

7.3.3.2 Example use: semi-formal process

The semi-formal process is typically used on small-to-medium projects of a non-critical nature (Figure 7.22).

Figure 7.22 shows the subset views that are realised when applying a semi-formal process. The activities in the process may be executed as follows:

- ‘Elicit requirements’. In this case, the requirements would come from sources that can be formally identified and be represented as the Source Elements in the Source Element View.
- ‘Identify context definition’. In this case, only the stakeholders are being considered, so the ‘System Context Definition View’ is missing.
- ‘Analyse requirements’. This forms a major activity in this process and involves creating a context for each of the stakeholders identified.
- ‘Define acceptance criteria’. For each of the use cases, a number of Validation Views are generated. At the semi-formal level, these are the stakeholder scenarios and describe the stakeholder interactions with the system.
- ‘Establish traceability’. This activity generates the Traceability Views. If the modelling is carried out properly, this should be a straightforward activity as all the traceability paths have already been identified in the model.
- ‘Review’. This may take the form of a formal review meeting where the ACRE views that have been generated are assessed and commented on.
- ‘Baseline’. At the semi-formal level, base lining is a very important activity.

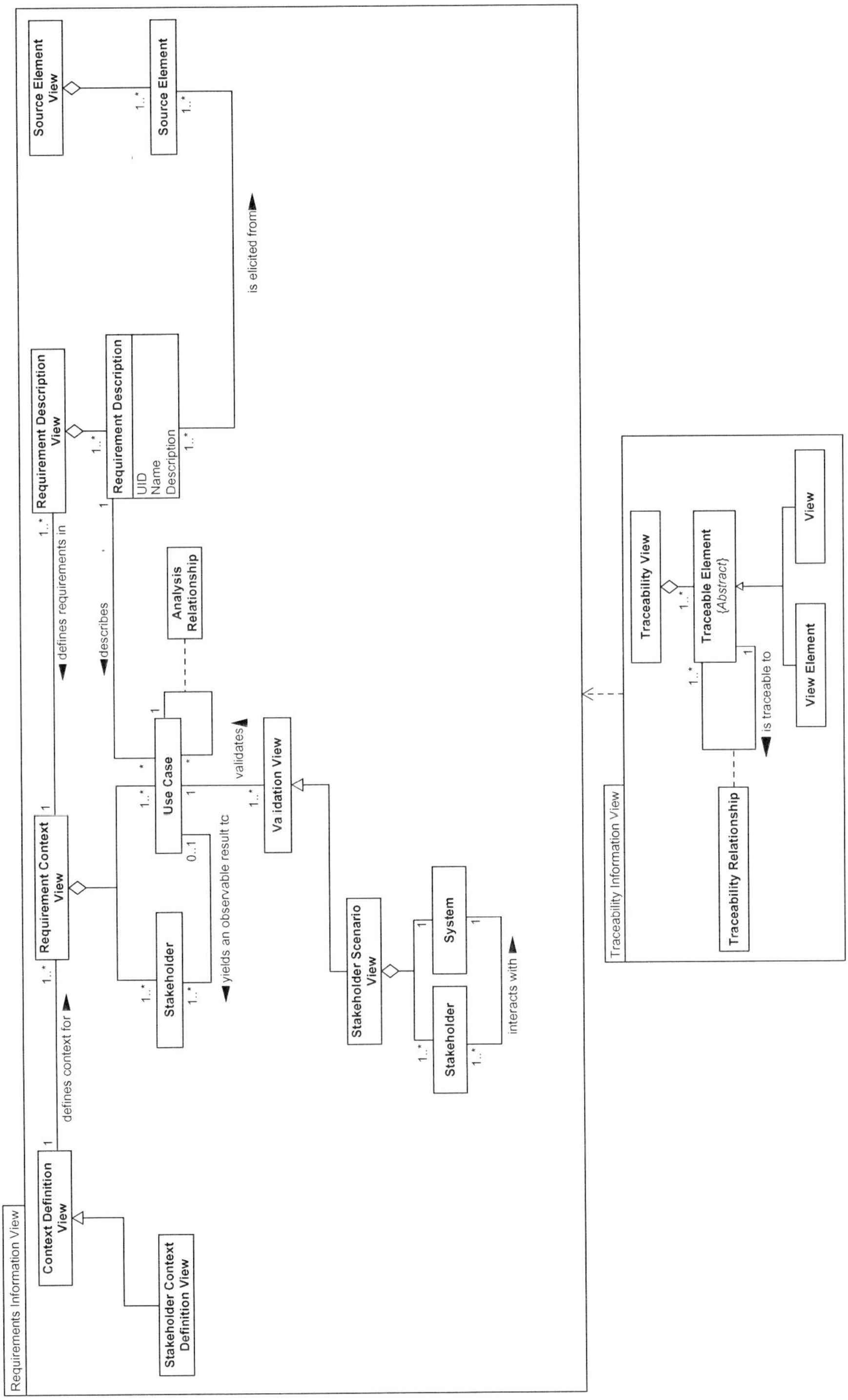

Figure 7.22 Example of a semi-formal process

The process here starts to look a lot more like the full ACRE process, with many of the views being generated.

7.3.3.3 Example use: formal process

The formal process is executed on process that is critical in some way, e.g. on safety-critical systems, mission-critical systems and so on. The formal process may also be used on long-term projects and projects that have a high cost associated with them.

Figure 7.23 shows the subset views that are realised when applying a semi-formal process. The activities in the process may be executed as follows:

- 'Elicit requirements'. In this case, this is a major activity. All requirement sources are formally identified and captured in the Source Element View. The requirements themselves now have some limitations on the way that they can be defined, as there is now the Definition Rule Set View to consider that defines rules for how the requirements can be described.
- 'Identify context definition'. In the formal process, this usually involves looking at both the Stakeholder Context Definition View and the System Context Definition View.
- 'Analyse requirements'. There is a large number of these views generated and, of course, the more views there are, the more analysis that needs to be carried out.
- 'Define acceptance criteria'. This is a massive undertaking, as all three levels of validation scenarios are considered.
- 'Establish traceability'. This activity generates the Traceability Views. If the modelling is carried out properly, this should be a straightforward and very time-consuming activity because all the traceability paths have already been identified in the model.
- 'Review'. This involves many formal reviews with different groups of stakeholders.
- 'Baseline'. Again, this becomes a very large exercise at this level.

The formal process represents all the views that can be considered in the ACRE process.

7.3.4 Deploying the process

So far, we have discussed the process itself and how the process can be tailored, depending on the level of scale or rigour required by the system. This is, of course, essential, but all of this great work can still come to nothing if the process is not deployed effectively. Deployment is not just a matter of providing people with a process 'manual', but is a complex area in itself where a number of requirements must be considered:

- *The process must be made available.* Each person who is to use the process must have access to it. In the past, process documentation and information has taken the form of large, printed and often very heavy documents, famous for

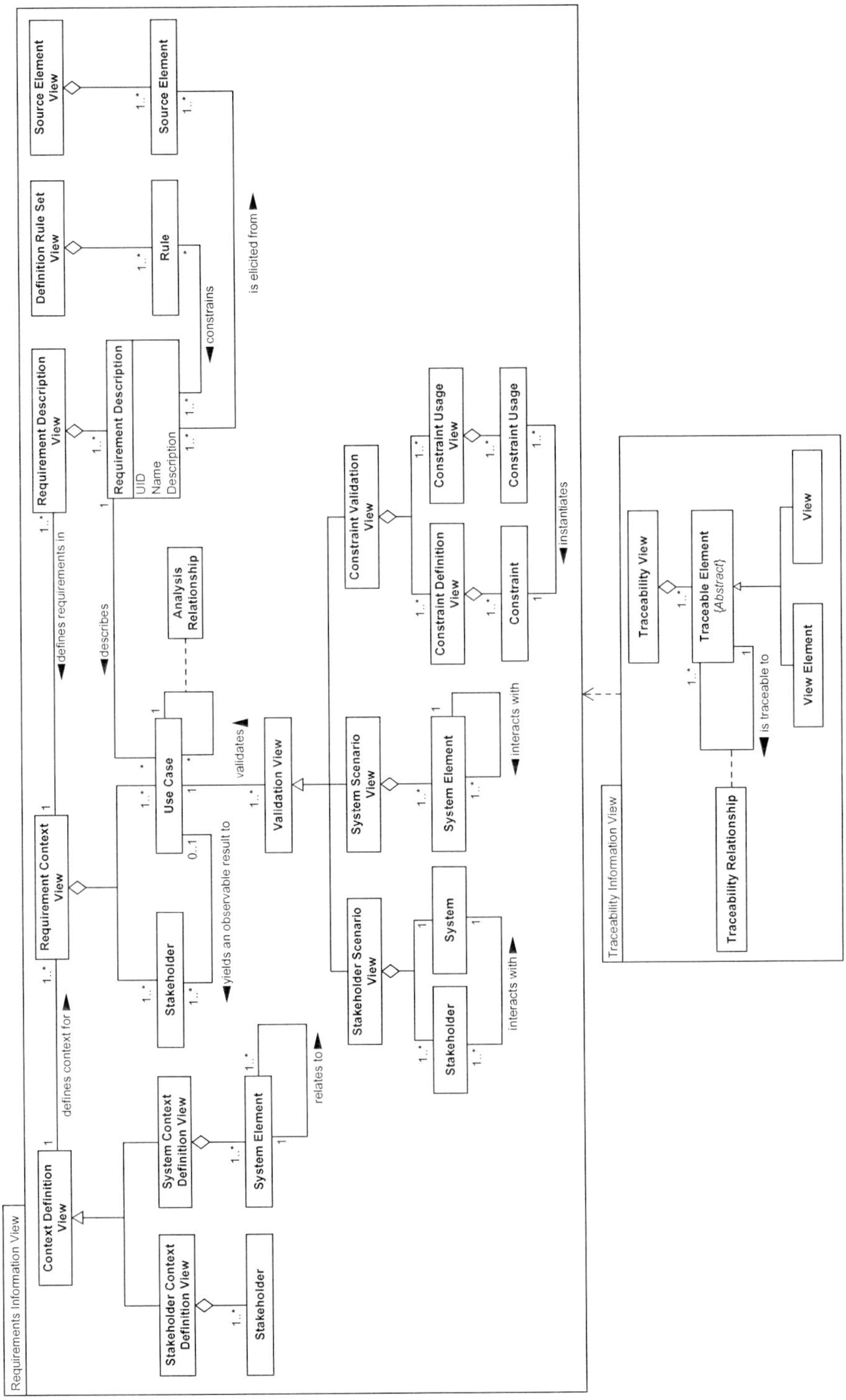

Figure 7.23 Example of a formal process

gathering dust on shelves. In today's business, web-based approaches (whether they are intranets or the Internet) are far more efficient ways to store, update and disseminate information amongst workers.

- *The medium used to convey the information must be one that is accessible to all relevant users.* Relating back to the previous point, the end users must be able to access the information easily. It may seem quite a safe assumption that, e.g. everyone has access to a web browser, but not everybody has access to the Internet. In many cases, this may be because of security issues, such as no Internet connection at all or a limited-access intranet. Another issue to consider is that if a third-party product is being used as part of the deployment, licensing will become an issue.
- *People must be aware that the process exists.* There is no point having the best process in the world if nobody knows anything about it. People must know that a process exists, and also the scope and context of the process.
- *The format of the presentation must be one that people can understand.* This is not necessarily the same as the SysML modelling that was used to engineer the processes. When developing any system, the techniques that are used to perform the development processes are not usually the same as the end product. Therefore, the processes may be developed using a structured approach (such as the 'seven views') and a standard modelling notation (such as the SysML), but the final output may look very different. Most people would be horrified to see the final process definition in SysML, and many prefer text descriptions and simplified flow charts.
- *The deployment should make people's lives easier.* It must be easy to access the information. In fact, it should involve the minimum of effort and be of less hassle than opening a book. This may seem a little extreme, but if the information is very easy to access, people will use it – if not, they will not use it.
- *The deployment should be a two-way activity in terms of feedback.* A process constantly evolves as time goes on, and an essential part of this is being able to obtain feedback from the end users of the process.
- *The deployment should be consistent with the processes.* This could be in terms of the project management or other processes that are being used. The deployment of the processes should be carried out in a structured and consistent fashion – in other words, by following a process.
- *The deployment should contribute to continuous process improvement, such as CMMI.* Any work involving processes is an ongoing activity if not for the simple fact that the world turns and things change; therefore, processes need to be checked on an ongoing basis to ensure that they are fit for the purpose.
- *The deployment itself should be part of a larger MBSE initiative.* This book is concerned with requirements engineering, but always within the context of a wider MBSE initiative.

The list presented here is not intended to be exhaustive but, rather, to stimulate thought about process deployment. This list can be used pragmatically when it comes to selecting tools for process deployment, which is discussed in the following section.

7.3.5 Summary of 'Process'

This section shows how the ACRE process can be used for real-life projects and situations. This entail the following:

- *The definition of the process.* The ACRE process is modelled using the 'seven views' approach and using SysML as the notation. Again, any notation or sufficiently rigorous approach may be used. This also include mapping the ACRE process to source standards; in this case ISO 15288.
- *The use of the process.* It is important that the process is flexible in terms of its scale and rigour, and three examples of how the ACRE process may be used have been shown, along with the views that would be necessary. Again, these three views are for information only, based on previous experience, and do not feel limited to them.
- *The deployment of the process.* The process may be very well defined, but if it is not deployed correctly, the process will fail. A number of key characteristics for process deployment have been discussed.

So far, the 'people' and 'process' aspects of the ACRE have been discussed, which leaves the 'tool' aspect, which is discussed section 7.5.

7.4 The 'Tool'

To realise the full benefits of MBSE, 'sharp' tools are essential. There are three main issues that are discussed in this section:

- *Considering the types of tools available.* Before any sort of tool selection is performed, it is important to identify the types of tools that may be applicable to model-based requirements engineering.
- *Considering how the tools are used.* This involves looking at each tool used at the main views in the process, and how the tools are used in conjunction with one another.
- *Considering tool selection.* When the tools have been identified and their usage is understood, it is important to carry out a tool selection exercise that looks at various aspects of each tools and measure them, and then use the results as a basis for deciding between tools.

Each of these three issues is discussed in detail in sections 7.41, 7.42 and 7.43 respectively.

7.4.1 Considerations for tool availability

There are many different types of tools that can be used for MBSE when performing any tool selection. For the purposes of this book, the following types of tools are considered:

- *Pen and paper system tools.* The so-called PAPS tools are available to anyone and everyone. They have the advantage of being free (if you own a writing instrument and writing surface) and eminently portable, and run on a number of different platforms – mainly paper and whiteboards. The PAPS tools are often overlooked or dismissed by many people, but they form a very important and necessary part of any professional systems engineer's toolkit. Regardless of the software-based tools available, PAPS should be the first tools touched by anyone until they have thrashed out their basic ideas and concepts. With the advantages in technology over the last few years, the PAPS tools are now no longer confined to their original pads of paper or whiteboards. The ubiquity of smart phones, smart boards and automated pen-and-paper pads means that the PAPS tools can truly form part of a MBSE toolset. The PAPS tools may be used to realise any or all of the views required by the ACRE approach to model-based requirements engineering.

- *Office-type tools*. The first set of software-based tools that must be considered is the standard office-base software that is available for free (or as good as free) in most organisations. Office tools, such as word processors spread sheets and presenting programmes may be used to realise any or all of the views required by the ACRE approach to model-based requirements engineering.
- *Drawing tools*. Most people have access to a basic drawing tool. This may be a specific drawing tool, or an integral part of a presenting programme or word processing application. When using established notations for the modelling of requirements (such as SysML), it actually means that just about any application that can draw basic shapes and show basic ASCII text characters can visualise any diagram in the notation. Drawing tools may be used to visualise any of the graphic-based views in the ACRE approach to model-based requirements engineering.
- *Requirements management tools*. These tools are usually quite simple in what they can do, yet very powerful in terms of their use and benefits. What these tools allow you to do is to *manage* and *describe* the requirements in a system and to *define relationships* between them. What these tools do not allow you to do is to *model the requirements* in a system. This is a simple yet, as it turns out, much misunderstood belief about what these tools do. Requirements management tools are often seen as a silver bullet for requirements engineering, so caution must be exercised that they are being used properly. Requirements management tools may be used to realise the Requirement Description View and all Traceability Views in the ACRE approach to model-based requirements engineering.
- *Modelling tools*. There is often some confusion between modelling tools and drawing tools, but the distinction is quite clear and simple. Modelling tools produce models and drawing tools produce pictures. The difference between modelling and drawing pictures is discussed elsewhere in this book. The capabilities associated with modelling tools differ enormously and these are often the deciding factors in the selection of tools. Some modelling tools are so powerful that they can be used to realise any and all of the views required by the ACRE approach to model-based requirements engineering.
- *Document generation tools*. These tools are usually part of, or are bolt-ons to, other types of tools that have been discussed so far. One of the benefits of MBSE is that it is an approach that can increase productivity and save vast

amounts of time. One of the ways that these savings can be realised is through effective use of the humble document generation tool. When the requirements process is understood, it is possible to define templates for various artefacts, all of which can be derived from the model. The document generation tools allow the user-friendly face of model-based requirements engineering (text!) to be generated from any model. Document generation tools are used to provide well-formatted and presented artefacts that are based on views in the ACRE approach to model-based requirements engineering.

- *Mathematical modelling tools*. When it comes to critical systems, formal mathematical-based modelling is required. The modelling tools that have been discussed so far have focussed on visual modelling, rather than on mathematical modelling; hence, this has been drawn out as a separate discussion point. Generally speaking, visual modelling tools are not good at mathematical modelling and, likewise, mathematical modelling tools are not very good as visual modelling, so in many cases, different tools are required. Mathematical-based modelling tools may be used to realise some of the Validation Views in ACRE approach to model-based requirements engineering.
- *Data transformation tools*. It should be clear from looking at the plethora of tools in this list that tools integration may very well be an issue for many people. This is because, in reality, most organisations use a toolset that is made up of different tools (this is discussed further later in this section). As a result of this, there is often a need to be able to use data in more than one tool that, typically, requires the use of a tool to transform data from a format used in one tool to a format used in another one.

Any or all of these tools may be used together to form a 'toolset' or 'tool chain'. There are several types of tool chain that should be considered:

- *Closed tool chain*. This term is used to refer to a set of tools that can only be used in conjunction with a limited set of other tools, usually from the same tool provider. Closed tool chains comprise tools that have bespoke or proprietary interfaces that cannot be accessed by every other tool, but only by a limited set of tools.
- *Open tool chain*. This tool is used to refer to a set of tools that can be used, potentially, with any other tool. Open tool chains comprise tools that have open interfaces that are defined according to some sort of accepted norm, or open standard, and that may be accessed by any other tool that complies with this standard.
- *Heterogeneous tool chains*. This term is used to refer to tools that exist separately and are not able to communicate with other tools without external help. Heterogeneous tool chains comprise tools that may be from different or the same manufacturers but that cannot be easily integrated when the first comes 'out of the box'.
- *Integrated tool chains*. This term is used to refer to a set of tools that exist separately but that can work together with no external help. Integrated tool chains comprise tools that may be from different or the same tool providers, but that work together when fresh 'out of the box'.

Just to further complicate matters, these different types of tool chains may be mixed. For example, it may be possible that a tool chain exists that is both closed and integrated, another that is open and integrated, another that is open and heterogeneous and so on. This often results in data transformation tools that are necessary to enable the integration of tools in these tool chains.

Now that the different types of tools have been discussed, it is still not possible to choose between them, as it is first necessary to look at how the tools are used within the process.

7.4.2 Use of tools with the process

This section presents some examples of how the various types of tool, toolsets and tool chains may be used to realise the ACRE approach to model-based requirements engineering. The examples shown here relate to the different types of processes used that have been discussed in the previous section, but this time the emphasis is on how the various types of tools may be used to realise the different views.

The focus in the discussion is not so much on which tools are *possible* to be used for each example, but which tools are the *most appropriate* to be used for each example. The use of tools may be considered as being analogous to the use of cooking equipment, in what is colloquially known as '*Pemberton's Cooking Analogy*'. This analogy equates the maturity of a person's cooking requirements to those of an organisation's modelling.

Consider the first example, where a young student has just made the brave step of moving away from home and must, possibly for the first time, start to fend for themselves and cook their own meals. It is not difficult to imagine that a student may meet all of their dietary requirements with a toaster, a microwave oven and a kettle. This is perfectly adequate for them and the situation that they find themselves in. To provide the student with a full range of pots, pans and cooking utensils would be a waste of time and effort as not only would the student not use them properly, but they would have no appreciation whatsoever for what sophisticated and impressive cuisine they could create with such a fine set of tools. This is analogous to an organisation which is just starting to implement MBSE and does not yet know how it will use the tools, or what its requirements are for their use. For them, it is fine to simply make use of existing office-based tools.

Consider now the same student who has graduated, started a career and now shares a flat or first house with a new partner. The classic 'beans on toast' is no longer sufficient for this new, more mature lifestyle, and it is necessary to expand the kitchen toolset to include some basic pots, pans and utensils. They will probably be cheap and of sufficient quality to get by in cooking. This situation is analogous to the situation where an organisation has reached the limits of office-based tools and will now be looking for more specific modelling tools.

Consider again the same student who has now matured into a world-class chef and understands many nuances of the culinary world and can truly appreciate the difference having a good set of tools can make. The chef now sees the benefit of the high-quality tools and the price now seems almost irrelevant as excellent result require excellent tools. This is the situation where an organisation has a high maturity of modelling and can see the value in sophisticated tools.

The point to note here is that different tools suit different people. A lot of this comes down to how mature the modelling in your organisation is and how close you are to achieving MBSE.

7.4.2.1 Example tool implementation: quick and dirty process

The 'quick and dirty' process is used when the system is either very simple or has a very limited set of timescales associated with it. Clearly, almost all of the tools discussed previously may be used to realise this approach but, when time is against you on a project, it may be that the quickest and (what is perceived as) the easiest-to-use tools are more suitable than tools that may actually be far more powerful.

Two examples are considered, the use of a PAPS tool and the use of office-based tools.

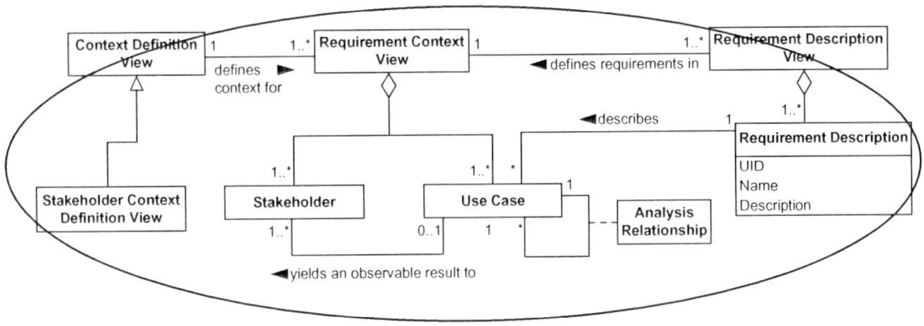

Figure 7.24 Quick and dirty tool implementation example – PAPS tool in a heterogeneous tool chain

Figure 7.24 shows that a PAPS tool may be used to realise all of the views that have been identified as essential for this project. Note that this comes under the broad banner of a heterogeneous tool chain, although in reality the interface issue is an artificial one due to the nature of a PAPS tool.

In some cases, all of these views may be created in a log book, as follows:

- The 'Requirement Description View' may be a simple hand-written list of high-level descriptions for the requirements.
- The 'Requirements Context View' may be a set of use case diagrams that have been drawn in a log book.
- The 'Context Definition View' may be a block definition diagram that has been written into a log book.

These views may then stay in the log book, may be transcribed into documents (see the next example) or may make use of digital photography or smart pen and pad to get the information into an electronic format that may be more easily communicated with the rest of the project stakeholders.

The most immediate advantage to the PAPS approach is that it is free and can be done almost anywhere. Also, provided that the person doing work is competent in the approach and the modelling, there is no need for them to be competent in the use of a specific tool. This approach is also very well suited to brainstorming sessions and is far more suitable than using a sophisticated tool, and it should be considered to be used before any tools are used.

Of course, the downside of the PAPS approach is quite steep because there is no automation and hence very few benefits of the MBSE approach in terms of time-saving activities. For example, all consistency checking must be carried out manually and all documents must be generated by hand (Figure 7.25).

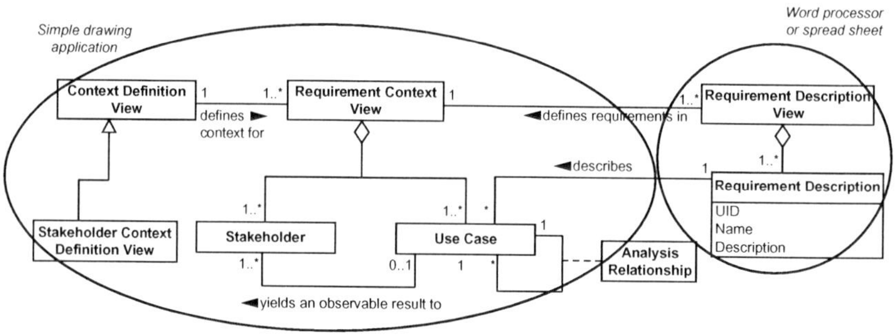

Figure 7.25 Quick and dirty tool implementation example – office-type tools

The example in Figure 7.25 shows how office-type tools may be used to realise the views required for this project. Note that depending on the office tools selected, this tool chain can potentially fit into any of the four categories described previously. These may be realised in the following way:

- The 'Requirement Description View' may be defined as a word-processed document that describes the requirements in a list or in a table. Another way would be to represent the requirements in a table in a spread sheet.
- The 'Requirement Context View' may be realised using use case diagrams in a simple drawing tool in a word processor or even in a presentation application.
- The 'Context Definition View' may be realised using a block definition diagram in a simple drawing tool in a word processor or even a presentation application.

In many cases, the tools here are used to transcribe information that was generated using a PAPS tool in order to get the views into easily manipulated formats that can be communicated to other stakeholders on the project.

The example shown here has the advantage of being quick and simple to implement. However, the downside of this approach is that the model, as it stands, is very difficult to manage and must be verified and validated entirely manually, which, of course, is both time-consuming and prone to error. This is a perfectly valid approach, but not one that realises all the true benefits of MBSE because there is no automation between the views and everything is done manually.

7.4.2.2 Example tool implementation: semi-formal process

The semi-formal variation of the process is used where there is a requirement for a certain amount of rigour on the project, or where the project has quite a long timescale. Two implementations are considered here, the first of which is an office-based tool implementation and the second uses bespoke modelling and management tools.

Figure 7.26 shows a possible implementation of the semi-formal process.

The example in Figure 7.26 shows how office-type tools may be used to realise the views required for this project. Note that depending on the office tools selected, this tool chain can potentially fit into any of the four categories described previously. These may be realised in the following way:

- The 'Source Element View' may be realised using almost any tool or file type. For example: emails, digital photos, scans of log books, word processor documents and so on.
- The 'Requirement Description View' may be defined as a word-processed document that describes the requirements in a list or in a table. Another way would be to represent the requirements in a table in a spread sheet.
- The 'Requirement Context View' may be realised using use case diagrams in a simple drawing tool in a word processor or even in a presentation application.
- The 'Context Definition View' may be realised using a block definition diagram in a simple drawing tool in a word processor or even a presentation application.
- The 'Validation Views' may be realised using sequence diagrams in a simple drawing tool or as text-based ordered lists in a word processor.
- The 'Traceability View' may be realised using tables in a spread sheet. By using a hypertext links in the spread sheet, it is also possible to start to provide simple navigation between the various office documents.

Note here that we are starting to see the beginnings of automation by using hypertext links. This is the tip of the iceberg in terms of the benefits that MBSE can offer, but it is a start point. Again, the more powerful the tools, the more benefits that are able to be realised by using the MBSE approach.

The next example uses more sophisticated tools and, therefore, we would expect to see more automation and, hence, more benefits than when using the office-based tools.

The example in Figure 7.27 shows how a proprietary toolset may be used to realise the views required for this project. Note that depending on the tools selected, this tool chain can potentially fit into any of the four categories described previously. For the purposes of this example, let us assume that this proprietary toolset consists of a modelling tool and a requirements management tool that are being used in conjunction with standard office-type tools. These may be realised in the following way:

- The 'Source Element View' may be realised using almost any tool or file type. For example, emails, digital photos, scans of log books, word processor documents and so on.

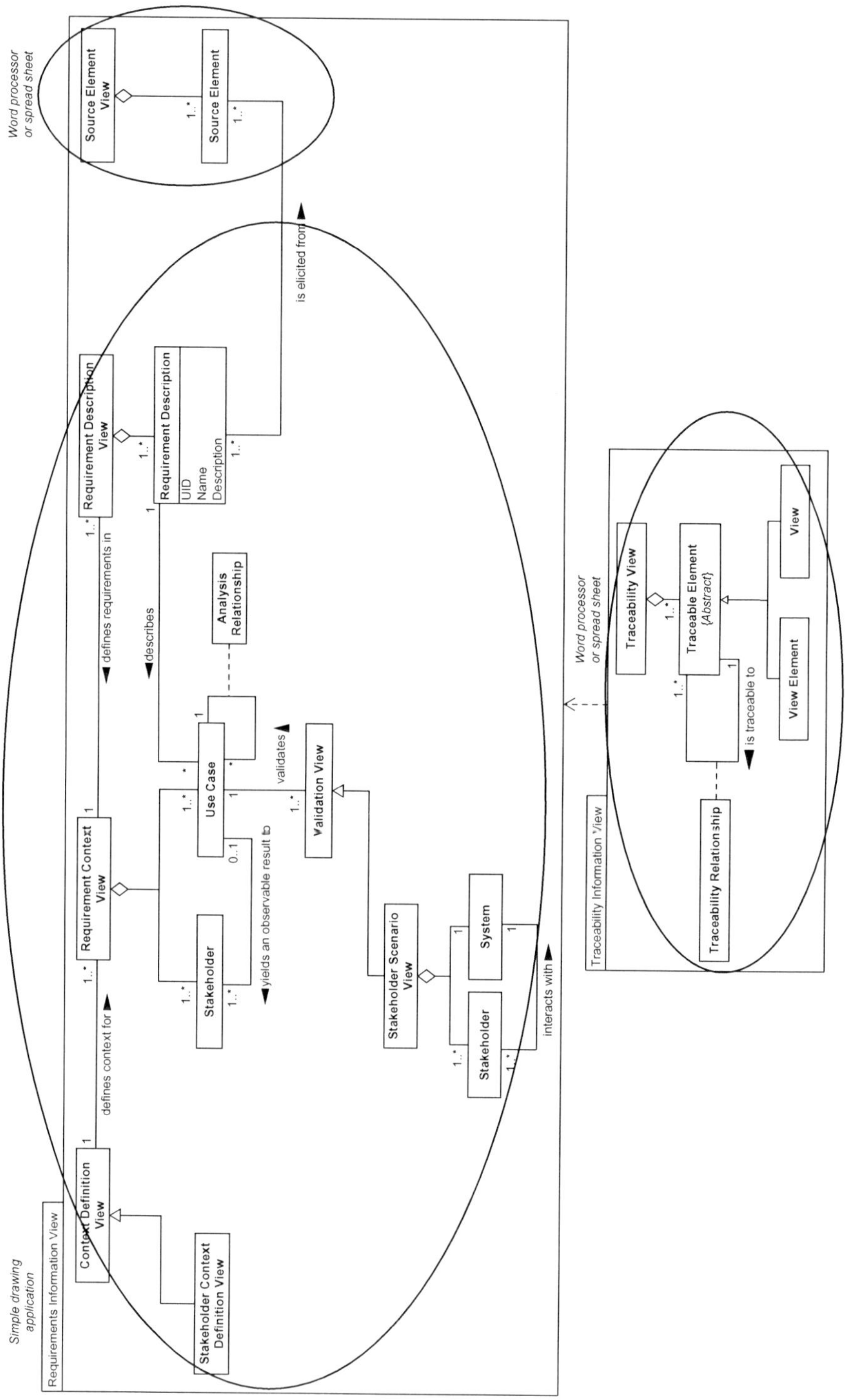

Figure 7.26 Semi-formal tool implementation – office-type tools

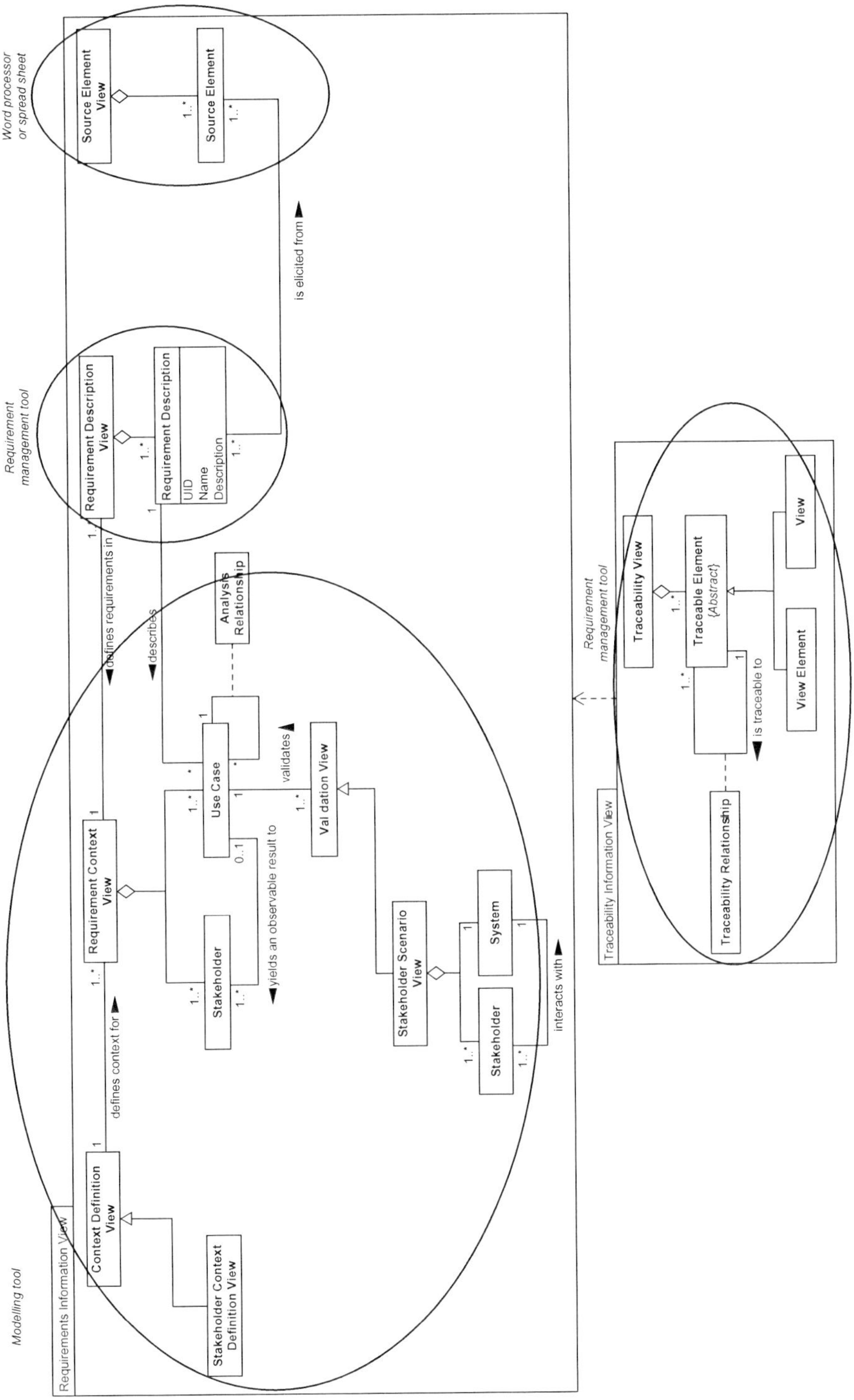

Figure 7.27 Semi-formal tool implementation – proprietary tools

- The 'Requirement Description View' may be defined in the requirements management tool. Most requirements management tools excel in this area and allow the user to define a set of attributes for each requirement. Each requirement here is text based, but it is also usually possible to paste in graphics and images, which will prove useful in some of the later views.
- The 'Requirement Context View' may be realised using use case diagrams in the modelling tool. This information may also be represented in the requirements tool, depending on the sophistication of the tool. At the very least, each use case diagram may be pasted into the management tool as an image. This may then be used for the basis of traceability in the 'Traceability View' later. If the tool is more powerful, there may be an automated link between the modelling tool and the requirements management tool, which would have a number of advantages, as discussed later in this section.
- The 'Context Definition View' may be realised using a block definition diagram using the modelling tool.
- The 'Validation Views' may be realised using sequence diagrams in a simple drawing tool or as text-based ordered lists in a word processor. Again, depending on the level of sophistication of the tools, there are advantages of using the proprietary toolset over the previously described office-based tools.
- The 'Traceability View' is realised using the requirements management tool. This is one of the main uses for the requirements management tool – establishing traceability. At the moment, these traceability links would be put into the requirements management tool manually, but, once established, can be used to perform impact analysis and other types of investigation.

Now that we are starting to use sharper tools, we will start to see more benefits. One of the main advantages to using a proprietary toolset is that it is now possible to automate the interactions between the two tools. The advantages are given below:

- First, the diagrams from the model and their representations in the management tool are linked, so that they can always be made consistent by simply pressing a button to update the information in the management tool from the modelling tool, or vice versa.
- Another advantage is that descriptions that are associated with elements in the model (such as actors or use cases) may then be used automatically in the requirements management tool, avoiding the need for tiresome 'cutting and pasting'.
- Both tools should allow documents, websites etc. that exist outside the tools to be linked to, which provide even more flexibility and power when it comes to traceability.
- The use of the two tools together provides rigour both in terms of the technical correctness of the model (in the modelling tool) and the management of the requirements artefacts (in the requirements management tool).

There are, however, several disadvantages of using this approach:

- Very often the integration or communication between tools is non-trivial and can entail quite a large piece of work. Of course, this should only need to be carried out

once, and then it can be reused as often as desired. This initial overhead should be taken into account when considering the semi-formal tool implementation.
- On a related note, tools from different providers may have different versions, so the configuration and version control of the tools must be taken into account.

The semi-formal process may be seen quite often in reality and can offer many benefits to a project or programme.

7.4.2.3 Example tool implementation: formal process

The formal process is used where the system is a mission-critical system or the project is being conducted on a very large scale.

On projects such as these, the initial cost of the toolset being used is usually less of an issue as it tends to be a very small part of the overall project budget. More importantly, the level of maturity of modelling of the project should(!) be higher and, therefore, the full benefits of a powerful toolset may be realised.

Two examples of implementation are looked at: the first with a heterogeneous tool chain and the second with an integrated tool chain.

Figure 7.28 shows how the formal process may be implemented using a heterogeneous tool chain. In this case, each tool exists in its own right and needs some effort to be put into the integration of the tools. Examples of tools that may be used here are a SysML modelling tool, a requirements management tool, simulation tools, data exchange tools and spread sheets.

One possible way to realise the views is as follows:

- The 'Source Element View' may be realised using almost any tool or file type. For example: e-mails, digital photos, scans of log books, word processor documents and so on.
- The 'Requirement Description View' may be defined in the requirements management tool or in a modelling tool that supports SysML requirements diagrams, or both.
- The 'Requirement Context View' may be realised using use-case diagrams in the modelling tool. This information may also be referenced or repeated in the requirements management tool.
- The 'Context Definition View' may be realised using a block definition diagram using the modelling tool.
- The 'Validation' Views may be realised using sequence diagrams in the modelling tool. In the case of the formal process then a simulation tool may be required to define the parametric constraints.
- The 'Traceability View' is realised using the requirements management tool or with a separate traceability tool, such as spread sheet or workbench-type environment.

With the use of more tools, then one would expect to see more benefits. In fact, the benefits and disadvantages are very similar to the ones discussed for the semi-formal process. The main benefits are automation and the ability to generate process artefacts. The main disadvantages are again related to integrating the various

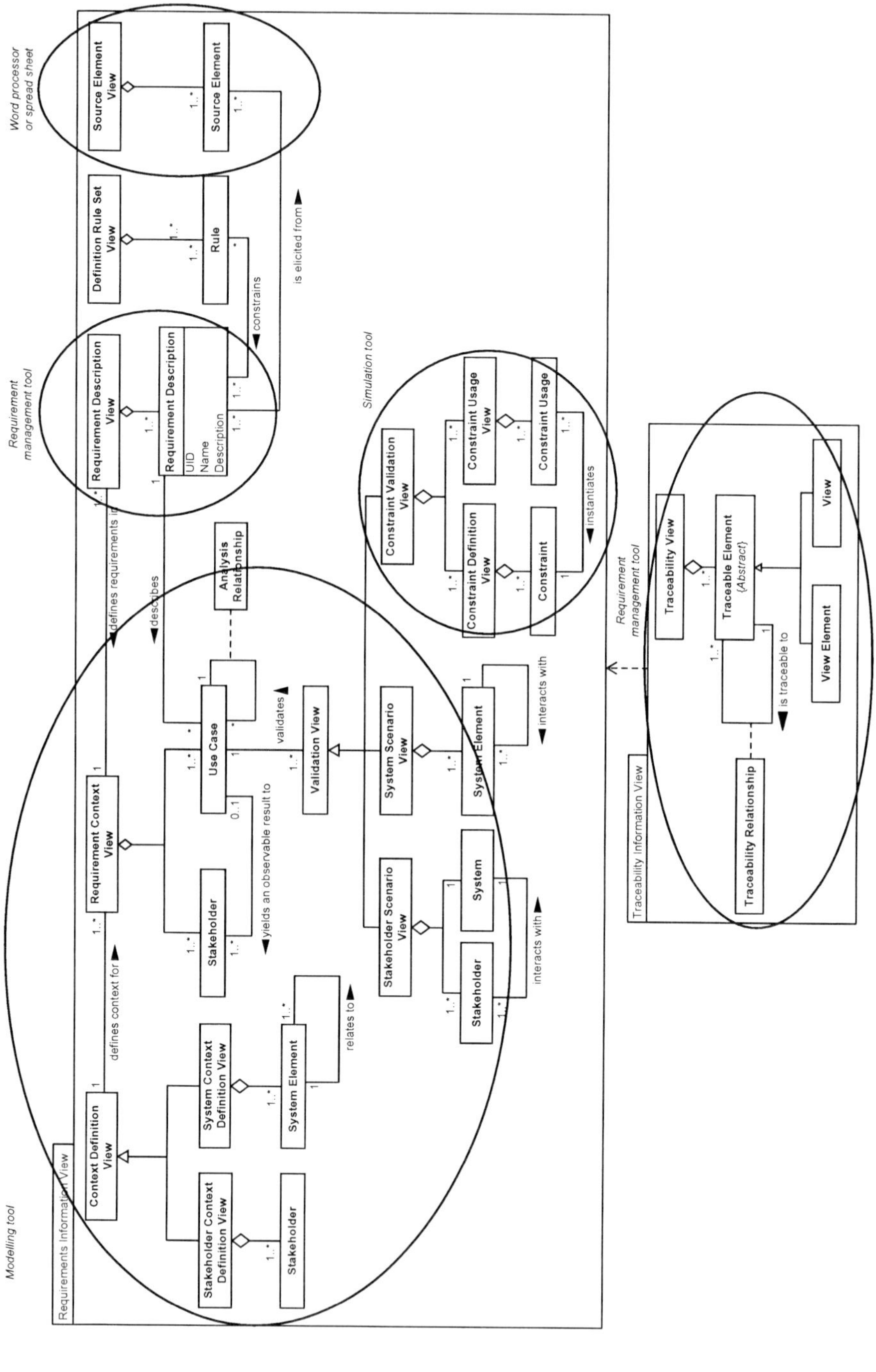

Figure 7.28 Formal tool implementation – heterogeneous tool chain

tools. These are problems that should be addressed with the next example of tool implementation (Figure 7.29).

The second example is that of the formal process implementation when using an integrated tool chain. The tools may come from different vendors but all the hard work of tool integration has already been carried out by the tool provider. At the heart of integrated tool chain is the modelling tool that provides interfaces to all of the other tools, such as requirements management tools, simulation tools, checking tools, and document generation tools.

One possible way of implementation of the views is as follows:

- The 'Source Element View' may be realised using almost any tool or file type. For example: emails, digital photos, scans of log books, word processor documents and so on. These will be linked back to elements within the model or may even have SysML blocks to represent them.
- The 'Requirement Description View' is defined in the modelling tool using SysML blocks. If anyone wants to see the same information in a requirements management tool, a button is pressed and the management modules are generated automatically. These may then be edited and then re-imported back into the model – the model is the master.
- The 'Requirement Context View' may be realised using use case diagrams in the modelling tool.
- The 'Context Definition View' may be realised using a block definition diagram using the modelling tool.
- The 'Validation Views may be realised using sequence diagrams in the modelling tool. The parametric constraints may be defined using SysML requirements diagrams in the modelling tool and then, at the push of button, they are sent to the simulation tool for formal validation.
- The 'Traceability View' is generated automatically from the model. When the modelling is carried out properly, traceability is inherent in the model and needs not be entered again.

The true benefits of MBSE start to become apparent now. Just in terms of automating artefacts, checking and simulation, the following benefits may be realised:

- *Automatic verification of the model.* This may take forms as there may be standard checks and bespoke checks. When using SysML, there are a number of standard consistency checks that should be carried out on any model. However, in addition to these checks, there may be bespoke checks associated with the process that is being followed. For example, a number of consistency checks have been identified for the ACRE process in Chapter 5, which can be automated. This automation of consistency checks is a massive timesaver. Performing consistency check manually takes a long time, is incredibly tedious and is very prone to human error. Just performing the consistency checks for the ACRE process can take anything between a few hours and a few days, depending on the size of the project. By automating this, the whole process takes anywhere from a few seconds to just under a minute.

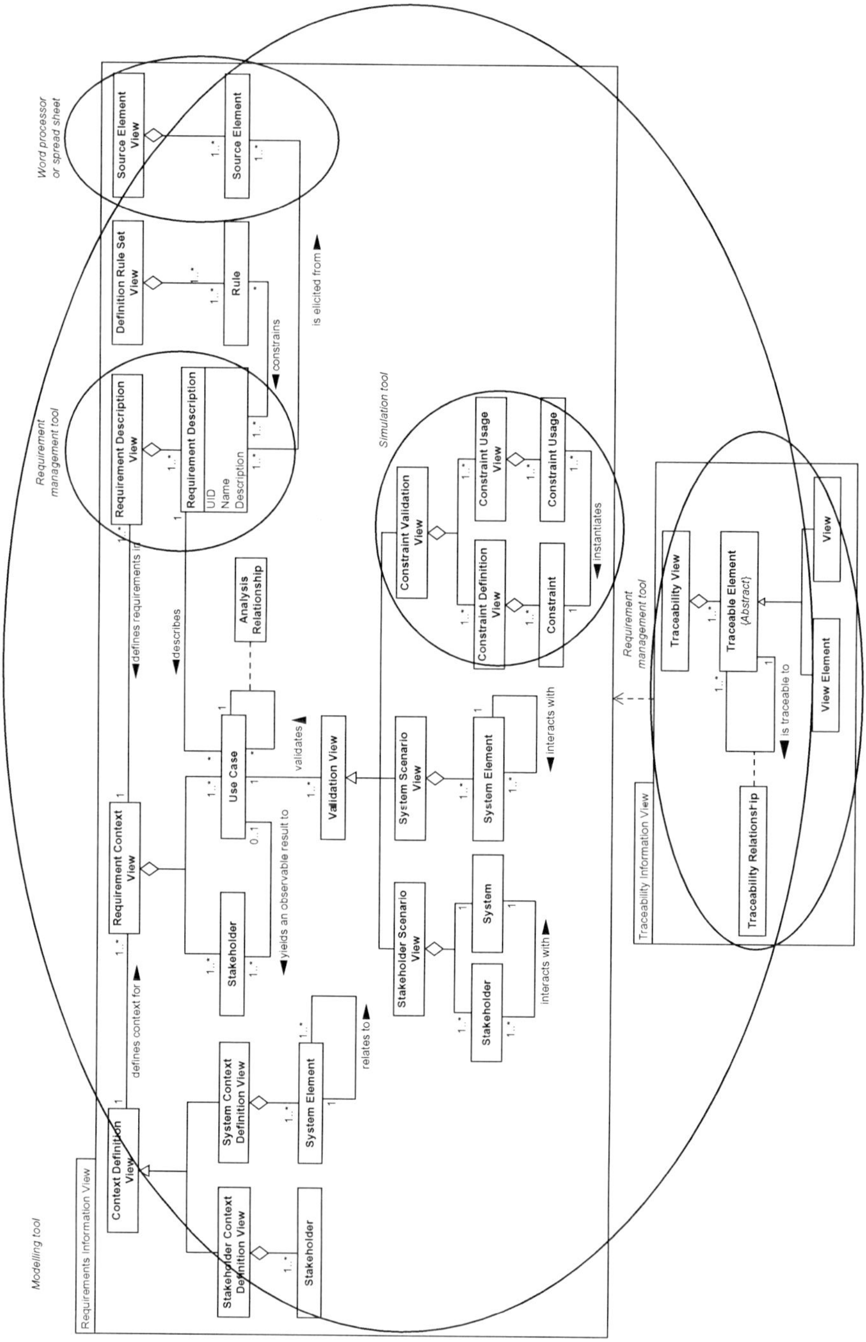

Figure 7.29 Formal process implementation – integrated tool chain

- *Automatic generation of documents.* The modelling allows the user to define document templates that may be used by standard word processors. Elements in the model are used to populate the document that is then produced as the word-processed document. This requires some effort to set up the templates, but once done, it can save a lot of time. Apart from saving time, it also means that the word-processed document becomes a 'throwaway' document in that if it ever needs to be changed then it is simply discarded. Any changes are made to the model and then the document is regenerated. This means that a document can be generated at any point in time and that it is guaranteed to be consistent with the model.
- *Automated interface with simulation tools.* SysML modelling tools are usually not very good at performing mathematical simulation as it is not what they are intended for. Therefore, a good interface between the modelling tool and the simulation tool can provide an excellent way to validate the parametric constraints that have been defined in the model.
- *Requirements management automation.* As stated previously, if the model is correct, traceability comes for free; there are many cases, however, where there is still a need for a separate requirements management tool, whether this is a true or perceived need. Again, the model is the master and the management tool information is simply automatically generated. Any change made in the management tool may then be imported back into the model, assuming that the interface is bi-directional.

Many of these points are revisited in Chapter 6 in the case study, where examples are given for these views.

7.4.3 Tool selection

The selection of tools for a project or an organisation can be the most important decision that is made on a project. Getting the right tool is very important, but getting the wrong tool is usually much worse than not getting any tool at all.

There are a number of factors that should be borne in mind when selecting tools, each of which is discussed in the following sections.

7.4.3.1 Providing modelling capability

This is probably the most obvious factor and one that most people will put a lot of effort into. The modelling capability describes the drawing, management and usability of a tool. This is aimed primarily at modelling (and some drawing) tools and should not be taken lightly or assumed. Many tools claim compliance with a specific modelling notation, e.g. SysML, when, in fact, they do not meet all the standards. This can be very important because it is not just a matter of how much of the standard is implemented, but also to what degree is the notational meta-model implemented in the tool. For example, if you want consistency checking as part of your tool, it needs to have a meta-model and not just a simple drawing package.

The compliance for the standard is relatively easy to quantify and the SysML standard has a full specification that can be used for the comparison. There are some aspects, however, that are not so easy to quantify, such as the usability.

The tool should be easy to use in that it should have an intuitive interface. Unfortunately, what one person thinks is perfectly intuitive (e.g. a programmer) may not seem at all intuitive to someone else (an end user). Therefore, it is often desirable to have a tool that has a configurable user interface.

Another aspect of usability that should be considered is the navigation of the model. How easy is this, are there search functions, can one diagram be automatically navigated to from another and so on.

7.4.3.2 Providing system functionality

This is a very pragmatic concern that describes the actual system requirements in terms of any hardware and software requirements along with networking capabilities. Some tools, particularly more powerful ones, come with quite heavy installation requirements. Also, this problem can be compounded when using multiple tools from multiple vendors.

7.4.3.3 Providing interoperability capability

This is key to MBSE and describes how a tool may be required to operate with other tools, technologies or clients. The previous sections have discussed how one of the biggest problems with tool chains is the integration of them. It is important to look for what data formats may be used as outputs for compatibility of tools, what interfaces are provided as standard, what interfaces may be programmed and so on.

7.4.3.4 Ensuring vendor's quality of service

This describes how to ensure that the vendors can meet all of your requirements. If you are working on a project that is in any way mission critical and expensive or that is scheduled to take a long time, it is important that good support is available from the tool provider. Also, it is important that the tool provider is still be in business when the project ends. Generally speaking, large companies have more chance of existing over the next 10 years, whereas a small provider may not. On the other hand, smaller companies tend to provide better, or certainly more personal and consistent support, as they can be for more pro-active and reactive in their support.

7.4.3.5 Being compatible with process model

This describes how the tool may be required to fit in with a particular approach to working, which may put additional requirements on the tool. A good example of this would be support for the ACRE process in the form of profiles, templates and so on.

For a more in-depth discussion on the selection of modelling tools, see Reference 6.

7.5 Conclusions

This chapter has discussed how the three aspects of MBSE (people, process and tools) may be implemented for real projects and systems. This has involved discussions on the following:

- How to ensure that people are competent, with example competencies and scopes.
- How to teach the ACRE approach with a teaching guide.
- A definition of the ACRE process using the 'seven views' approach to process modelling.
- Examples of three ways to implement the ACRE process ranging from a 'quick and dirty' process, to a 'semi-formal' process right up to a 'formal process'.
- The types of tool that may be considered.
- Examples of how tools may be used for the different types of process execution.
- How to select tools.

The main aim of this chapter has been to see how the process can be implemented in real-life situations and considered the pragmatic issues of using the ACRE process.

References

1. Bloom B.S. (ed.). *Taxonomy of Educational Objectives: The Classification of Educational Goals*. Susan Fauer; 1956. pp. 201–207
2. International Council on Systems Engineering (INCOSE). 'INCOSE Competencies Framework, Issue 2.0'. International Council on Systems Engineering (INCOSE); November 2006
3. Holt J., Perry S. *A Pragmatic Guide to Competency: Tools, Frameworks and Assessment*. London, UK. BCS Publishing; 2011
4. Holt J. *A Pragmatic Guide to Process Modelling*. London, UK. 2nd edn. BCS Publishing; 2009
5. *BS ISO/IEC 15288:2002 Systems Engineering – Systems Life Cycle Processes*. London, UK. ISO Publishing; 2002
6. Holt J. *UML for Systems Engineering – Watching the Wheels*. London, UK. 2nd edn. IET; 2005
7. Anderson L.W., Krathwohl D.R., Airasian P.W., Cruikshank K.A., Mayer R. E., Pintrich P.R., et al. (eds.). *A Taxonomy for Learning, Teaching, and Assessing – A Revision of Bloom's Taxonomy of Educational Objectives*. Addison Wesley Longman; 2001
8. Holt J., Perry S. *SysML for Systems Engineering*. London, UK. IET; 2008

Appendix A
Summary of notation

During the process of stepwise refinement, a notation which is natural to the problem in hand should be used as long as possible.

Niklaus Wirth (1934–)

A.1 Introduction

This appendix provides a summary of the systems engineering modelling language (SysML) notation used throughout this book. Although the SysML contains nine types of diagram, only five are used in this book. For each of these five SysML diagram types, three diagrams are given:

- A simplified meta-model diagram for that diagram type
- The notation used on that diagram type
- An example of that diagram type

This appendix is intended to be a summary of the information found in Chapter 3 and therefore each diagram is presented without comment. For more information on the SysML and the five diagrams used in this book, see Chapter 3 and Reference 1.

A.2 Block definition diagrams

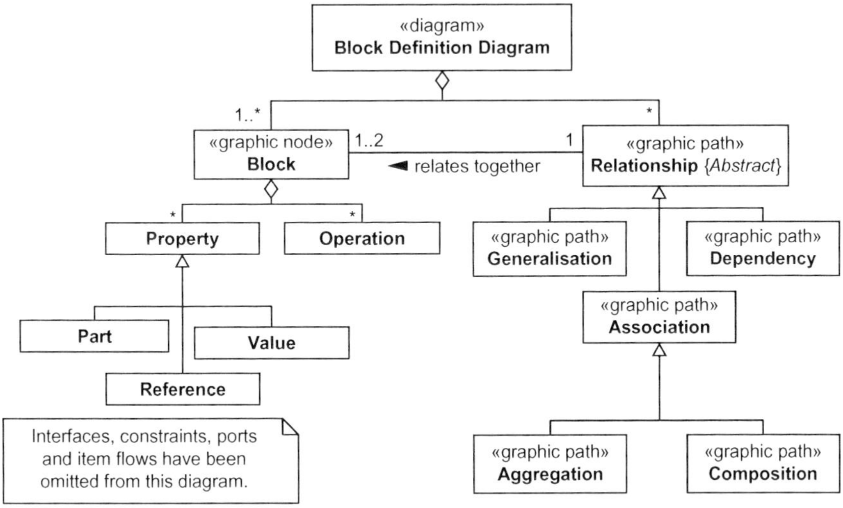

Figure A.1 Simplified meta-model for the block definition diagram

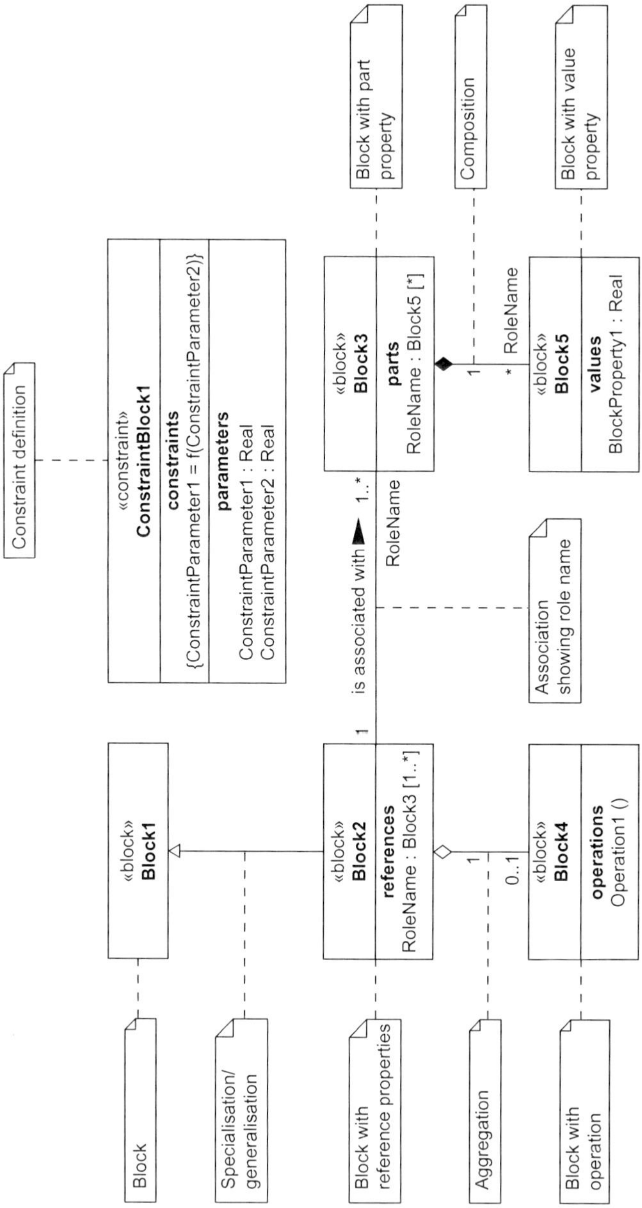

Figure A.2 Graphical symbols for elements in a block definition diagram

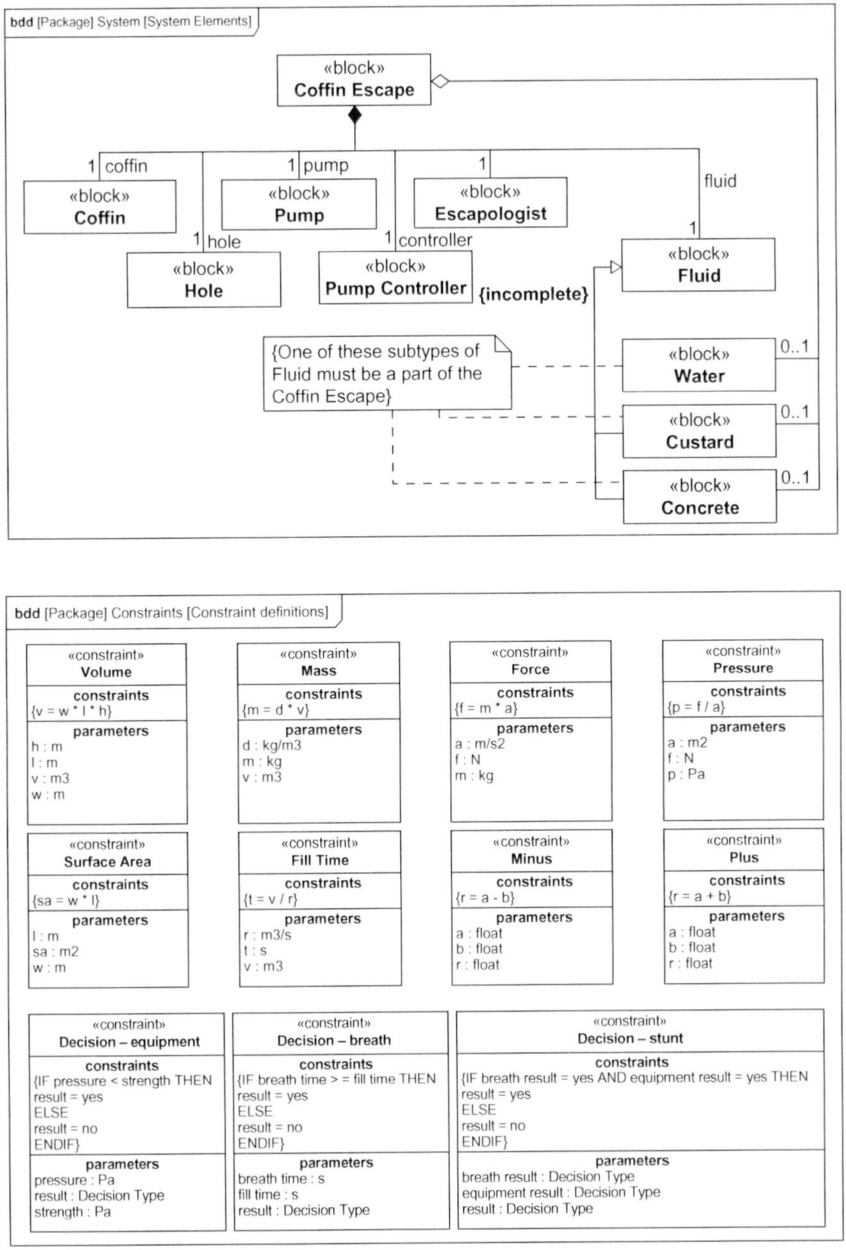

Figure A.3 Example block definition diagrams, including one showing definition of constraints

A.3 Requirement diagrams

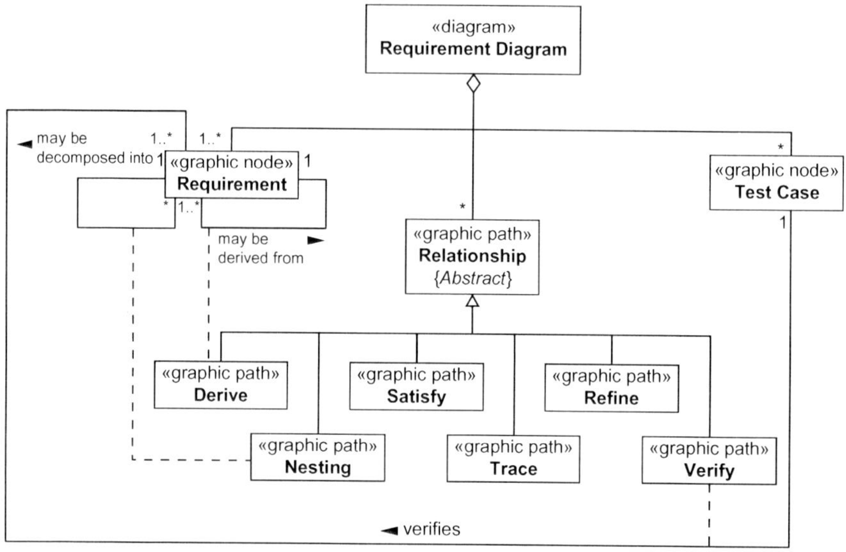

Figure A.4 Simplified meta-model for the requirement diagram

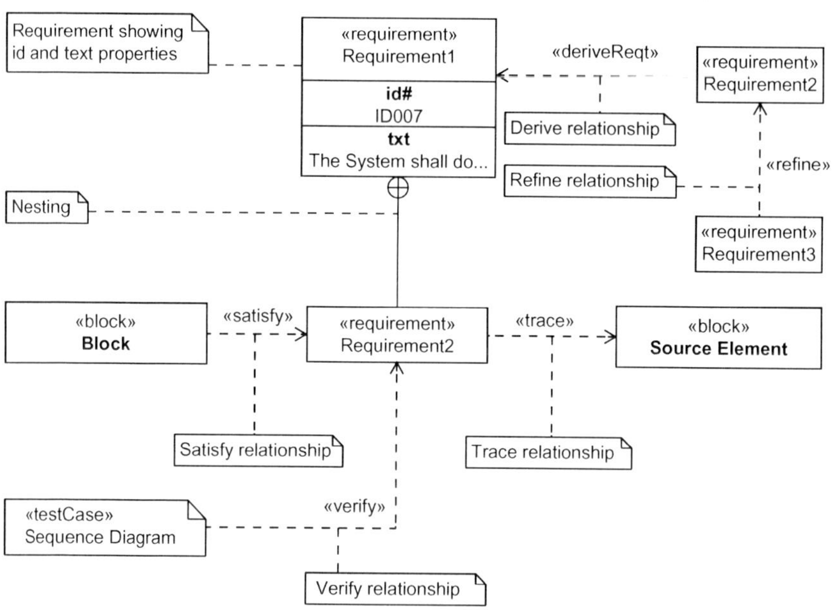

Figure A.5 Graphical symbols for elements in a requirement diagram

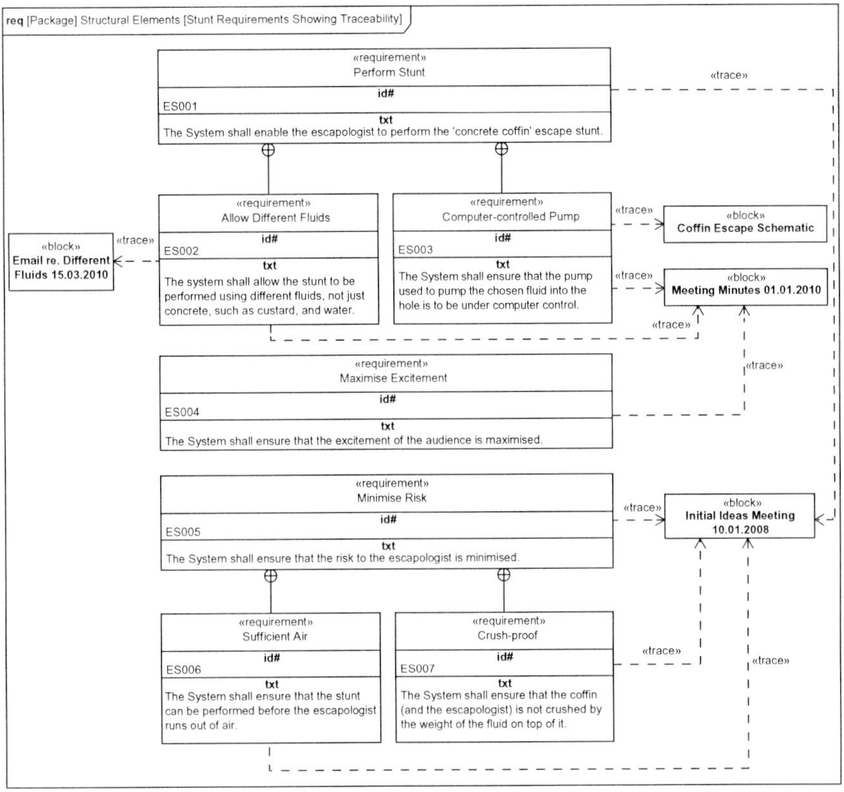

Figure A.6 Example requirement diagram

A.4 Parametric constraint diagrams

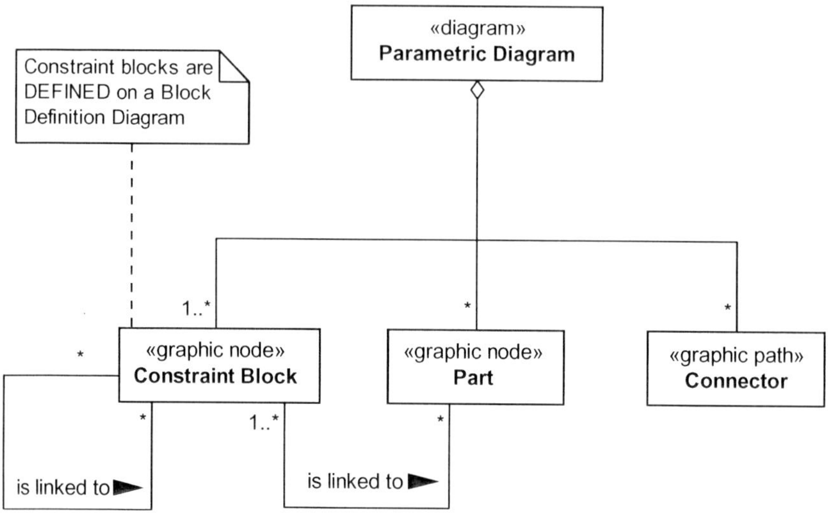

Figure A.7 Simplified meta-model for the parametric constraint diagram

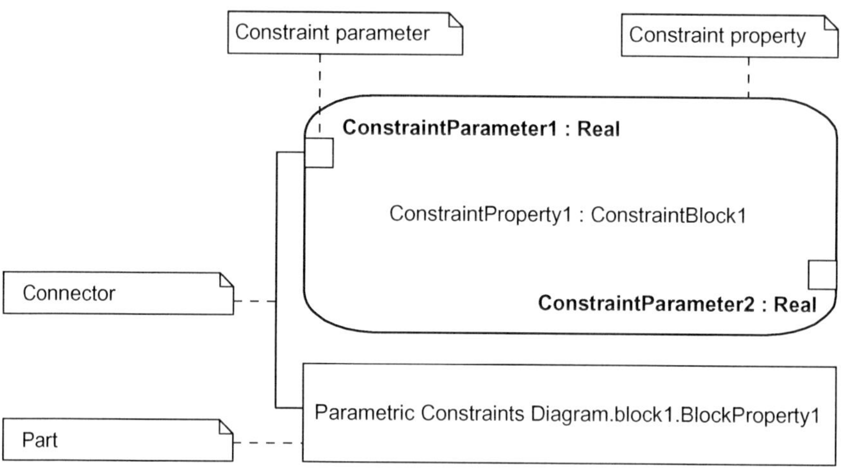

Figure A.8 Graphical symbols for elements in a parametric constraint diagram

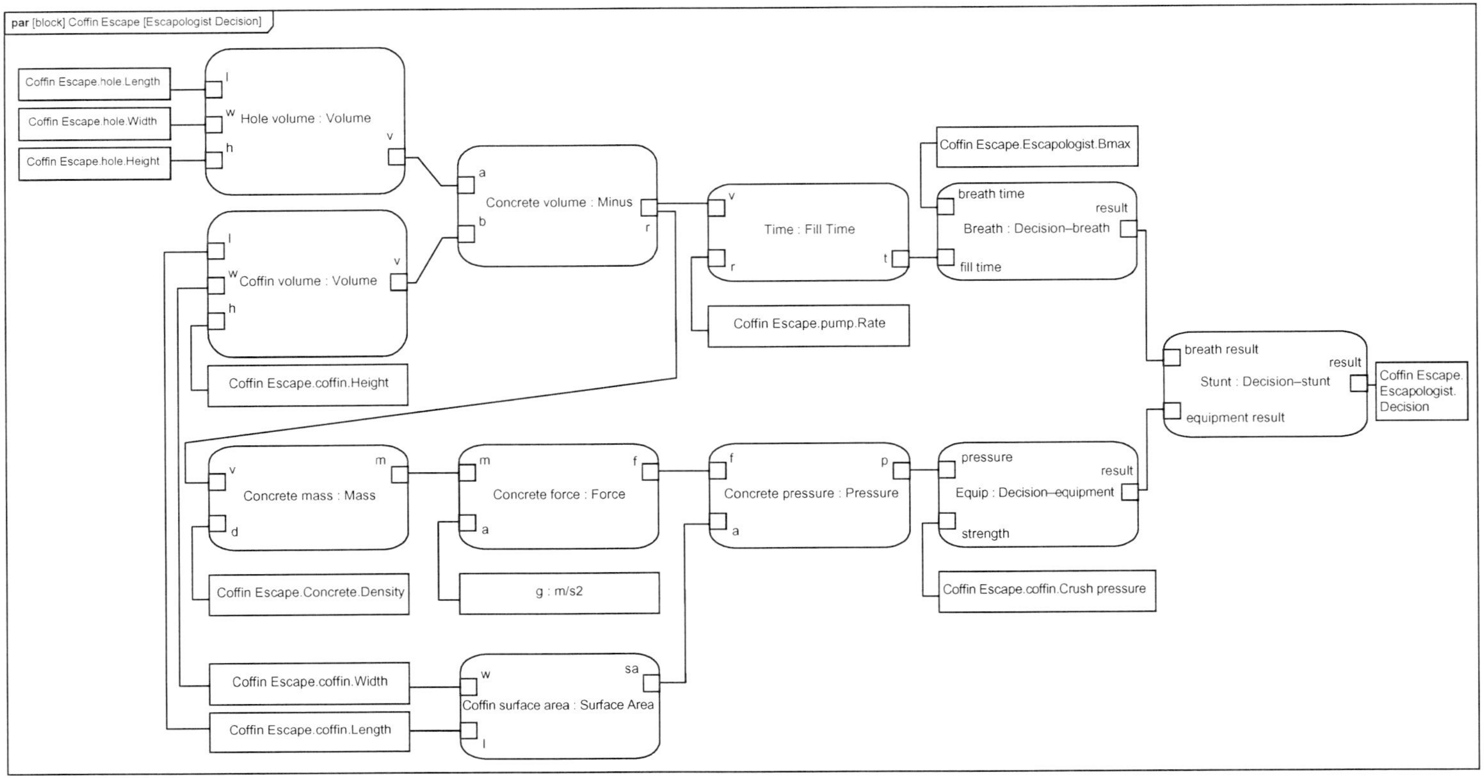

Figure A.9 Example parametric constraint diagram

A.5 Use case diagrams

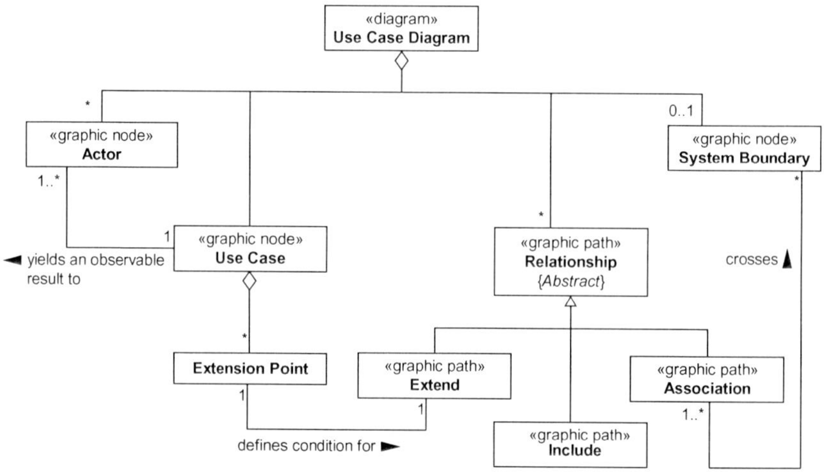

Figure A.10 Simplified meta-model for the use case diagram

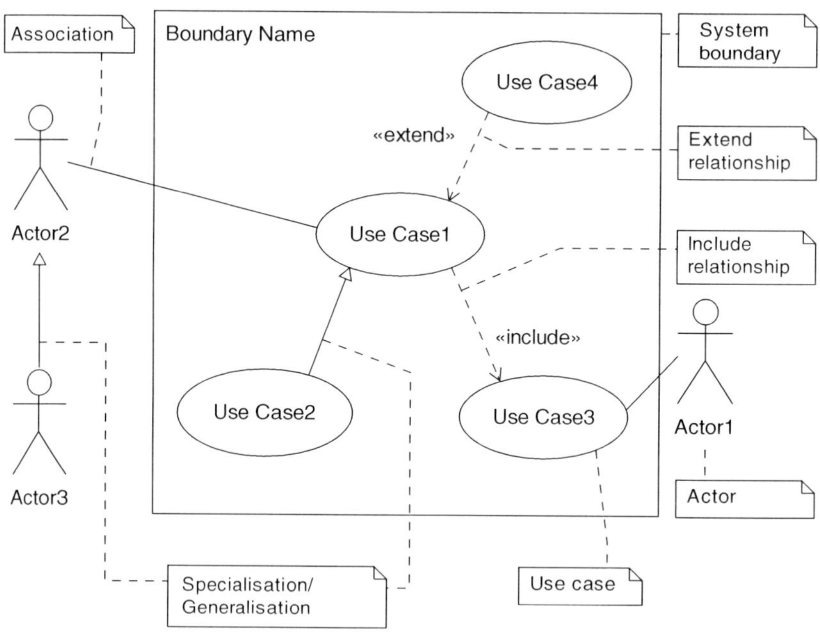

Figure A.11 Graphical symbols for elements in a use case diagram

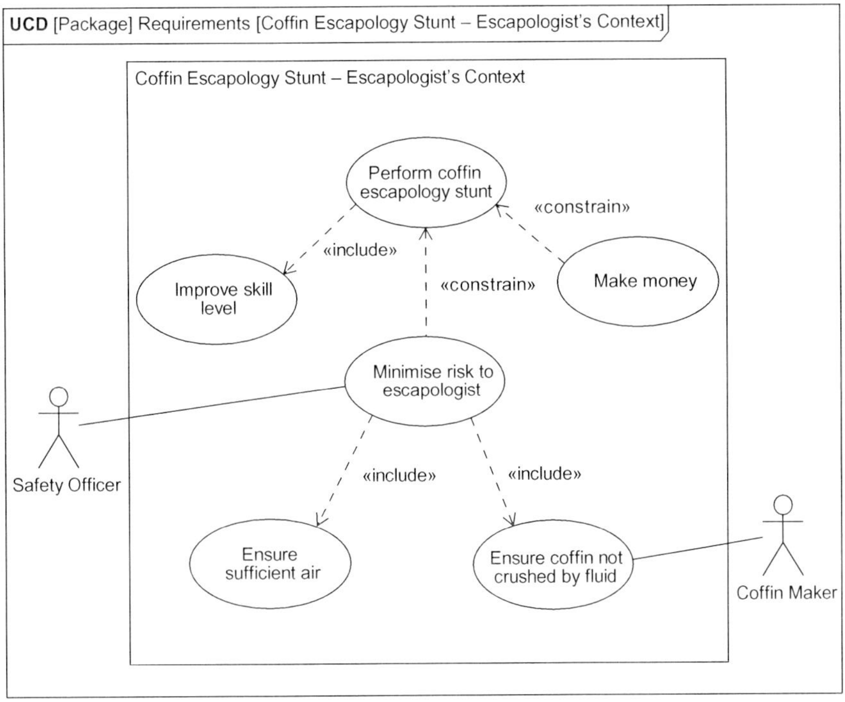

Figure A.12 Example use case diagram

A.6 Sequence diagrams

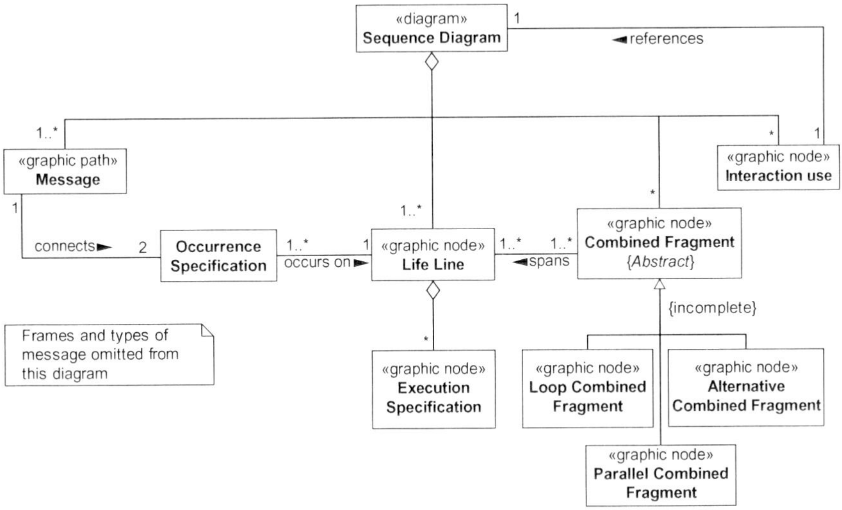

Figure A.13 Simplified meta-model for the sequence diagram

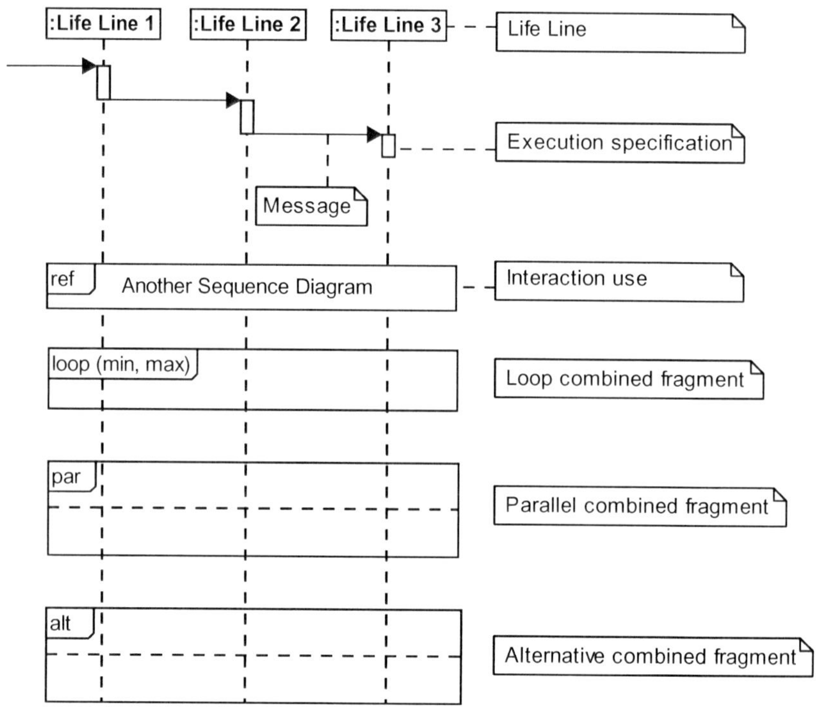

Figure A.14 Graphical symbols for elements in a sequence diagram

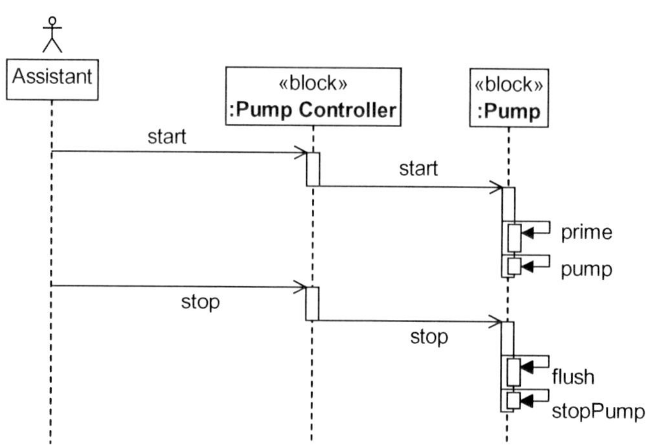

Figure A.15 Example sequence diagram

Reference

1. Holt J., Perry S. *SysML for Systems Engineering*. IET: London; 2008

Appendix B
Advanced sequence diagram notation

Step by step one goes very far.

Proverb

B.1 Introduction

This appendix gives a summary of some of the advanced notations that can be used with SysML sequence diagrams that allow the following to be shown:

- Parallel behaviour
- References to other sequence diagrams
- Alternative steps
- Loops

The notation to allow these constructs are known as *combined fragments*. For full details, see Reference 1. Examples of the notation for each of these combined fragments showing how they are used are given in the following sections.

B.2 Showing parallel processing

Parallel paths can be shown in sequence diagrams using a *parallel combined fragment*. Each parallel path appears in a separate compartment within the combined fragment frame. The parallel compartments are divided by a dashed line, and the combined fragment uses the keyword *par*.

Figure B.1 shows a sequence diagram with two parallel combined fragments, each of which has two parallel regions. The first parallel combined fragment shows the 'Begin stunt' message being sent from the 'Set up' lifeline to the 'Start' lifeline at the same time as the 'Start' lifeline sends the 'Begin stunt' message to the 'Escape' lifeline. Similarly, the second parallel combined fragment shows the 'Start escape' message being sent between the 'Start' and 'Escape' lifelines at the same time that it is sent between the 'Escape' and 'Monitor' lifelines.

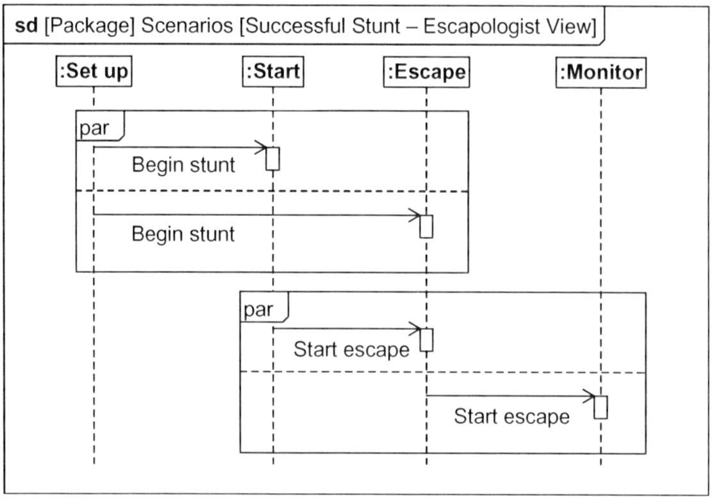

Figure B.1 Sequence diagram showing a parallel combined fragment

B.3 Referencing other diagrams

Often, when modelling scenarios, common behaviour is observed. Rather than repeating this behaviour on every sequence diagram that needs it, SysML allows other sequence diagrams to be referenced to allow reuse of scenarios.

For example, say that we have some common functionality that we want to show on multiple scenarios. First, we model this using a sequence diagram. An example is shown in Figure B.2.

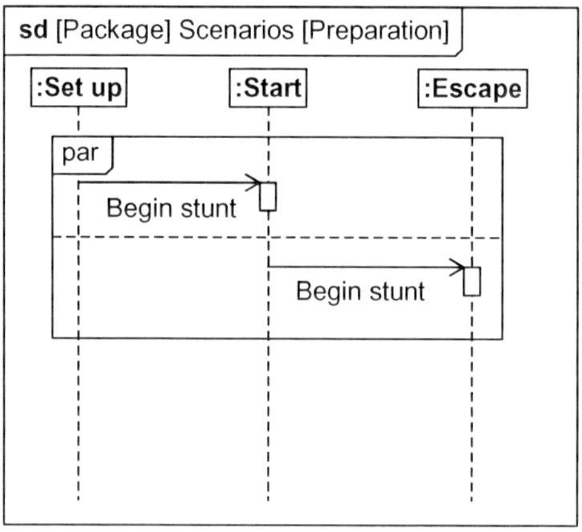

Figure B.2 Sequence diagram defining common functionality to be referenced

This functionality can then be reused on another sequence diagram using a *reference combined fragment*. Each referenced scenario appears in a separate frame with the keyword *ref*, as shown in Figure B.3.

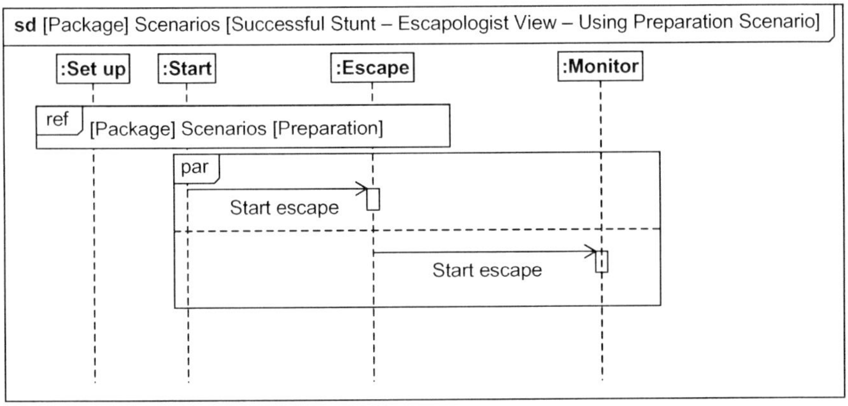

Figure B.3 Sequence diagram showing the use of a reference combined fragment

The lifelines that appear in the sequence diagram referenced must appear on the referencing diagram, and the reference combined fragment must be placed over those lifelines as in Figure B.3.

B.4 Showing alternatives

Sometimes two or more scenarios are so similar that showing alternative paths on a single diagram rather than one per diagram is desirable. SysML allows scenarios to be modelled in this way using *alternative combined fragments*.

This consists of a frame with the keyword *alt* that is divided into separate compartments, one for each alternative, by dashed lines. Each compartment should have a *guard condition* that indicates the conditions under which that alternative is executed. The absence of a guard condition implies a true condition. The guard condition *else* can be used to indicate a condition that is true if no other guard conditions are true. Although there is nothing in SysML to prevent the use of guard conditions where more than one can evaluate to true, this leads to a non-deterministic sequence diagram and is to be avoided. An example of a sequence diagram showing two alternatives is shown in Figure B.4.

B.5 Showing loops

The final combined fragment to be considered allows looping behaviour to be shown. The *looping combined fragment* is shown using a frame with the keyword *loop*. The keyword may be accompanied by a repetition count specifying a *minimum* and *maximum* count as well as a guard condition. The loop is executed while

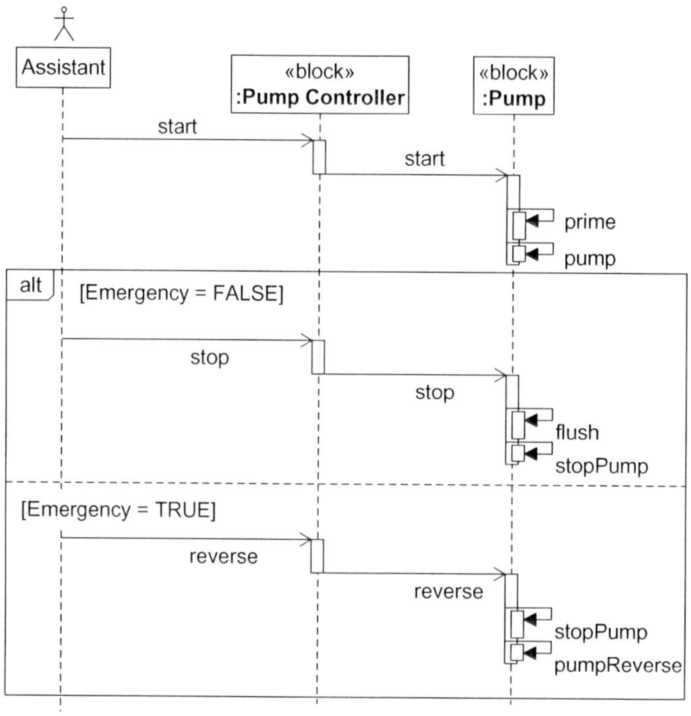

Figure B.4 Sequence diagram showing the use of the alternative combined fragment

the guard is true, but at least the minimum count, irrespective of the guard, and never more than the maximum count.

The syntax for loop counts is the following:

- loop minimum = 0, unlimited maximum
- loop(repeat) minimum = maximum = repeat
- loop(min, max) minimum and maximum specified, min ≤ max

An example sequence diagram showing a loop combined fragment is shown in Figure B.5.

The diagram shows a loop with no repetition count (which is the same as a *loop forever*), and a guard that indicates that the loop is to continue while the coffin escape stunt is not complete.

There are many other types of combined fragment defined, but the four discussed here are the most often used. For details of the other types of combined fragment, such as the *break* or *opt* fragments, see Reference 1.

In addition, there is nothing to prevent the nesting of combined fragments. For example, a *loop* may have a *par* fragment inside it, with *ref* fragments and perhaps even *alts* in each parallel region. Remember, though, that one of the key aims of

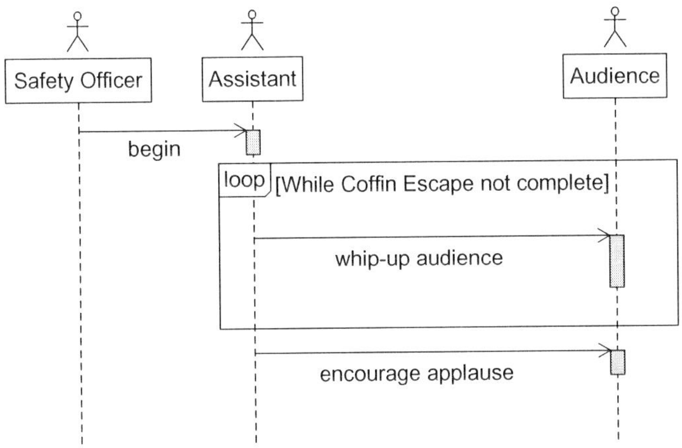

Figure B.5 Sequence diagram showing the use of a loop combined fragment

modelling is to *improve* the communication of complex ideas and such diagrams, whereas valid SysML should be used with caution as it can rapidly become very difficult to understand and make the communication worse rather than better.

Reference

1. Rumbaugh J., Jacobson I., Booch G. *The Unified Modeling Language Reference Manual*. 2nd edn. Boston, MA: Addison-Wesley; 2005

Appendix C

A seven views summary of the UCAM processes

> Don't get involved in partial problems, but always take flight to where there is a free view over the whole single great problem, even if this view is still not a clear one.
>
> *Ludwig Wittgenstein (1889–1951)*

C.1 Introduction

Chapter 7 has discussed the importance of having competent people as an essential part of a model-based approach to requirements engineering. To assess the competency of staff, it introduced the concepts of competency assessments and frameworks that can be tailored, developed and deployed using a technique such as the universal competency assessment model (UCAM), a model- and process-based approach to competency assessment.

A full discussion of UCAM is beyond the scope of this book and the reader is directed to Reference 1. The UCAM itself is modelled using a UML/SysML-based technique informally known as the 'seven views' approach. Again, a full discussion of the seven views approach is beyond the scope of this book, and Reference 2 gives full details.

However, this appendix gives a very brief overview of each of the seven views from Reference 2 together with a model of the UCAM processes expressed using the seven views approach. No explanation of the UCAM diagrams is given here; the full description can be found in Reference 1.

C.2 Overview of the seven views approach

The 'seven views' approach is a model-based technique for modelling business processes and is described fully in Reference 2. It defines seven inter-related views of any process and posits that the seven views are the minimum that are needed to ensure a process, or set of processes, is both completely and consistently specified. The seven views and the relationships between them are shown in Figure C.1.

The diagram shows each of the seven views and also shows, using stereotypes (the words enclosed in angle brackets, e.g. «use-case diagram»), how each of the concepts described in the seven views approach may be realised using the SysML.

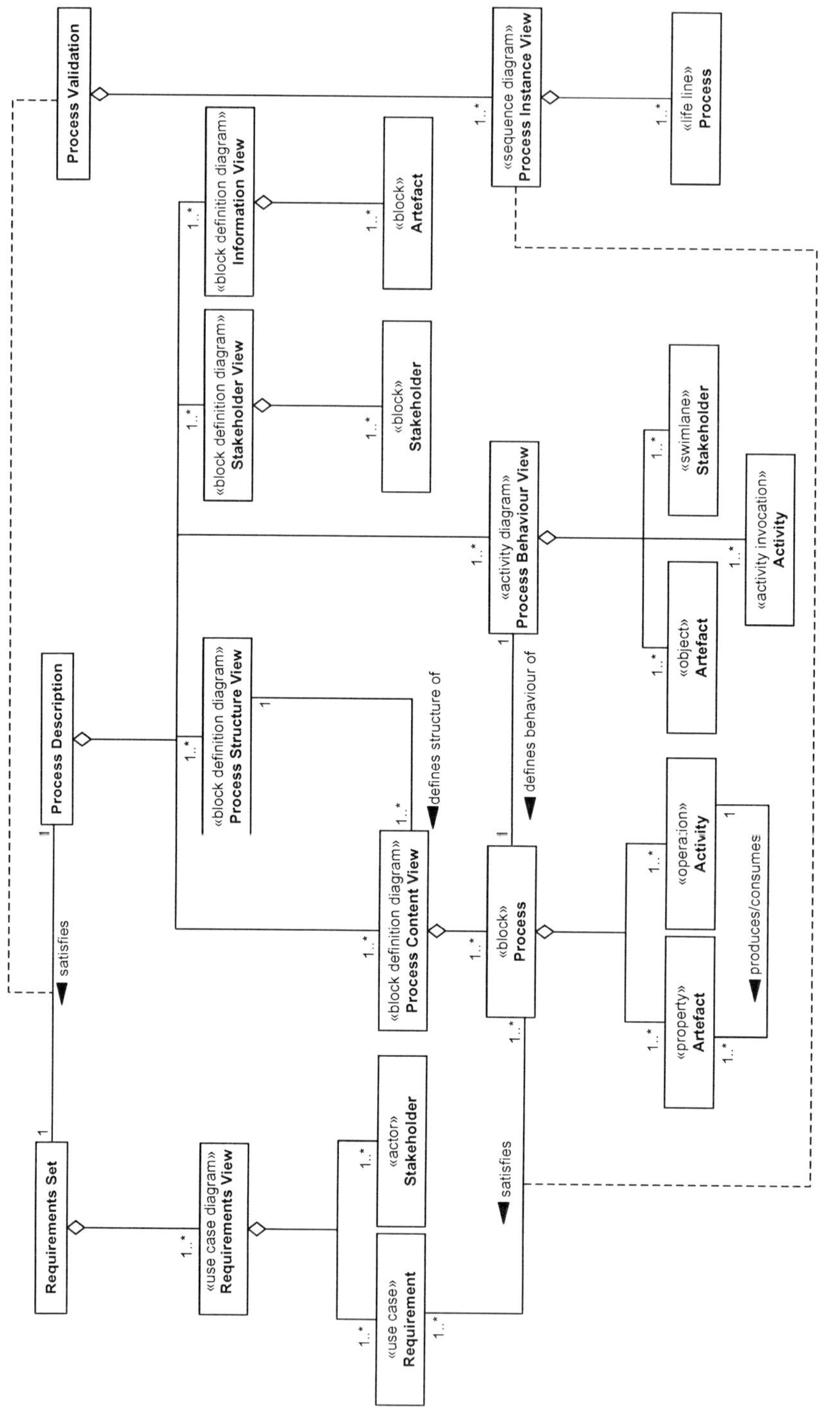

Figure C.1 The seven views and their relationships

It also shows how the seven views are related to the important concepts of the requirements for a group of processes (the 'Requirements Set' element in the diagram), the description of a set of processes (the 'Process Description' element) and the need to validate a group of processes (the 'Process Validation' element). Each of the seven views is very briefly described here.

C.2.1 The Requirements View

The Requirements View (RV) specifies the overall aims of a process or set of processes and can, if desired, be represented by more than one view. Often an RV is produced from the point of view of a particular stakeholder or a group of stakeholders.

The RV is essential for validation of the processes to ensure that they continue to fulfil the needs of the organisation. Requirements for processes do change and so the RV needs to be checked periodically to ensure that any changes to the requirements have been captured, allowing the processes to be changed, if necessary, to address the changed requirements.

C.2.2 The Stakeholder View

The Stakeholder View (SV) identifies the stakeholder roles that have an interest in the processes being defined. It presents stakeholders in a classification hierarchy and allows additional relationships, such as managerial responsibility, to be added. The stakeholders appearing on the SV must be consistent with those shown on the RVs.

C.2.3 The Process Structure View

The Process Structure View (PSV) specifies the structure of concepts and the terminology used when defining processes. For example, when one executes a process, does one carry out 'activities' or 'tasks' and are 'artefacts' or 'work products' generated? The PSV defines this vocabulary in order to ensure the consistency in terminology used.

If many different processes have to be mapped to each other, the PSVs for each set of processes form the basis for this process mapping, allowing the terminology used in one process model to be related to the terminology used in another.

C.2.4 The Process Content View

The Process Content View (PCV) identifies the processes available, showing the activities carried out and the artefacts produced and consumed. It may show general associations and dependencies between processes.

It is important to understand that the PCV only *identifies* processes. It does *not* show how they are carried out.

C.2.5 The Process Behaviour View

The Process Behaviour View (PBV) shows how an individual process behaves, and each process identified on the PCVs should have a PBV that defines its behaviour. A PBV shows the order of activities within a process, the flow of information through the process (i.e., the flow of artefacts around the process) and the responsibilities, in terms of stakeholder roles, for carrying out the activities.

The activities and artefacts shown on a PBV must be consistent with those shown for the process on a PCV, and the stakeholder roles indicating responsibility must appear on both the SVs and RVs.

C.2.6 The Information View

The Information View (IV) identifies all the artefacts produced or consumed by a process, showing the relationships between them. IVs can be created at both high and low levels. High-level IVs simply identify artefacts and relationships, whereas low level IVs show the detailed structure and content of individual artefacts.

C.2.7 The Process Instance View

The Process Instance View (PIV) shows instances of processes and the stakeholders involved in order to validate the processes by relating the execution of a sequence of processes back to the requirements for the processes. The PIVs along with the RVs ensure that the processes are fit for purpose and that all the requirements for the processes are met.

C.3 The Seven Views of UCAM

The UCAM approach is a model- and process-based technique for understanding and developing competency frameworks and for conducting competency assessments. It is described fully in Reference 1. This section presents a seven views model of the UCAM processes. Note that the «block» stereotype has been omitted from the various block definition diagrams.

C.3.1 The Requirements View

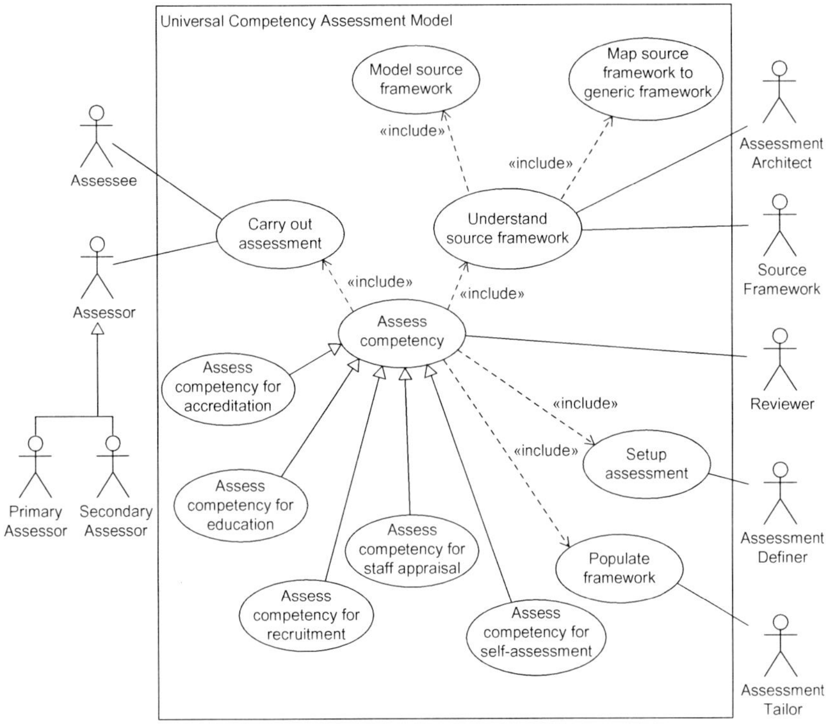

Figure C.2 The RV for the UCAM processes

C.3.2 The Stakeholder View

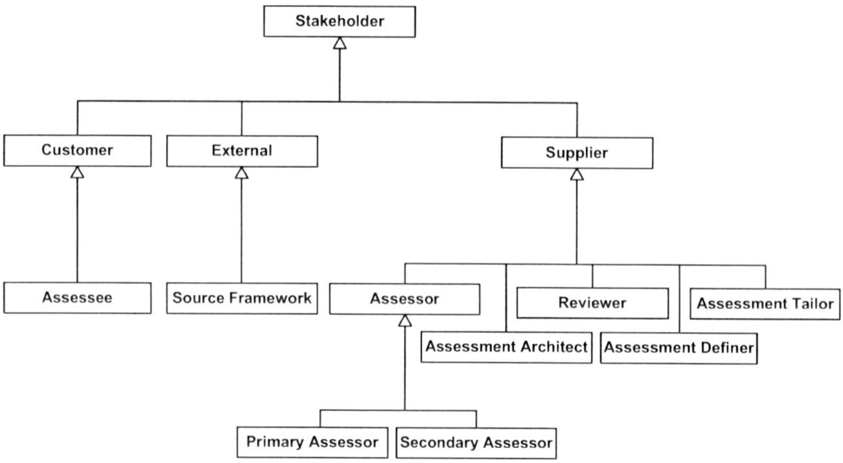

Figure C.3 The SV for the UCAM processes

C.3.3 The Process Structure View

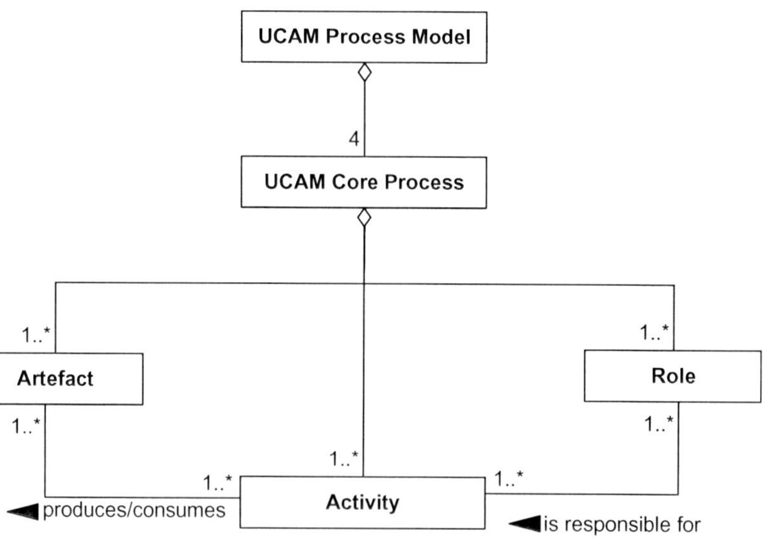

Figure C.4 The PSV for the UCAM processes

C.3.4 The Process Content Views

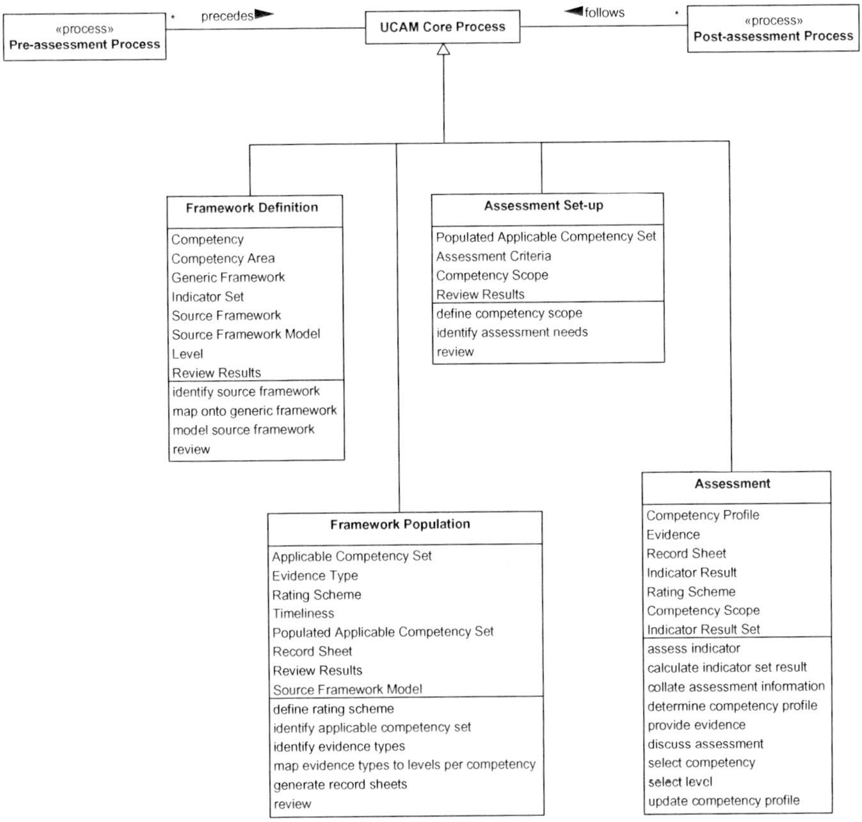

Figure C.5 The PCV for the UCAM processes

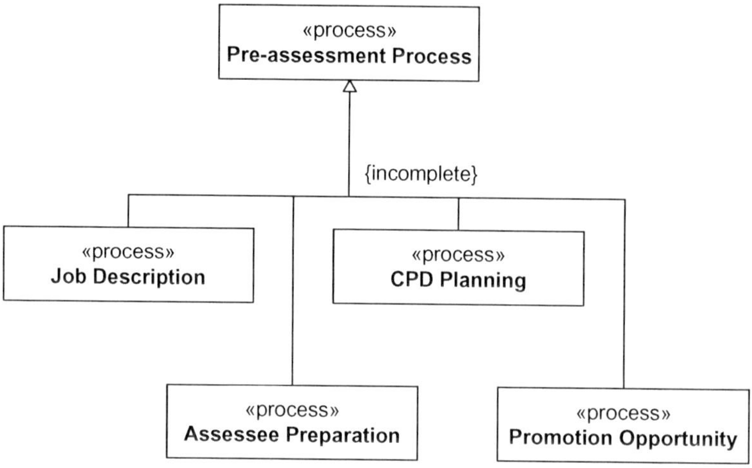

Figure C.6 Partial PCV for suggested pre-assessment processes

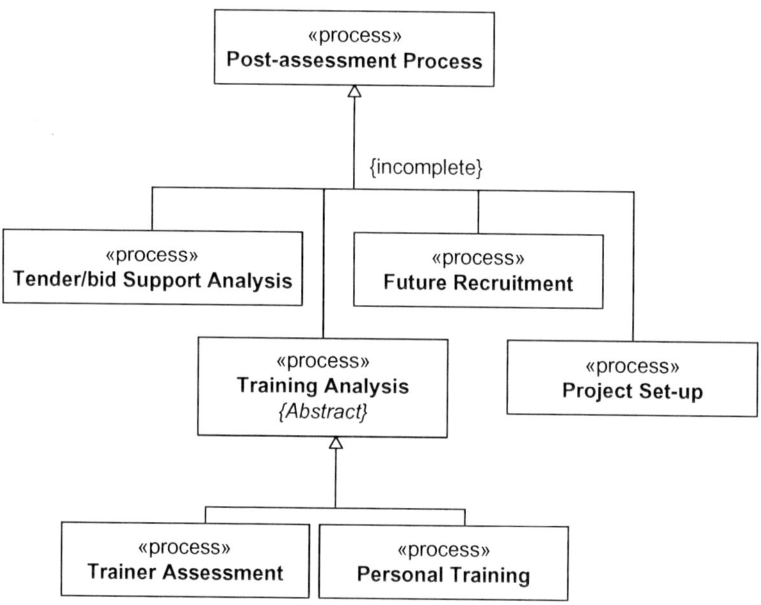

Figure C.7 Partial PCV for suggested post-assessment processes

C.3.5 The Process Behaviour Views

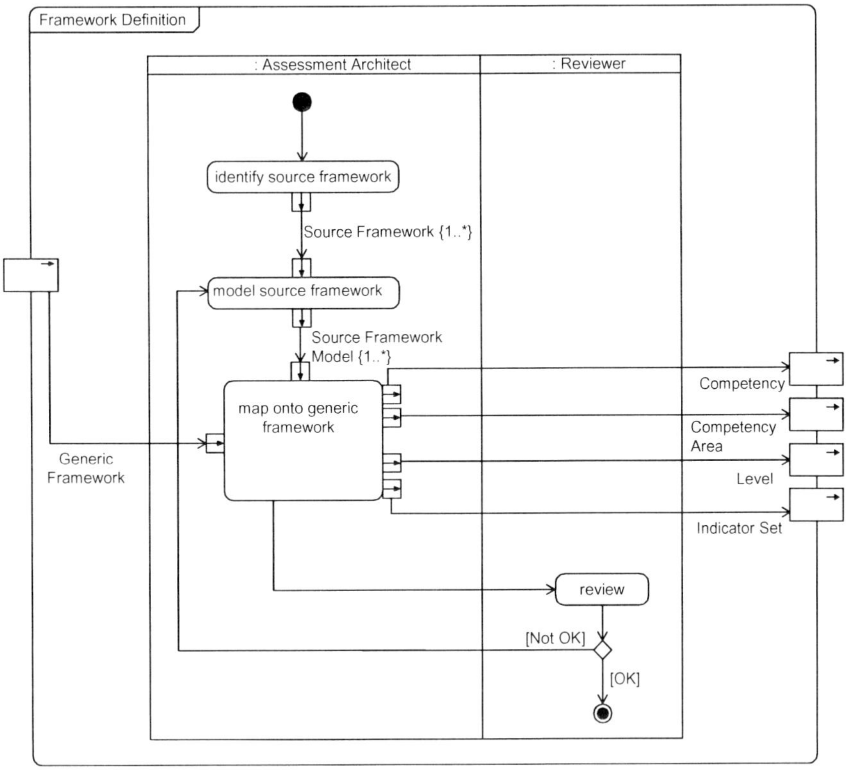

Figure C.8 The PBV for the UCAM 'Framework Definition' process

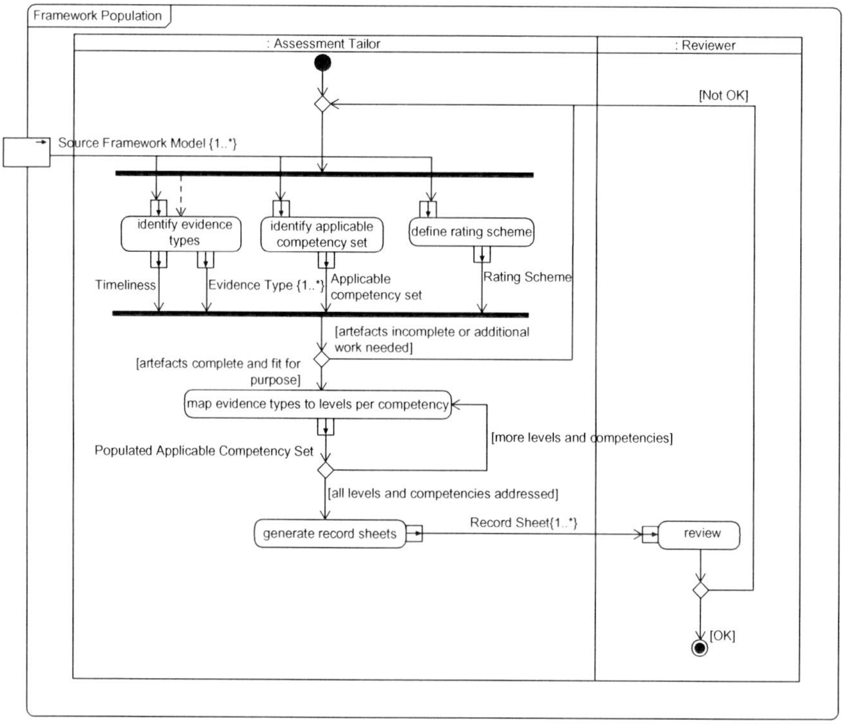

Figure C.9 The PBV for the UCAM 'Framework Population' process

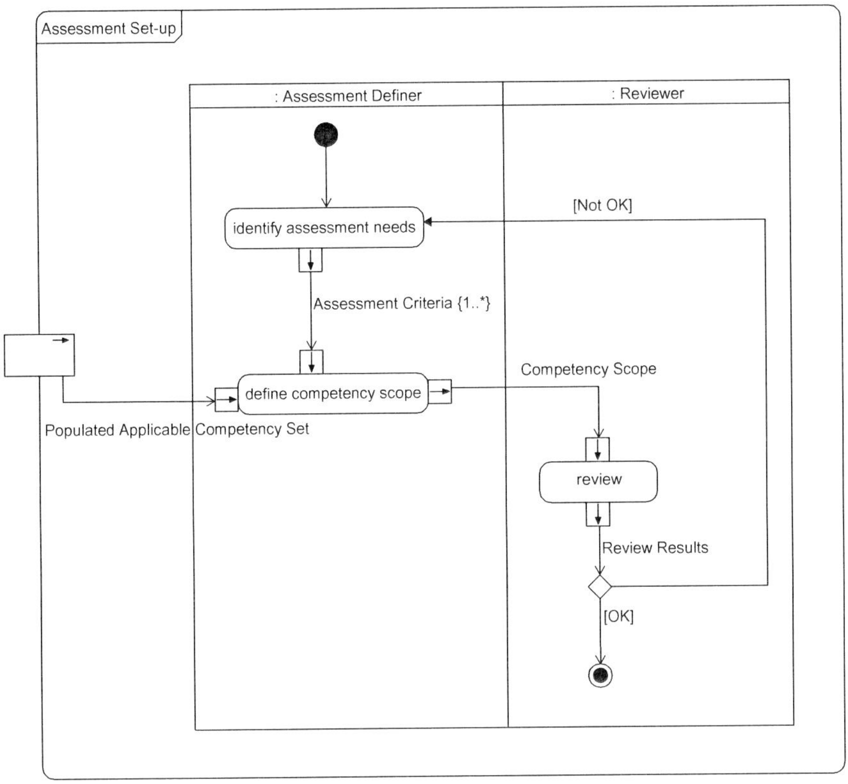

Figure C.10 The PBV for the UCAM 'Assessment Set-up' process

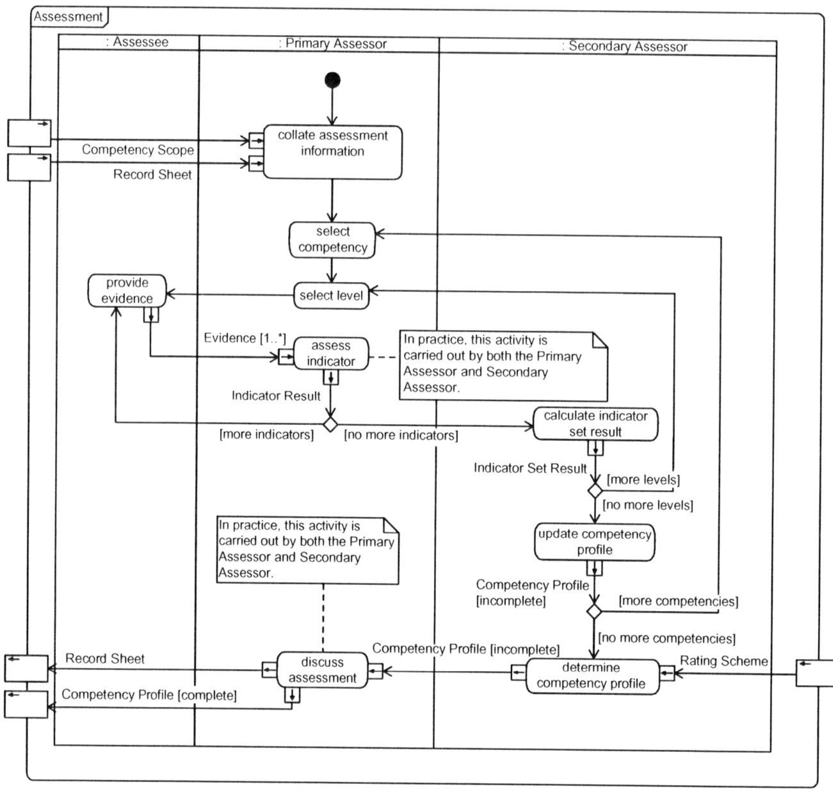

Figure C.11 The PBV for the UCAM 'Assessment' process

C.3.6 The Information Views

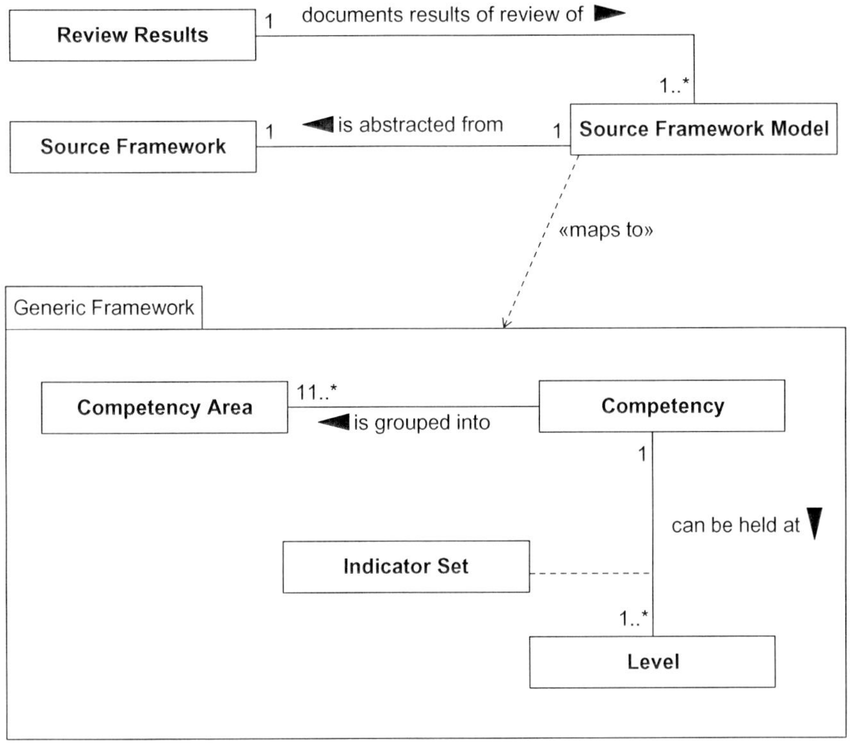

Figure C.12 The IV for the UCAM 'Process Definition' process

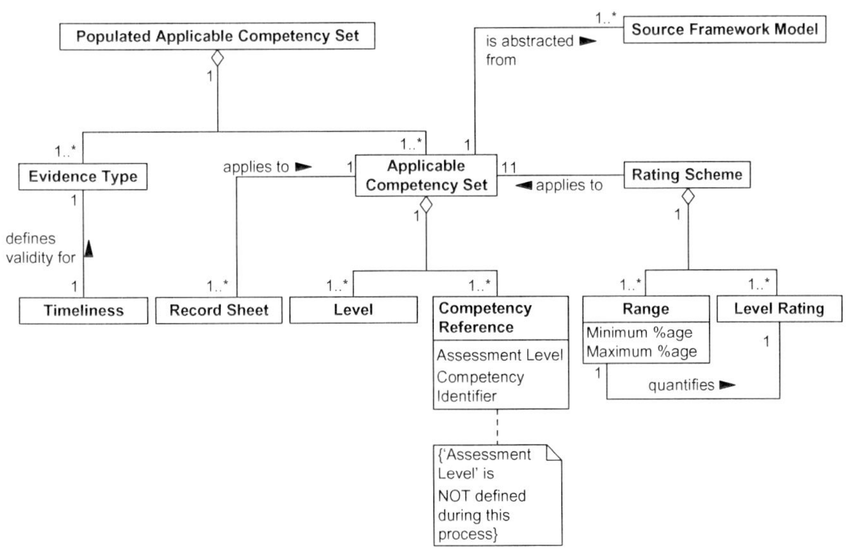

Figure C.13 The IV for the UCAM 'Framework Population' process

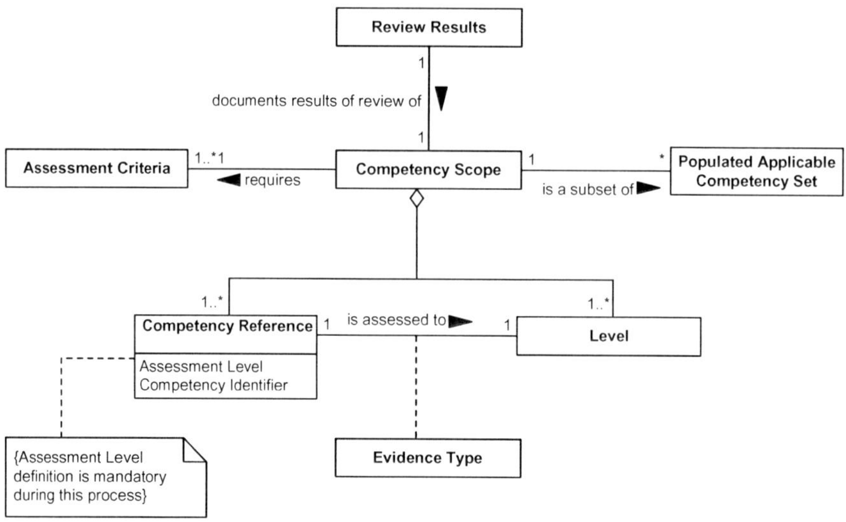

Figure C.14 The IV for the UCAM 'Assessment Set-up' process

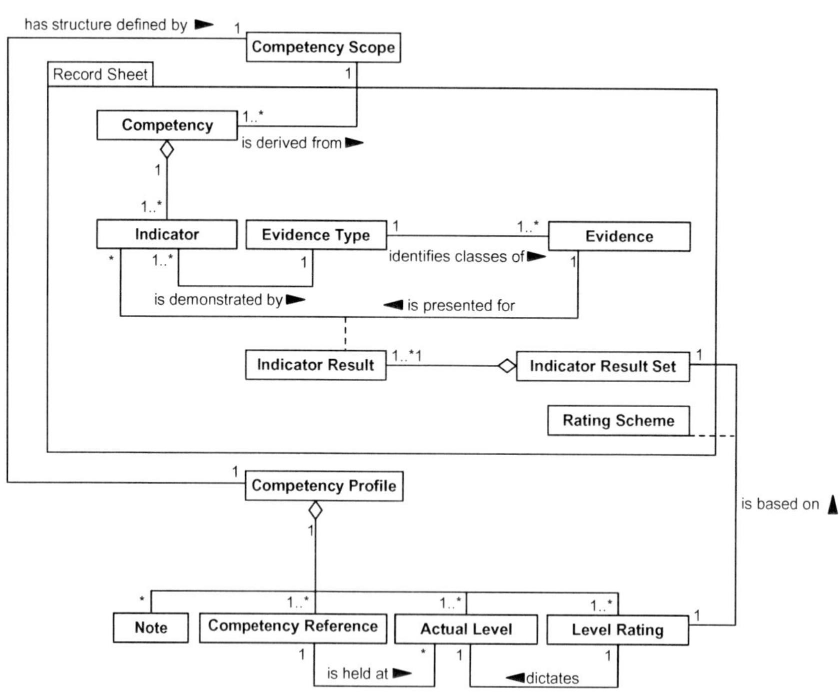

Figure C.15 The IV for the UCAM 'Assessment' process

C.3.7 The Process Instance Views

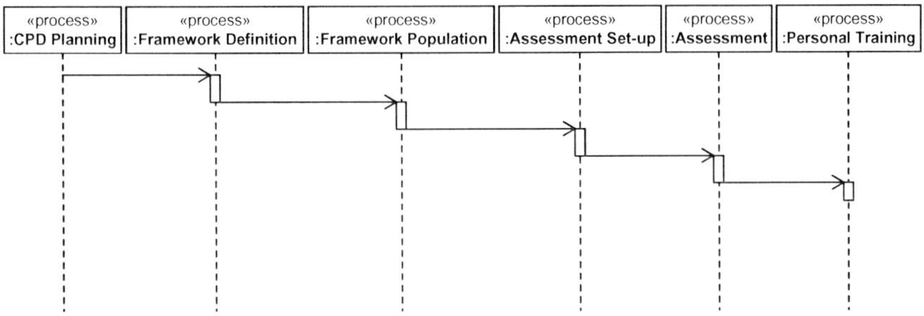

Figure C.16 Example PIV for UCAM processes showing process execution for self-assessment

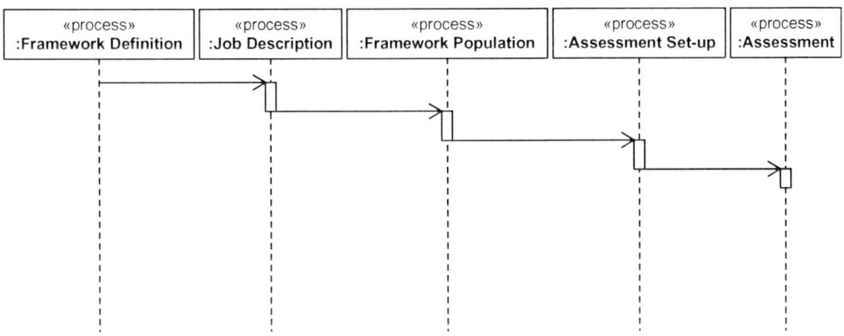

Figure C.17 Example PIV for UCAM processes showing process execution for recruitment

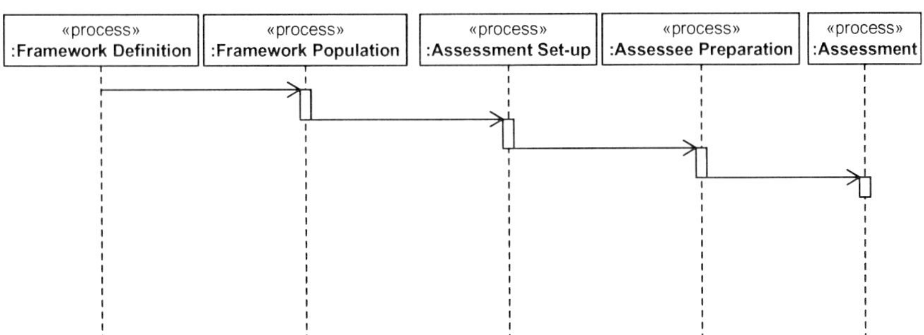

Figure C.18 Example PIV for UCAM processes showing process execution for appraisals

References

1. Holt J., Perry S. *A Pragmatic Guide to Competency: Tools, Frameworks and Assessment*. BCS, the Chartered Institute for IT: Swindon; 2011
2. Holt J. *A Pragmatic Guide to Business Process Modelling*. 2nd edn. BCS, the Chartered Institute for IT: Swindon; 2009

Index

Page numbers followed by *f* and *t* indicate figures and tables, respectively.